Wanderers in space

exploration and discovery in the solar system

Wanderers in space

exploration and discovery in the solar system

Kenneth R. Lang
Charles A. Whitney

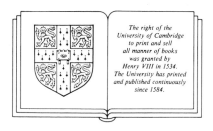

Cambridge University Press

Cambridge
New York   Port Chester   Melbourne   Sydney

Published by the Press Syndicate of the University of Cambridge
The Pitt Building, Trumpington Street, Cambridge CB2 1RP
40 West 20th Street, New York NY 10011–4211, USA
10 Stamford Road, Oakleigh, Melbourne 3166, Australia

First published 1991

Printed in Great Britain by Butler & Tanner Ltd, Frome

British Library cataloguing in publication data
Lang, Kenneth R.
Wanderers in space: exploration and discovery in the solar system.
1. Solar system
I. Title   II. Whitney, Charles A. (Charles Allen), *1929–*
523.2

Library of Congress cataloguing in publication data
Lang, Kenneth R.
Wanderers in space: exploration and discovery in the solar system/
    Kenneth R. Lang, Charles A. Whitney.   p.   cm.
Includes bibliographical references.
1. Solar system.   I. Whitney, Charles Allen.   II. Title.
QB501.L26   1991
523.2—dc20   90—1473   CIP

ISBN 0 521 24976 7 Hardback
ISBN 0 521 42252 3 Paperback

To Marcella and Jane

Contents

Figures

Chapter 3

Chapter 4

Preface

.

The moons and planets have changed from mere points of light to fascinating, diverse worlds. No two of them are exactly alike. These worlds have been discovered and explored through the space-age extension of our senses with giant telescopes and the inquisitive eyes of remote spacecraft.

Spacecraft have visited all the planets known to ancient peoples. Human beings have visited the Moon, and robot spacecraft have landed on Venus and Mars. Here we present the results of this captivating voyage of discovery, recording more than two decades of extraordinary accomplishments.

This book includes numerous photos from spacecraft and diagrams, as well as a few works of modern art. Each illustration has been chosen for its artistic value and for the new insights it offers. Altogether, they provide the best available metaphors and images of the previously invisible worlds.

Our voyage begins with the Moon, a stepping stone to the planets, examining its battered surface under a sky that remains black in broad daylight. There is no sound, weather, water nor life on the Moon, and there apparently never has been. The lunar world may provide clues about the Earth's early history, but we are still not even sure where the Moon came from.

We then proceed to Mercury, which looks like the Moon on the outside and resembles the Earth on the inside. A space-age glimpse of Venus' invisible, cloud-covered surface then reveals a torrid world that is hot enough to melt lead and vaporize former oceans.

In contrast to the other worlds, our water planet, Earth, is teeming with life. We inhabit the only place in the vast solar system with abundant liquid water and free oxygen in the air. Ancient continents slide over the globe, colliding and coalescing with each other like floating islands, as ocean floors well up from inside the Earth and remain in eternal youth. Life on this restless world is protected by a thin membrane of air that is being constantly modified by life itself, as well as by a magnetic cocoon that occasionally lights up like a cosmic neon sign.

Robot spacecraft have revealed exotic and unexpected details on Mars, including towering volcanoes, deep canyons, broad arroyos, and evidence of former floods.

After traversing the asteroid belt, and noting that a colliding asteroid may have wiped out the dinosaurs, we continue with the kaleidoscopic results of the Voyager missions to Jupiter, Saturn, Uranus and Neptune. Here we find cyclonic storms, liquid hydrogen, and helium rain. Diverse rings surround all four of these planets – Jupiter's single ring, Saturn's huge rings of ice, Uranus' dark, narrow hoops, and Neptune's clumpy rings. The currently-active volcanoes on one of Jupiter's satellites have turned the satellite inside out, and one of Saturn's satellites has a nitrogen-rich atmosphere. Uranus is tipped on its side and its magnetic field has a peculiar tilt, while its satellites show elaborate surface detail. Neptune's weather may be driven by internal heat, while icy volcanoes may be erupting on its satellite, Triton. Pluto is a tiny world of rock and ice, and it has an oversized moon.

Our voyage continues to the icy comets. We discuss space-age photographs of Halley's dark nucleus and speculate on the invisible comet cloud that extends halfway to the nearest star.

The asteroids and comets are relics of the formation of the solar system, and they provide vital clues about its birth. This birth is our last topic, and it includes a discussion of the possibility of invisible worlds orbiting other stars.

We are grateful to the numerous experts who have read individual chapters and commented on their accuracy and completeness. They include Jayne Aubele, Alan P. Boss, John C. Brandt, Joseph A. Burns, Alastair G.W. Cameron, Michael H. Carr, Clark R. Chapman, Lawrence Colin, Barney J. Conratl, Armand H. Delsemme, Stanley F. Dermott, Larry W. Esposito, Owen Gingerich, Lawrence Grossman, Robert M. Haberle, Norman H. Horowitz, William Hubbard, D.M. Hunten, Torrence V. Johnson, Stephen M. Larson, Myron Lecar, Conway Leovy, John S. Lewis, Brian Marsden, Ursula Marvin, David Morrison, Gordon H. Pettengill, Gerald Schubert, Zedenek Sekanina, Conway W. Snyder, Larry Taylor, Stuart Ross Taylor, Joseph Veverka and Fred Whipple.

The title for this book was suggested by Peter Sturrock over drinks in a Chinese restaurant.

The Astrologers of Life. Two silhouetted figures search the heavens in this 1947 painting by Rufino Tamayo. It may represent modern man's attempts to understand the Universe. A comet and a full Moon illuminate the azure blue sky. The geometric diagrams in the foreground could portray stellar or planetary configurations. A red radio tower in the background sends out signals, perhaps to civilizations on other worlds. (Courtesy of Sotheby Parke Bernet Inc., New York, © 1985.)

1 Worlds in motion

1.1 Astronomy, the oldest and newest science

When we stroll under the starlit sky and look upward at familiar constellations, such as the Pleiades, Orion, the Great Bear, or Cassiopeia, we are taking part in an age-old drama. We are the actors, our planet Earth provides the stage, and we proclaim the stars and planets as the backdrop for our story.

The names of the constellations remind us that the ancient story-tellers and astrologers were profiling human history and writing a tribute to the human spirit when they translated their mythology to the patterns of the sky. The writers of the *Book of Job* put the names of these constellations into the very mouth of God when they described God's challenge to the arrogance of Job:

> Can you bind the chains of the Pleiades,
>   or loose the cords of Orion?
> Can you lead forth the Maz'zaroth in their season,
>   or can you guide the Bear with its children?

Those writers were, in a sense, the novelists of their day. And if we read carefully, we may notice that the astronomers of today – the spiritual grandchildren of the ancient astrologers and mathematicians – are writing about human achievement and insight when they describe new discoveries. Few stories could be more glorious than the story of our breaking the bonds of gravity and penetrating the misty barrier of our atmosphere – a penetration with body and eye and, above all, with the eye of the mind.

The "Maz'zaroth" mentioned in the quotation from Job were the constellations of the zodiac that mark the Sun's yearly path about the sky. Many of these constellations depict animals. In the circle of figures we can find a crab, lion, scorpion, goat, fish, and bull. They remind us of the time – even before the advent of astrology – when Mother Nature was seen as a living force, a nurturing friend who lived among the animals.

The roots of astronomy were planted by astrologers who recognized that the Sun, Moon and planets were not whimsical and did not wander like foraging animals across the sky, but followed an intricate and knowable pattern. These astrologers began to track the Sun, whether for religious reasons or for forecasting the outcome of a battle, or for designating the days to plant the new crops. Their meticulous records provided the first data for the later development of models of the heavens. The ancient Greek models of the Universe were, for the most part, constructed by mathematicians who felt that the underlying structure of the Universe was essentially mathematical and physical, rather than human or animalistic.

The line between myth and science in the ancient practice of astrology–astronomy was a fine one. Observations of the sky revealed countless patterns of space and time, patterns that permitted reliable predictions of the seasons and that could be used to mark the passage of the decades. Many ancient buildings appear to be aligned according to the passages of the celestial bodies, from the stone pillars of Stonehenge (Fig. 1.1) to the walls, windows, and corridors of ancient observatories in Egypt, India, China, and Central America. The Sun, Moon, and planets were observed, not only because they permitted keeping track of the passage of time, but also because many priests insisted the heavens held the powers of nature and ruled the lives of humans.

Figure 1.1. Stonehenge. Sunrise is framed by the ancient stones of Stonehenge in southern England. This monument was used to find midsummer and midwinter four thousand years ago – before the invention of writing and the calendar. The Sun rises at different points on the horizon during the year, reaching its most northerly rising on midsummer day (summer solstice on June 21). After this, the rising point of the Sun moves south along the horizon until it reaches its most southerly rising on midwinter day (winter solstice on December 22). An observer located at the center of the main circle of stones at Stonehenge watched midsummer sunrise over a marker stone located outside the circle; midwinter sunrise and sunset were framed by other stones within the circle. (Courtesy of Owen Gingerich.)

Because of this mixed heritage, astronomy as we know it today is a young science. It emerged from astrology, by adopting the methods of physics and by rejecting many of the old traditions. Astronomers today have rejected the notion of a direct, day-to-day influence of the planets on human activity. Such notions do not stand up when they are tested closely.

Of course, the work of modern astronomers is built on ancient sightings of the Sun, Moon, and planets, whose motions are now understood in terms of the theory of gravitation. The development of this theory was a necessary prelude to space flight, so we shall begin our story with a brief introduction to the moving worlds about us.

1.2 Sun dance

Figure 1.2. Curved shadow of Earth. This multiple-exposure photograph of a total lunar eclipse reveals the curved shape of the Earth's shadow, regarded by ancient Greek astronomers as evidence that the Earth is a sphere. Only a spherical body will cast the same circular shadow on the Moon during different eclipses. The Moon turns blood-red during its total eclipse. This photograph was taken by Akira Fujii during the lunar eclipse of December 30, 1982.

(a) The spherical Earth

Many of the ancients thought the Earth was flat, and it may actually appear flat to us as we stroll across its surface. However, in Aristotle's time (384–322 B.C.) the curvature of the Earth had already been inferred from the shape of its shadow. During a lunar eclipse, the shadow of the Earth moves across the Moon's face. This shadow is always round, regardless of the orientation of the Earth, and only a spherical body can cast a round shadow in all orientations (Fig. 1.2). Of course, the spherical Earth has been revealed in photographs from space (Fig. 1.3).

Figure 1.3. Earth from space. When seen from space, the Earth shows shadowed, crescent shapes like the Moon. The dividing line between day and night marks the twilight zone. Because it is summer in the northern hemisphere, there is more sunlight in the north (top) than the south (bottom). This photograph was taken in July 1969 when the Apollo 11 spacecraft was 180 000 kilometers away from the Earth, travelling on its way to land the first men on the Moon. Most of Africa and Asia are visible, but night has already fallen in India (right). As the Earth spins, Africa will next enter night. (Courtesy of NASA.)

Figure 1.4. Multiplying your sunsets. Because the Earth is round, you may see more than one sunset in an afternoon. Stand at the west side of a hill, near its foot, and when the Sun has set, run uphill. You will see the Sun reappear for a moment. The same may be achieved by running downhill at sunrise.

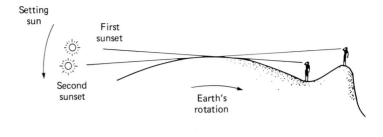

You can convince yourself that the surface of the ocean is curved by watching a ship disappear over the horizon; first the hull and then the mast sink from view. If you climb a nearby bluff, you can see the ship again as you peer over the curve of the horizon.

The roundness of the Earth also permits you to see more than one sunset (Fig. 1.4).

(b) Sun time

The daily westward motion of the Sun across the sky is the basis of the sundial, which tells time by the Sun. The sundial may be considered to be a clock with just one moving part, the Earth – or the Sun, if you prefer to think of it that way. The design of a sundial can be a fascinating geometrical puzzle, but there is also a very simple procedure for constructing an accurate sundial without any calculations. The first step is to find a level piece of open ground on which the Sun can shine all day. Next, cut a strong, thin stick several feet long with a point at both ends. Jam it straight down into the ground so the tip of its shadow falls on smooth ground. Then cut a series of short sticks to be used for locating the tip of the shadow each hour, or each quarter hour. Mark each stick with the time it is supposed to represent and sit down with a good book or a friend.

As the Sun moves across the sky and the shadow moves across the ground, jam a small stick into the tip of the shadow each hour, or each quarter hour. If the upright stick, called a gnomon, is long enough, you will be able to detect the motion of its shadow in less than a minute. If the stick is 2 meters tall, the shadow of the tip will move about one-third its height, or 60 centimeters, in an hour. This is equivalent to one centimeter per minute. By contrast, the Eiffel Tower is 300 meters high, and the tip of its shadow rushes across the ground at a rate of 1.5 meters per minute due to the rotation of the Earth.

(c) The seasons

At noon each day, the Sun reaches its highest point in the sky, and it will be close to the north–south line. If you watch the sundial all year, you will notice that the Sun's height (indicated by the length of the stick shadow) at noon varies with the season. For viewers in the northern hemisphere, summer officially begins when the Sun has reached its most northerly excursion. At that season, it rises highest and traces its longest arc across the sky (Fig. 1.5).

During a year, the Sun appears to be carried north and south because its annual path is tilted 23.5 degrees from the Earth's equatorial plane. This path is called the ecliptic, because that is where eclipses can occur. In either hemisphere, the greatest sunward tilt occurs in summer when the Sun is more nearly overhead and its rays strike the surface more directly (Fig. 1.6).

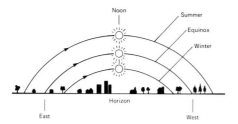

Figure 1.5. Sun's motion across the sky looking south. The Sun's maximum height in the sky and its rising and setting points on the horizon change with the season. The Sun rises exactly east and sets exactly west on the vernal (spring) equinox and on the autumnal equinox. In the summer, the Sun rises in the northeast, reaches its highest maximum height, and stays up longest. The Sun rises in the southeast and remains low in the winter when the days are shortest.

(d) Night and day

One half the Earth is bathed in sunlight at each moment, and the circle that divides the globe in light and dark halves is the twilight region. If we could look down onto the north pole as the Earth rotates, we would see that the rotation is counter-clockwise. (This direction of motion, called ''prograde,'' is also the direction in which the Earth and other planets move about the Sun.) Every point moves eastward in a circle that carries it alternately into darkness and sunlight (Fig. 1.7).

Twice each year, the Sun crosses the equator. On these days, it rises in the east at 6:00 a.m. (local time) and sets in the west at 6:00 p.m. (local time). These are the times of equinoxes when day and night are of equal length no matter where you are on Earth. The dates are the first day of spring (around March 21 in the northern hemisphere) and the first day of autumn (around September 21 in the north). At other seasons of the year, the times and directions of sunrise and sunset will depend on your latitude as well as the date.

Two other dates are important: the solstices, when the Sun is farthest from the Earth's equator. On June 21, the Sun is farthest north and days in the northern hemisphere are longest; on December 21, the Sun is farthest south, and days in the southern hemisphere are longest.

(e) Sunset colors

At sunset, the dazzling white image of the Sun fades to bright red, because the blue light has been scattered out of the sunbeams by its great path through the air. (The light that has been removed from the sunbeams goes to create the bright skies beyond the western horizon.) After the last beam of sunlight vanishes, the western horizon continues to glow bright yellow but the eastern horizon quickly darkens. This darkness in the east grows upward from the horizon as the Earth's shadow rises into the air.

When the Sun has sunk 4 degrees below the horizon (about 20 minutes

Figure 1.6. The seasons. As the Earth orbits the Sun, its rotational axis always points in the same direction (towards Polaris), but the northern and southern hemispheres are tilted toward or away from the Sun by varying amounts. This variable tilt produces the seasons by changing the angle at which the Sun's rays strike different parts of the Earth's surface. The greatest sunward tilt occurs in the summer when the Sun is more nearly overhead and the Sun's rays strike the surface most directly. In the winter, the relevant hemisphere is tilted away from the Sun and the Sun's rays obliquely strike the surface. When it is winter in one hemisphere it is summer in the other one. (Notice that the radii of the Earth, the Sun and the Earth's orbit are not drawn to scale.)

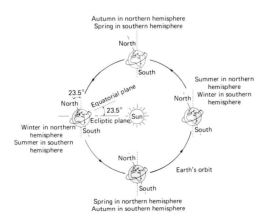

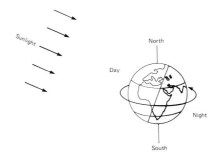

Figure 1.7. Night and day. The Earth rotates with respect to the Sun once every 24 hours, causing the sequence of night and day. Each point on the Earth's surface moves in a circular track parallel to the equator; each track spends a different time in the Sun depending on the season. This drawing depicts summer in the northern hemisphere and winter in the southern hemisphere. Because the northern part of the Earth's rotational axis is tipped towards the Sun, circular tracks in the northern hemisphere spend a longer time in the Sun than southern ones. On the first day of spring or fall (the equinoxes), all circular tracks are half in sunlight and the Sun stays up the same amount of time in both hemispheres.

after sunset at moderate and low latitudes), one of the most interesting parts of sunset has begun. The western sky, at a point halfway to the zenith, becomes suffused with a bright glow of rose-red light. It rapidly spreads, and then within 5 or 10 minutes it fades as quickly as it had appeared. This is called the "purple glow" because it combines the blue background of the sky with the red light of the vanished Sun. It comes from thin layers of cloud or dust that are still illuminated by the sunlight high in the atmosphere (Fig. 1.8). It can be particularly strong during the first few weeks and months after an intense volcanic eruption.

1.3 Motion of the Moon

(a) The inconstant Moon

The Moon's daily motion westward from horizon to horizon is caused by the eastward rotation of the Earth, and its monthly circuit eastward against the background stars is caused by its orbital motion about the Earth. Once each month, the Moon comes nearly in line with the Sun, and vanishes into the glare of daylight for several days. This is the time of new Moon.

If we look into the western twilight with a pair of binoculars one or two evenings after the new Moon, we may catch a glimpse of the Moon's thin crescent pointing to the Sun beneath the horizon. The horns of the crescent are balanced on either side of a line from the center of the Moon to the center of the Sun. In a few minutes the pale crescent follows the Sun into the thick air of the horizon and is lost. If we turn to look toward the east, we will see the sky darken as the shadow of the Earth climbs toward the zenith and night replaces day. In a half hour or so, the first stars become visible.

On the next night, we have an easier time finding the Moon. It is slightly higher, farther from the Sun's light and the crescent is thicker so it stands out better. Each successive night, the Moon's crescent will become thicker as more of the Moon's face becomes illuminated by the Sun (Fig. 1.9).

According to Vachel Lindsay's poem,

> The Moon's the North Wind's cooky.
> He bites it, day by day,
> Until there's but a rim of scraps
> That crumble all away.
>
> The South Wind is a baker.
> He kneads clouds in his den,
> And bakes a crisp new Moon that . . . greedy
> North . . . Wind . . . eats . . . again!

Figure 1.8. Blue skies, red sunsets and the purple glow. Our air is a colorless gas, but the sky is usually blue and sunsets are red. The incident sunlight contains all colors, but air molecules scatter blue light more strongly than red light. When the Sun is overhead, the light that reaches us is mostly scattered sunlight, and this causes the sky to appear blue. When the Sun sets, its rays pass through a maximum amount of atmosphere. Most of the blue light is scattered out before it reaches us, and light from the setting Sun therefore becomes strongly reddened. (Dust also helps to redden sunsets.) Shortly after the Sun has set, the upper atmosphere is illuminated by red sunlight which is scattered down into the twilight zone. This causes the western sky to glow with an intense purple or pink light.

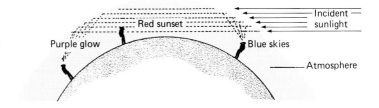

| Waxing crescent | First quarter | Waxing gibbous | Full Moon | Waning gibbous | Third quarter | Waning crescent |

Figure 1.9. The appearance of the Moon. The term crescent is applied to the Moon's shape when it is less than half-lit; it is called gibbous when it is more than half-lit but not yet fully illuminated. During the monthly cycle, the Moon waxes (grows) from crescent to gibbous, and then after full Moon, it wanes (decreases) to a crescent again. (Lick Observatory Photograph.)

The Moon may be the "North Wind's cooky" as the poem says, but its shape is close to a sphere and we see more of it in sunshine as it moves around the Earth, from new to full (Fig. 1.10).

By the third or fourth night after new Moon, we can see that the Moon is all there, because the crescent is extended into a full circle by a faint glow often called the "old Moon in the new Moon's arms" (Fig. 1.11). This glow is earthshine produced by sunlight that has bounced off the Earth, and it remains visible for several nights. Then the Moon's crescent becomes so wide and bright that the earthshine is lost in its glare.

If we could transport ourselves to the face of the Moon around the time of new Moon and look up toward the Earth, we would see it hanging fully-lit in a black sky (Fig. 1.12). Its light would be 84 times brighter than full Moon, because it has 14 times the surface area and is 6 times more reflective than the Moon. We would have enough light to read by and to pick our way across the rough lunar surface. The Moon's soil at our feet would glow in the light of the earthshine.

At the end of the first week after new Moon, the Moon's face would have become a semi-circle. This is the time of dichotomy, when the face is divided into two equal parts by the edge of the shadow, called the terminator. This is the first quarter, and by now the moonlight has become so bright that the earthshine is almost impossible to see.

Figure 1.10. Phases of the Moon. Light from the Sun illuminates one half of the Moon, while the other half is dark. As the Moon orbits the Earth, we see varying amounts of its illuminated surface. The phases seen by an observer on Earth (below) correspond to the numbered points along the lunar orbit. The period from new Moon to new Moon is 29.53 days, the synodical month. As the Earth completes its daily rotation, all night-time observers see the same phase of the Moon. The Moon completes one rotation during one complete revolution around the Earth, so that the same side keeps facing the Earth at all times, but the Sun illuminates first one part of the Moon's face and then another.

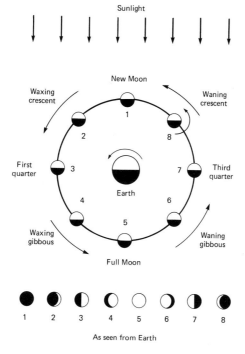

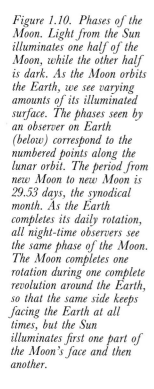

Figure 1.11. Venus and crescent Moon with earthshine. The horns of the crescent Moon point towards Venus a few hours after the planet passed behind the Moon. These relatively rare occultations always involve a crescent Moon, since Venus never strays far from the Sun in the sky. Faint earthshine fills the globe of the Moon. The earthshine is produced by sunlight that has bounced off the Earth. (Courtesy of Johnny Horne, photographer.)

Two weeks after new Moon, we see full Moon rise in the east at sunset. It remains in the sky for the entire night and sets at sunrise. Near full Moon, the sunlight comes over our shoulders on its way to the Moon. For a few days the Moon is extraordinarily bright and most stars are lost in its glare. Examining the Moon's face with binoculars we see bright streaks called lunar rays emanating from a few bright craters and curving part way around the Moon. Most of the craters are difficult to see, because there are no shadows, but we easily see the large bright and dark areas that form the so-called "Man in the Moon" or the "Man and Woman" in the Moon.

Carefully looking at features on the Moon's face night after night, we notice that they do not move. They are stationary because the Moon keeps its same face toward the Earth through the entire month – and it has done so for uncounted millions of years. The Moon keeps the same face toward the Earth because it rotates on its own axis in exactly the time it takes to orbit the Earth.

One more week brings third quarter, when the face of the Moon is again divided in halves by the shadow terminator, bright on the east and dark on the west.

In Shakespeare's *Romeo and Juliet*, Act 2, Scene 2, Romeo swears his love for Juliet, referring to the silvery cast of moonlight,

> Lady, by yonder blessed Moon I vow
> That tips with silver all these fruit-tree tops –

This appearance is an illusion caused by our inability to see colors in faint light. In the next line of the play, the astronomically astute Juliet protests:

> O, Swear not by the Moon, the inconstant Moon,
> That monthly changes in her circled orb,
> Lest that thy love prove likewise variable.

Toward the end of the fourth week, the old Moon becomes a thin crescent pointing the way to the Sun. We can catch the earthshine if we are up before sunrise and find the Moon in the eastern twilight. The cycle of the month is completed when the Moon vanishes in the glare of the Sun. It remains invisible from Earth for about three days before re-emerging as new Moon once more. (The motions are summarized in Table 1.1)

Figure 1.12. Earth-rise. This view from the Apollo 11 spacecraft shows the Earth rising above the Moon's horizon. The foreground moonscape is Smyth's Sea, just barely visible from Earth as a flat lava bed, or mare, at the extreme eastern edge of the Moon (Courtesy of NASA).

Table 1.1. *Moon's cycle during the month*

Phase	Rises at	Sets at
New	Sunrise	Sunset
First quarter	Noon	Midnight
Full	Sunset	Sunrise
Third quarter	Midnight	Noon

Because the plane of the Moon's orbit is tilted relative to the Earth's equator, the "inconstant Moon" also swings north and south each month. This north and south motion affects the Moon's height in the sky and the times of moonrise and moonset.

The winter full Moon rises much higher in the sky than the summer full Moon. This is contrary to the behavior of the noonday Sun, which is high in the summer and low in the winter. The difference occurs because the full Moon always lies nearly opposite the Sun. Hence, the Moon's north and south motion is complimentary to the Sun's motion, and the *summer* full Moon is *low* in the southern sky while the *winter* full Moon is *high* in the sky.

(b) Eclipses of the Moon and Sun

Once or twice in a typical year, the Moon's eastward orbital motion carries it through the Earth's shadow. This is an eclipse of the Moon, and it can be seen from half of the Earth, where the full Moon is in the sky. Figure 1.13 shows the Sun–Earth–Moon system (not to scale) and indicates the Earth's shadow at the time of a lunar eclipse. There are two regions in the shadow: the umbral region where the Sun appears totally hidden, and the penumbral region where it is partially hidden. The umbral shadow is darker, and it is in the shape of a narrow cone pointing away from the Sun. The Moon is a deep red when in the umbral shadow of the Earth (Fig. 1.14). Ancient Hebrew writers often used this appearance as a metaphor to describe the end of the world. For instance, the prophet Joel declared that the Lord

will give portents in the heavens and on the Earth, blood and fire and columns of smoke. The Sun shall be turned to darkness, and the Moon to blood . . .

An eclipse of the Sun occurs when the Moon appears to hide a portion of the Sun and the Moon's shadow falls on the Earth (Fig. 1.15). A total eclipse occurs along a relatively narrow region of the Earth, where the Sun is completely hidden. At other regions of the Earth, the Sun will be partially eclipsed.

The Moon's orbit about the Earth is slightly elongated, and when the Moon and Earth are at their average separation, the tip of the Moon's umbra reaches almost precisely to the center of the Earth. In such a situation, total eclipses of the Sun are possible inside a narrow track drawn by the umbral shadow as it sweeps across the Earth's surface (Fig. 1.16).

Figure 1.13. Lunar eclipse. During a lunar eclipse the initially full Moon passes through the Earth's shadow. A total lunar eclipse occurs when the entire Moon moves into the umbra. Because no portion of the Sun's surface can be seen from the umbra, it is the darkest part of the Earth's shadow. Only part of the Sun's surface is blocked out in the larger penumbra. A partial lunar eclipse occurs when the Moon's orbit takes it only partially through the umbra or only through the penumbra.

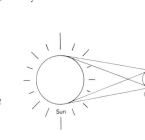

Figure 1.14. The blood red Moon. If the Earth had no atmosphere, the Moon would disappear in darkness during a total lunar eclipse. As shown here, the Moon actually becomes dark red for an hour or so. This is because the Moon is illuminated by sunlight that is bent part way around the Earth and is reddened by passing through the Earth's atmosphere, just as the Sun is reddened at sunset. If the Earth is heavily clouded, the sunlight is obstructed and the Moon is particularly dark during a lunar eclipse. (Courtesy of Eric Mandon who took this photograph with the 14 centimeter (5.5 inch) refractor of the Observatoire Populaire de Rouen during the lunar eclipse of 16 September 1978.)

Figure 1.15. Solar eclipse. During a solar eclipse, the Moon casts its shadow upon the Earth. No portion of the Sun's surface can be seen from the umbral region of the Moon's shadow, but the Sun is only partially blocked in the penumbral region. A total solar eclipse is only observed from the tip of the umbra. It traces a narrow path across the Earth's surface.

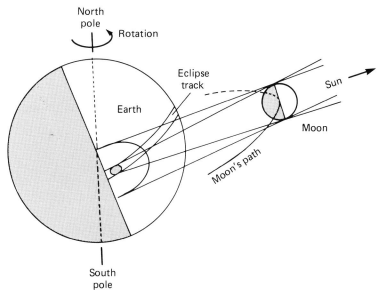

When the Moon is at a distant part of its orbit, the umbra does not quite reach the surface of the Earth. In this case, the Moon appears smaller than the Sun, and a bright ring of the Sun's disc can be seen around the edge of the Moon. This is an annular eclipse, and it has none of the darkness and excitement of a total eclipse. An annular eclipse of the Sun is similar to a partial eclipse, because the Sun's disc is blindingly bright. In contrast, during a total eclipse the bright solar disc is hidden from view and we can glimpse the Sun's tenuous outer atmosphere. Eclipse tracks for the interval 1985–95 are shown in Figure 1.17.

Figure 1.16. Total eclipse of the Sun. A multiple-exposure photograph of a total eclipse of the Sun. The circular form eclipsing the Sun is the Moon. Because the Moon and the Sun have nearly the same angular extent, the Moon blocks out most of the Sun's light during a total solar eclipse. This photograph was taken by Akira Fujii on February 16, 1980.

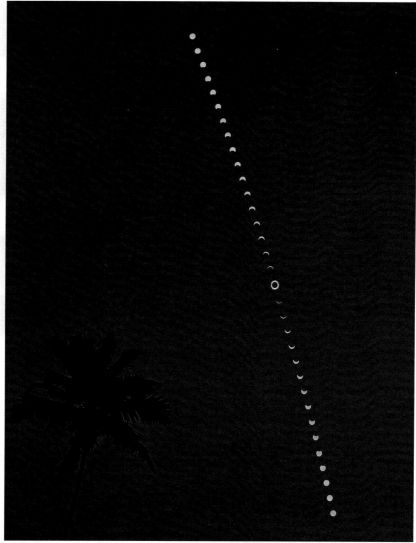

Figure 1.17. Solar eclipse tracks. The central paths of 16 solar eclipses between 1985 and 1995 are shown. Shaded paths represent total eclipses; open ones are annular (the October 1986 and March 1987 eclipses are annular total). Ellipses show the instantaneous shape of the zone of annularity or totality at various points along each eclipse track. The cross near the center of each path marks the point of maximum eclipse.

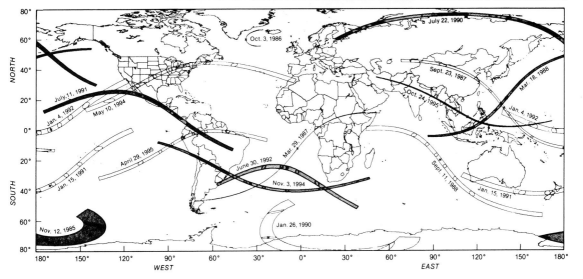

1.4 Understanding the motions of the planets

(a) Paths of the planets on the sky

The stars seem firmly fixed on the celestial sphere, but the ancients noted seven objects that wandered in a generally eastward direction: Sun, Moon, Mercury, Venus, Mars, Jupiter, and Saturn. Each wandering object was associated with one day of the week, and the present sequence of days of the week is associated with the Sun, Moon, Mercury, Venus, Jupiter, Mars, Saturn. In modern English, the connection between this list and the names of Sunday, Monday and Saturday are obvious, but the others have been adopted from a variety of sources.

Even the earliest sky-watchers must have noticed that all the wanderers were confined to a narrow track around the sky. This track is now known as the belt of the "zodiac," from an ancient Greek word for animal. The center of the belt is the Sun's yearly path, and its narrowness (about 9 degrees on each side of the ecliptic) is a sign that the planets move almost like marbles on a table because the planes of their orbits are closely aligned with each other.

When the outer planets (Mars, Jupiter, Saturn, Uranus, Neptune, and Pluto) come opposite the Sun and nearest to Earth, so they shine brightly in the midnight sky, their eastward motion is interrupted briefly. Each planet then moves backward toward the west for a few months before resuming its eastward movement. These backward loops are called retrogrades, and the retrograde of Mars is the most prominent because Mars is the nearest of the outer planets (Fig. 1.18). Ancient and modern explanations of this motion differ in their perspective on the solar system.

A few of the ancient Greeks believed that the Sun marked the center of the planetary system, but the Earth-centered system captivated the minds of most astrologers and astronomers until the 17th century. Geocentric and

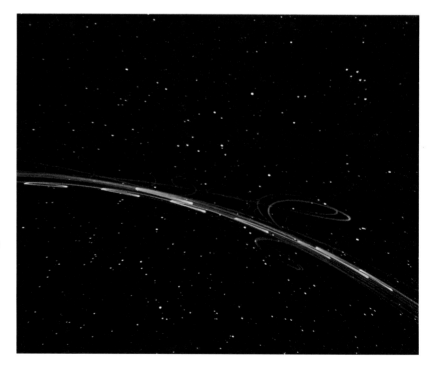

Figure 1.18. Retrograde loops. This photograph shows the movements of the planets as seen from the Earth. When they appear closest to the Sun and therefore brightest, the outer planets appear to stop in their orbits, then reverse direction before continuing on – a phenomenon called retrograde motion by modern astronomers. Here the apparent movements of Mars, Jupiter and Saturn have been recorded against the background stars at the Munich Planetarium. (Courtesy of Erich Lessing/Magnum.)

Figure 1.20. Galileo's observations of the Moon. Galileo Galilei's drawings based on his telescopic observations of the Moon and reproduced in his Sidereus Nuncius, *or* Starry Messenger, *in 1610. Galileo was an excellent draftsman, enabling him to detect and show the shadowed craters for the first time. Thomas Harriot made similar telescopic observations at about the same time, but he did not recognize the craters.*

heliocentric models were equally capable of predicting the observed positions of the planets, and most early astronomers did not feel that their models should behave as objects behave on Earth (Fig. 1.19). Most of the theorizing about astronomy in ancient Greece started from the notion that the sky is perfect and the planetary motions must conform to the ideal model of a circle. Claudius Ptolemy (around AD 140) was able to show that the observed motions of the planets – retrogrades and all – could be imitated quite well by properly compounding them from circles upon circles, or epicycles.

In the 16th century, Nicholas Copernicus (1473–1543) placed the Sun at the center of the planetary system and laid the groundwork for the "Copernican Revolution," *De revolutionibus orbium coelestium* ("On the Revolutions of the Heavenly Spheres," 1543). But Copernicus did not provide any new data on the positions of the planets; this was accomplished a short time later by Tycho Brahe. Also, Copernicus did not propose any new laws of motion; this was first attempted by Johannes Kepler (1571–1630), who painstakingly analysed Tycho's observations. Nor did Copernicus provide any new observational data that might argue for the Sun-centered model; this was the achievement of Galileo Galilei. But Copernicus' book became the symbol of a new view of the heavens, a view that was ultimately to unite the Earth and the planets in the domain of terrestrial physics.

The importance of Copernicus' book was in the attitude it implied toward the building of models, not in the details of his method for computing the planetary positions. He opened the way toward treating planetary models as explanatory theories and not merely computational devices. Models were not merely to be built up step by step until they fitted the observations, but they were to be based on a set of unifying laws that governed the mechanics of planetary motion.

(b) *Galileo and his* Starry Messenger

One of the most fascinating and lively books in astronomy was *Starry Messenger*, written by Galileo Galilei (1564–1642) and published in 1610. In it, Galileo described the discoveries he had made by turning a telescope to the heavens for the first time. His work brought the sky down to Earth, so to speak, and opened a new age in astronomy – an age of detailed observations

Figure 1.19. The First Step. This painting by Frantisek Kupka seems to convey the difficulties in understanding the apparent motions of the planets, or perhaps the process by which they were formed. (Courtesy of the Museum of Modern Art, New York.)

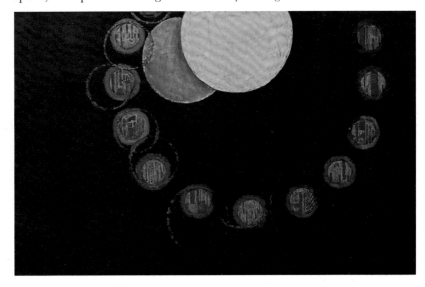

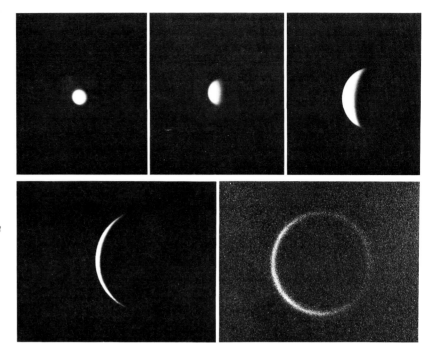

Figure 1.21. The phases of Venus. When fully illuminated, Venus looks small and far away; its apparent size is about 7 times larger when the crescents are narrowest and Venus is nearest to the Earth. After observing similar phases with his small telescope in 1610, Galileo wrote "Cynthiae figuras aemulatur mater amorum," or "The mother of love (Venus) emulates the figure of Cynthia (moon)." The phases and variations in the apparent size of Venus played an important role in convincing Galileo that the planets revolve around the Sun. (Lowell Observatory photograph.)

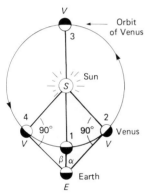

Figure 1.22. Determining the size of Venus' circular orbit. The planet Venus has a nearly circular orbit (eccentricity only 0.006) with a radius SV. Observations at inferior conjunction (1) and maximum elongation (2 or 4) give the angles α and β, from which it is possible to construct the right-angled triangles EVS and determine the lengths of the lines SV and SE. The ratio SV/SE is Venus' distance in astronomical units (A.U.). The average distance between the Earth and the Sun is exactly 1 A.U. by definition.

of the objects themselves, and not merely their motions among the stars. The Sun, Moon, and planets suddenly became physical objects, with irregularities, spots, and moons of their own. They were no longer the "perfect" celestial jewels imagined by the ancients. Galileo was delighted by his discoveries, because they seemed to give him ammunition against the classical ideas of Aristotle and the model of Ptolemy.

Figure 1.20 shows some of Galileo's sketches based on his own telescopic observations of the Moon. With his rudimentary telescope (a 4-centimeter refractor), Galileo was also able to discover the phases of Venus (Fig. 1.21) and the four large moons that circle Jupiter, resembling a miniature solar system. These observations fostered a new attitude towards the sky and ultimately led to a new set of laws for describing the motions of the planets, as well as those of the stars and galaxies.

(c) Constructing the orbits of the planets

If we could view the solar system from the direction of the northern constellation Draco far outside the zodiac, we would see that the orbits are nearly circular and the planets move counter-clockwise about the Sun. Using the known periods of the planets as well as some simple and direct observations from Earth, we can construct the shapes and sizes of their orbits relative to the size of the Earth's orbit. Doing so, we retrace Johannes Kepler's steps on his way to the discovery of his laws of motion.

Two different strategies are needed to construct the planetary orbits: one for the inner planets Mercury and Venus, and one for the outer planets. Kepler's methods made no assumptions about the orbits, but we shall eliminate unimportant complications by assuming that the Earth's orbit is nearly circular and that the planetary orbits are in a single plane.

Inner planets

Venus is the simplest case. Figure 1.22 shows a view looking down on the orbit of Venus from the north pole of the ecliptic and we can use it to

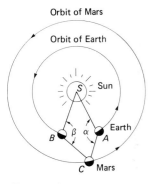

Figure 1.23. Finding the orbit of Mars. The method uses the fact that Mars returns to the same point of its orbit after one sidereal period of 687 days. Observations from the Earth on two dates separated by this interval determine the angles α and β. They are the angles between the Sun and Mars as seen from the Earth. These two angles can be used to construct the Earth–Mars lines AC and BC; their intersection is a point C on Mars' orbit. Repeating for other pairs of dates separated by the same interval will permit plotting the entire orbit.

interpret our observations of Venus. Three positions are marked: (1) inferior conjunction, (2) greatest westward elongation, and (3) superior conjunction. (The angular distance of Venus from the Sun as viewed from Earth at any instant is called its elongation.) The angles α and β are Venus' angular distances from the Sun at greatest elongations.

A few years' observations show that α and β are always the same. That is, Venus is always the same distance from the Sun at times of greatest elongation. This means that the orbit is circular and the greatest elongations are moments when the Sun–Venus line is perpendicular to the Earth–Venus line. The configuration at greatest elongation can therefore be modeled by a right-angled triangle with α at one of the vertices. This triangle is shaded in Figure 1.22. The hypotenuse is the Earth–Sun line, and the side opposite the angle α is the Sun–Venus line. The length of this side gives the relative radius of Venus' orbit.

Mercury is more of a challenge because its orbit is not circular, as we infer from the fact that it is closer or farther from the Sun at different moments of greatest elongation. We could ignore this eccentric behavior and get a crude answer by applying the same methods as used for Venus. But if we wish to see the shape of the orbit, each moment of greatest elongation must be plotted separately.

Outer planets

Now we determine the orbits of the outer planets. One method will suffice for all, and we will take Mars as the model, using the fact that Mars returns to the same location in space after one of its sidereal years, as shown in Figure 1.23. When the Earth is at point *A*, we observe the Sun–Earth–Mars angle, α, and then wait until one Martian sidereal year has passed. Suppose the Earth is at point *B* at that time; Mars will be back where it was, so if we measure a new Sun–Earth–Mars angle, β, and lay out the positions of the Earth at the two days, we can draw lines that must intersect at Mars. Then we choose another pair of points, *A'* and *B'*, separated by one Martian sidereal year, and repeat the process. (Of course, the points *A* and *A'* need not be separated by a long interval. Only the intervals *A–B* and *A'–B'* are critical.)

These are similar to the methods used by Kepler to trace out the paths of the planets. We now turn to the laws he developed to describe his results.

1.5 Harmony of the worlds

In the hope of developing a new and more accurate description of the solar system, the Danish astronomer Tycho Brahe (1546–1601) amassed a great number of observations of the planets. This was before the days of telescopes, and he used ingenious measuring instruments that resemble huge gunsights with graduated circles. He never succeeded in reconciling these observations with a planetary theory, but the mathematician Johannes Kepler was so bothered by the remaining discrepancies that he rejected the idea of circles and epicycles. He began to search for a description in terms of non-circular shapes, and the physical origin of these curves was of secondary importance for him at the outset.

Kepler tried egg-shaped ovals but rejected them, and he then found that the orbits can be described by ellipses with the Sun at one focus. This ultimately became known as Kepler's first law of planetary motion, but his contemporaries were puzzled by this discovery and many of them, even Galileo, ignored the idea. What had ellipses to do with the planets? And, even if it were granted that the planets moved in elliptical paths, how would this knowledge help predict where the planets would be found?

To the first question, Kepler had no answer; ellipses were merely a

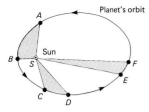

Figure 1.24. Kepler's first and second laws. Kepler's first law states that the orbit of a planet about the Sun is an ellipse with the Sun at one focus. According to Kepler's second law, the line joining a planet to the Sun sweeps out equal areas in equal times. This is also known as the law of equal areas. It is represented by the equality of the three shaded areas ABS, CDS *and* EFS. *It takes as long to travel from* A *to* B *as from* C *to* D *and from* E *to* F. *A planet moves most rapidly when it is nearest the Sun (at perihelion); a planet's slowest motion occurs when it is farthest from the Sun (at aphelion).*

geometrical tool for him. But to the second he did. He found that, as each planet moved about its orbit, it moved faster when it was closer to the Sun. In fact, he was able to state the relationship in a precise mathematical form that can be explained with the help of Figure 1.24. Imagine a line drawn from the Sun to a planet. As the planet swings about its elliptical path, the line (which will increase and decrease in length) sweeps out a surface. Kepler's second law states that each planet sweeps out the area of its orbit at a constant rate. This is also known as the "law of equal areas." During the three equal time intervals shown in Figure 1.24, the planet moves through different arcs because its orbital speed changes, but the areas swept out are identical.

When the planet is closest to the Sun it must be moving most rapidly. Its closest point is called the perihelion; and its most distant point is the aphelion, where the planet moves most slowly. The distance between perihelion and aphelion is the major axis of the orbital ellipse. Half that distance is called the semi-major axis and designated with the letter, *a*. The semi-major axis is very nearly the average distance of the planet from the Sun.

The first two laws are purely geometrical and they describe the shape of the orbit. Kepler's third law concerns the rate at which each planet sweeps around its orbit. It is also known as the harmonic law and it states that the squares of the planetary periods are in proportion to the cubes of their average distances from the Sun. The orbital speed decreases with increasing distance, so more distant planets have longer periods and they move more slowly. Figure 1.25 shows the relationship between period and orbital size for the major planets and for the brighter moons of Jupiter. Note that both sets of bodies follow a line with the same slope, as required by Kepler's third law. Once the period and the average distance are known, the position in the planet's orbit may be computed for any date, given its position on any other date.

The existence of such regularities led Kepler to the idea of an "animating principle" at work between the Sun and the planets, that is, a force that might be considered to accelerate the planets. The motion of falling bodies on Earth, and the concept of acceleration, had been studied in the 14th century, but Kepler, toward the end of the 16th century, was evidently the first to apply the ideas of force and acceleration to the motions of the planets. He compared the planets to great stones and suggested that the motions of the planets, and of objects on Earth, were united by a single type of force emanating from the Sun, and he supposed (incorrectly, as it turned

Figure 1.25. Kepler's third law. The orbital periods of the planets are plotted against their semi-major axes, a, *using a logarithmic scale. The straight line that connects the points has a slope 3/2, thereby verifying Kepler's third law that states that the square of the orbital periods increase with the cubes of the planets' distances. This type of relation applies to any set of bodies in elliptical orbits, including Jupiter's four largest satellites.*

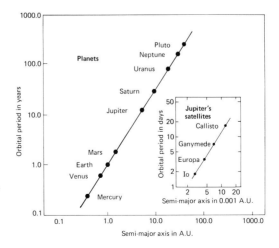

out) that the Sun's force varied inversely with distance from the Sun. The correct law, that gravitational attraction varies inversely with the square of the distance, was proposed by Isaac Newton a century later.

Kepler's three laws of planetary motion were firmly based in observations and they have remained as lasting monuments to his search for concise descriptive laws. His work represents a transitional stage in astronomy. In his day, the idea of unifying physical laws was in the air, but the formulation of such laws was not yet based on an adequate description of motion and force. It was left to Galileo, and other scientists of the 17th century, to perform experiments on accelerated motion and find the key.

Focus 1A An ancient clue to the Sun's great distance

Aristarchus of Samos (310–230 B.C.) devised an ingenious method for estimating the distance to the Sun. Although correct in principle, the method gave incorrect results because the Sun is so far away. Today, it serves as an indicator of the great distance from Earth to Sun.

The principle is shown in the figure, which is a view looking down on the Moon's orbit at the time of first and third quarters. The diagram is drawn out of scale, with the Sun much too close, in order to make the idea clearer. The trick is to catch the Moon when its face is exactly half-lit by the Sun – the instant of first or third quarter. At such moments, an observer on the Moon would see the Sun and Earth separated by an angle of exactly 90 degrees. As seen from the Earth, the Sun and Moon would be separated by slightly less than 90 degrees, an angle that we will write as $90-\beta$. The small angle β depends on the Sun's distance, and it is the angle between the Earth and Moon as seen from the Sun. (The sum of the angles in a triangle is 180 degrees.)

Aristarchus realized that the Moon should take longer to go from first quarter to third quarter than it does from third to first. By measuring the time difference, he hoped to determine the angle, β. The distances of the Sun and Moon, D_S and D_M, respectively, would be related by the equation $D_M/D_S = \sin\beta$.

Aristarchus (incorrectly) thought he had found $\beta = 3$ degrees, indicating that the Sun is about 20 times as far as the Moon. The true angle is much smaller (about 0.15 degrees) and would require measuring the instant of first quarter with an error less than 5 minutes of time. We now know that this is hopelessly difficult, given the rugged surface of the Moon, but Aristarchus' method did succeed in showing that the Sun is far beyond the Moon.

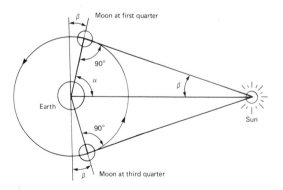

Kepler and Galileo are the two scientists who actually started the scientific revolution, whose seed had been planted by Copernicus – a revolution that culminated at the end of the 17th century with the work of Newton, to be described later in this chapter.

1.6 Finding the Sun's distance

By the middle of the 17th century, Kepler's description of the planetary motions had been accepted as giving a scale-model of the relative distances of the planets from the Sun. Astronomers knew that the Sun is much farther than the Moon (see Focus 1A, An ancient clue to the Sun's great distance), but they still lacked reliable values for the actual distance – the true scale of the model. They realized that if just one of the distances could be measured in kilometers, they could express the Earth–Sun distance (called the astronomical unit, A.U.) in kilometers and the problem of scale would be solved.

Since that time, the scale has been measured in several ways, and the methods can be divided into two categories:

1. Determine the circumference of the Earth's orbit in kilometers and divide by 2π.

    The circumference can be found by measuring the Earth's orbital speed in kilometers per second and multiplying by the number of seconds in a year. The speed is 29.8 kilometers per second, and at this rate a trip around the Earth would require 22 minutes. Multiplying the speed by the number of seconds in a year and dividing by 2π, we find:

    Average distance from Earth to Sun = 150 million kilometers
    = 1 astronomical unit.

    The distance from the Earth to Sun is almost 4000 times the distance around the Earth, or 400 times the distance to the Moon.

2. Determine the distance to a planet or asteroid and then express the Earth's orbital radius as a multiple of that distance.

    This second method is like measuring the mileage between any two cities on a map and dividing this by the number of inches on the map to find the scale in miles per inch. It was first achieved by triangulation of the nearby planets and a few asteroids from distant locations on the Earth. More recently, it has been achieved by bouncing radar signals off the nearby planets and multiplying half the round-trip time by the speed of light to compute the distance. (See Focus 1B, Radar-ranging to Venus.)

The best current value for the astronomical unit is 149 597 870 kilometers. With this value, the true distances and the average orbital speeds of the planets may be computed, with the results shown in Appendix A1. The Dutch physicist, Christiaan Huygens, in his *Celestial Worlds Discover'd* (London: 1698), described the scale of the planetary system in terms of the flight of a bullet. He imagined a bullet to travel 600 feet in a time equal to the "pulse of an artery." Such a bullet would spend 25 years in its passage from the Earth to the Sun. To make the journey from Jupiter to the Sun would require 125 years, and from Saturn to the Sun, 250 years. If we suppose a "pulse" to equal one second of time, modern values for the distances would be 26, 135, and 247 years, respectively, and the agreement is good.

1.7 Newton's apple and universal gravitation

According to tradition, Isaac Newton (1643–1727) was sitting under an apple tree when an apple fell next to him on the grass. This reminded him that the power of gravity seems undiminished even at the top of the highest mountains, and he wondered whether it might not reach to the Moon. Perhaps the Moon's motion in a circular orbit could also be thought of as a falling toward the Earth.

To compute the strength of the Earth's gravity at the Moon, Newton compared the Moon's acceleration with that of the apple. He assumed that the Moon moves in a circle and showed that the apple's acceleration was about 4000 times the acceleration felt by the Moon. Newton noticed that this ratio nearly equals the square of the ratio of the Moon's distance from the center of the Earth divided by the apple's distance:

$$(400\,000/6378)^2 = 3933$$

This equality led him to conclude that the acceleration falls off with the square of the distance.

Focus 1B Radar-ranging to Venus

Accurate distances to the nearby planets have been determined by sending radio pulses from Earth and timing their return a few minutes later. The figure shows the emission of a pulse toward Venus; when it bounces from Venus it spreads over the sky and we receive only a small fraction of the original signal, delayed by the round-trip travel time. If T is the round-trip time and c is the speed of light, the total distance travelled is cT and the distance to Venus is $cT/2$. For Venus, the round-trip time is 4.6 minutes at inferior conjunction and increases to 28.7 minutes at superior conjunction.

In computing the distance, there is a small correction for the relative motion of the planets and for the bending of light in the Sun's gravitational field. The accurate calculation of the gravitational correction is described by Einstein's general theory of relativity, and measurement of the correction has permitted verifying the theory. (See Focus 1C.)

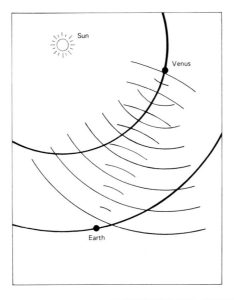

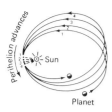

Figure 1.26. Precession of Mercury's perihelion. As illustrated in this exaggerated diagram, the perihelion of Mercury is slowly rotating ahead of the point predicted by Newton's theory of gravitation. (A planet's perihelion is the point in its orbit that is closest to the Sun.) This was at first explained by the gravitational tug of an unknown planet called Vulcan, but we now know that Vulcan does not exist. Mercury's anomalous motion was eventually explained by Einstein's new theory of gravity in which the Sun's curvature of space makes the planet move in a slowly revolving ellipse.

Figure 1.27. Weightlessness in orbit. The first untethered space walk, on February 7, 1984. Bruce McCandless II, a mission specialist, wears a 300-pound Manned Maneuvering Unit (MMU) with 24 nitrogen gas thrusters and a 35 mm camera. The MMU permits motion in space where the sensation of gravity has vanished. (Courtesy of NASA.)

The results of Newton's genius may be paraphrased as follows: the gravitational acceleration of one body by another is proportional to the mass of the attracting body (in this case the Earth). If we describe the quantity of acceleration as the ratio of force to the accelerated mass, we obtain the following expression for the force of gravity imposed by the Earth on the apple, or the Moon:

$$F_{\text{gravity}} = GMm/R^2$$

In this expression, G is the universal gravitational constant, M is the Earth's mass, m is the mass of the apple or the Moon, and R is distance from the Earth's center. This expression can be used to derive Kepler's harmonic law. Newton supposed that this force acted between all pairs of bodies in the Universe, and he made the great leap to a law that we may express as follows:

Universal gravitation

Every object in the Universe attracts every other object with a force that is directed along the line joining the two objects; the force is proportional to the product of their masses and inversely proportional to the square of their separation.

Of all the known forces of nature, gravitation exerts its influence to the greatest distance. It is unimportant over the short distances between atoms in a crystal, but the gravitational effect of stars and galaxies can be felt across the Universe. This behavior can be traced to two causes. In the first place, gravitational force decreases relatively slowly – with the inverse square of the distance – and this law gives gravitation a much greater range than the forces that hold together the nuclei of atoms, or the quantum forces that prevent atomic electrons from falling into the nuclei. In the second place, gravitation has no positive and negative polarity. This is in contrast to electricity, in which the repulsive and attractive forces among like and unlike charges can cancel each other. The effect of this cancellation is to shield distant atoms from each other's electrical forces. As far as is known, there is no gravitational repulsion between like masses, so there is no gravitational shielding, and every atom in the Universe feels the attraction of every other atom.

These are the reasons why the force of gravity plays the central role in governing the orbits of planets, controlling the flight of quasars, producing black holes, and defining the overall structure of the Universe.

1.8 A new theory of gravity

(a) Troubles with Mercury

For nearly two and a half centuries, the solar system appeared to behave according to the laws of Newton. The paths of planets, asteroids, and even comets appeared to be predictable with great precision, and many astronomers devoted their lives to unravelling the intricate motions of the solar system.

But there were problems with Mercury's motion. Newton's laws failed to provide the expected connection between old and new measures of the planet's position, and a detailed analysis showed that the orientation of Mercury's elliptical orbit was rotating in space at a rate of 43 arc-seconds per century – slightly greater than predicted (Fig. 1.26). In effect, the point of closest approach to the Sun, the perihelion, advances too rapidly.

The French mathematician, Urbain Jean Joseph Leverrier (1811–1877), detected the unexplained advance of Mercury's perihelion, and he

Table 1.2.

Newton's theory	Einstein's theory
1. Mass produces a force called gravity.	1. Mass distorts space–time, and the resulting motion imitates the action of a force.
2. Gravity acts instantly at all distances.	2. Gravitational effects propagate at the speed of light.

attributed it to an unknown planet orbiting the Sun inside Mercury's orbit. Such a planet would be difficult to find in the glare of the Sun, but it might be seen as a dark spot whenever it crossed the Sun's face. And, indeed, astronomers soon reported having seen the planet crossing the Sun, so Leverrier announced a new planet, which he named Vulcan. But the announcement was premature; the objects were merely sunspots, moving across the surface of the Sun as it rotated with a period of about 27 days. Vulcan was never found, and the discrepancy was later found to have a different cause.

(b) Curved geometry unites space, time and matter

In approaching the problem of gravitation and planetary motion, Albert Einstein (1879–1955) started from what he called the principle of equivalence:

The effects of gravity and an accelerated reference frame are equivalent in all respects.

The principle of equivalence explains why astronauts experience weightlessness in orbit (Fig. 1.27). When a spacecraft is falling freely through space, everything inside acts as though gravity were eliminated.

Einstein also adopted the principle of covariance. This postulate states that the laws governing light and matter will hold equally well in an accelerated frame of reference. This viewpoint led him, not merely to a revision of Newton's law of gravitation, but to a fundamentally new relationship between matter and space. The result was his general theory of relativity, one of the greatest accomplishments of modern science.

According to the general theory of relativity, space is distorted in the neighborhood of matter, and this distortion is the cause of gravity. In the absence of matter, space is not distorted and is described by the geometry developed by the ancient mathematician, Euclid. In the presence of matter, space becomes curved and it must be described by non-Euclidean geometry (Fig. 1.28).

The result is a gravitational field that departs slightly from the exact inverse-square law of Newton. This departure produces planetary orbits that are not exactly elliptical. Instead of returning to its starting point to form a closed ellipse after one orbital period, the planet moves slightly ahead in a winding path that can be described as a rotating ellipse. This produces an advance of perihelion, and the amount predicted by Einstein for Mercury was 43 arc-seconds per century – exactly the observed amount. Because the amount of space curvature produced by the Sun falls off with increasing distance, the perihelion advances for the other planets are much smaller than Mercury's.

We may summarize some of the differences between Newton's and Einstein's theories of gravity as in Table 1.2.

Figure 1.28. Space curvature. A massive object creates a curved indentation upon the flat Euclidean space that describes a world which is without matter. Notice that the amount of space curvature is greatest in the regions near the object, while further away the effect is lessened.

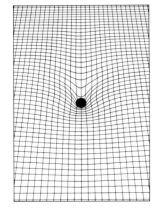

Focus 1C The weight of light

Near a massive object, the departures from Newton's laws can be quite spectacular. For example, when light passes near a star, its path is bent by the curvature of space. The star acts as a huge weak, gravitational lens and background stars are slightly displaced. Astronomers of the 18th century, treating particles of light as though they were "bullets" obeying Newton's laws, had computed that the Sun's gravity ought to deflect a light beam. But the light bending predicted by Einstein was twice the value predicted by Newton's laws, and when Einstein's value (1.75 arc-seconds for a Sun-grazing light ray) was confirmed by observations of a solar eclipse in 1919, he became an international hero.

Subsequent measurements seemed less definite, but a dramatic improvement in precision was achieved by radio interferometry of quasars, which are powerful radio sources with sharp, starlike images. These measures have confirmed Einstein's prediction to better than one percent.

Even greater precision has been obtained with radio signals from planets and spacecraft in our own solar system, because the Sun gravitationally alters the space around it and produces a slight extra delay in the passage of a radio signal. The delay occurs when radio waves travel along the curved path near the Sun. As illustrated below, this extra delay can be detected in radar echoes returned from Venus, but the measurement requires an extremely precise clock. The delay caused by the curvature of space around the Sun is only two ten-millionths of the total round-trip travel time of 1000 seconds! It becomes largest when the line of sight passes closest to the Sun. (Figure courtesy Irwin Shapiro.)

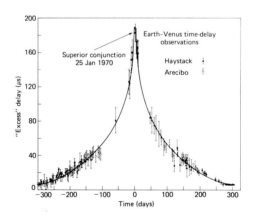

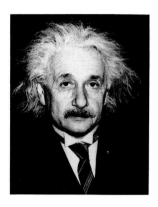

According to Einstein's theory, the gravity of any massive body ought to bend the paths of light rays (see Focus 1C, The weight of light). This is one of the famous predictions that have subsequently been observed (Fig. 1.29). So, Einstein's theory of gravitation is needed to explain details of gravitational interaction, including the motions of spacecraft. This brings us to the topic of this book, the space-age exploration of the solar system.

Figure 1.29. Albert Einstein. His prediction that the Sun would bend light by twice the amount predicted using Newton's theory was confirmed during eclipse expeditions in 1919. (Courtesy of the Bettmann Archive.)

Spaceshot of Moon. This photograph of the Moon was taken from the Apollo 11 spacecraft during its voyage toward our satellite in 1969. The large, dark circular area at the top is Mare Crisium. To the right of Mare Crisium is Mare Smythii, which is barely visible at the eastern edge of the Moon when viewed from the Earth. The western part of the far side of the Moon is seen in the right side of this photograph. The crater with a bright ray system at the upper right is named Giordano Bruno. At the lower right one can observe the basin Tsiolkovsky with its light-colored island. (Courtesy of NASA.)

2 The Moon: stepping stone to the planets

Because the Moon has no atmosphere, its sky remains pitch black in broad daylight and there is no sound or weather on the Moon.
Most of the features we now see on the Moon have been there for more than 3 billion years.
During its youth, the Moon was covered by a global sea of molten lava, but now it is covered by a layer of fine, powdery Moon dust.
There is no water or life on the Moon, and there apparently never was any.
The Moon acts as a brake on the Earth's rotation, causing the length of the day to steadily increase and the Moon to move away from the Earth.

Figure 2.1. The full Moon. The near side of the Moon that is always turned toward the Earth. This Earth-based view of the full Moon enhances the contrast between the dark maria and the bright rayed craters. The region near the Moon's south pole (lower center) is dominated by the magnificent rays of ejecta emanating from the relatively young crater Tycho. The dark circular Mare Imbrium is prominent in the northwest (upper left), immediately above the bright rays of craters Copernicus and Kepler (middle left). The dark circular Mare Serenitatis lies to the east (right) of Imbrium. (Lick Observatory photograph.)

Ocean tides have profoundly influenced life on Earth; for some animals they provide a key step in the reproduction process. Ironically, the Moon itself is a barren world that has remained almost unaltered for the past 3 billion years – except for an occasional impact and the formation of a new crater. It preserves a record of the early days in our neighborhood and gives mute evidence for a rain of planetesimals that may have struck the Earth as well. When the astronauts visited the Moon during the 1970s they hoped to find the key to the Moon's origin. Their voyages provided new perspectives, and this is the story told in this chapter.

2.1 The face of the Moon

(a) The view from a distance

The only features visible on the face of the Moon with the naked eye are the large, irregular areas of light and dark material (Fig. 2.1). Although it appears bright in contrast to the night sky, the Moon's face is darker than most rocks on Earth, as can be seen in a photograph of the landscape if the Moon is in the sky (Fig. 2.2).

Because the Moon's rotation is synchronized with its orbital motion, Earth-bound observers always see the same hemisphere of the Moon. We call this the near side of the Moon, in contrast to the invisible far side. You can demonstrate synchronous rotation by holding a ball at arm's length and slowly turning around. As your body completes one rotation you always see the same side of the ball, but the ball has completed one rotation while revolving once about your body.

When Galileo Galilei turned his primitive telescope on the Moon in 1609, he discovered that the dark patches were smooth and level, resembling seas (see Fig. 1.20). He called them *maria*, the Latin word for seas (*mare*, pronounced "MAHrey," is the singular for sea), although today we know there is no water on the Moon, and there apparently never has been. When he looked closely at the division between light and shadow – day and night – on the globe of the Moon, Galileo discovered that the dividing line was ragged and that he could see high mountain peaks casting long pointed shadows. These mountains were mostly confined to the brighter regions he called *terrae* (lands), now known as the highlands, because they are higher than nearby maria (Fig. 2.3). This was the first clear evidence that the Moon was not the perfectly smooth crystalline sphere that had been proclaimed in the writings of Aristotle.

Galileo wrote that

the surface of the Moon is not perfectly smooth, free from inequalities and exactly spherical, as a large body of philosophers considers with regard to the Moon and other bodies, but on the contrary, it is full of inequalities, uneven, full of hollows and protuberances, just like the surface of the Earth itself, which is varied everywhere by lofty mountains and deep valleys.

Galileo was also a person of measurements, and he devised a simple geometrical way to estimate the heights of lunar mountains from the lengths of the shadows they cast on the Moon's surface. He derived mountain heights that are comparable to those on Earth, but because the Moon is much smaller than the Earth, it is, relatively speaking, a very mountainous world.

The most distinctive lunar features are the circular craters that closely

Figure 2.2. Moon and Half Dome. Ansel Adams' beautiful photograph implies that the Moon's face is darker than most rocks on Earth. Although the full Moon appears bright in contrast to the night sky, the lunar surface has a gray pallor and reflects only 7 percent of the incident sunlight. (Courtesy of the Ansel Adams Publishing Rights Trust.)

pepper the Moon's face (Fig. 2.4). These craters are just beyond the limit of naked eye visibility, but a pair of binoculars will reveal a few of the larger ones. And around the time of full Moon, a pair of binoculars will also show bright streaks that radiate from several craters like the spokes of a wheel. These are the lunar rays, and they were produced by the debris of crater-formation. Some of the rays go more than one-quarter of the way around the Moon (Fig. 2.5).

The lunar shadows are as dark as the night sky and they give our first clue that the Moon has no atmosphere. Shadows on Earth are illuminated by the light of the Earth's atmosphere – the sky light, which on a clear day provides about one tenth the light of the direct Sun. On the Moon, there is only star light in addition to the direct light from the Sun, so the shadows are darker.

We can also find other evidence for the lack of an atmosphere on the Moon. The earliest was the abrupt vanishing of stars behind the edge of the Moon during lunar "occultations." (To *occult* means to hide.) It is in sharp contrast to the gradual fading of the Sun, stars, and planets when they set behind the horizon on Earth. If the Moon had an atmosphere, the starlight would gradually dim during a lunar occultation, but this vanishing actually takes less than one second. The Moon has insufficient gravity to retain an atmosphere.

Because there is no atmosphere on the Moon, there are no colorful sunsets, the sky is pitch black in broad daylight, no sounds relieve the eternal silence, and there are no clouds or weather on the Moon.

Figure 2.3. Smooth maria and rough highlands. This astronaut's view of the western edge of Mare Serenitatis (Sea of Serenity) shows the boundary between the smooth, lava-filled mare plain (right) and the rugged, heavily-cratered highlands (left). Ridges on the mare surface, sometimes called wrinkle ridges, have formed parallel to the edge of Serenitatis basin. The basin was created at the same time as the highlands during an intensive bombardment 4.0 billion years ago. Lava subsequently flowed out from inside the Moon about 3.5 billion years ago, filling the basin and creating Mare Serenitatis. (Courtesy of NASA.)

*Figure 2.4. Craters
Copernicus and Reinhold.
Bright rays of ejecta radiate
outward from the crater
Copernicus near the lunar
horizon. It is one of the
youngest large craters on the
near side of the Moon with
an estimated age of 900
million years and a diameter
of 93 kilometers. The craters
in the foreground are
Reinhold A and B.
(Courtesy of NASA.)*

With even a small telescope, new types of detail can be seen on the Moon; narrow ranges of mountains marking the submerged rim of a huge crater, an occasional sunlit peak standing in the midst of a shadowed crater, long sinuous "rilles" resembling ditches a kilometer wide and hundreds of kilometers long, craters whose rims overlap and obscure others, light and dark craters, central peaks and multiple rims, clusters of secondary craters evidently produced by debris from a large impact. The highlands are so thoroughly saturated with craters that no flat regions remain; a new crater would only be noticed by the fresh brighter material it turns up (Fig. 2.6). On closer examination, even the maria can be seen to have small craters here and there.

(b) Origin of the lunar craters

Many of the craters have fairly flat floors accented by a central peak. These peaks led to the idea that the craters were formed by volcanoes, but the resemblance to volcanoes is weak (see Focus 2A, Lunar craters – volcanoes or cosmic bombs?). The lunar craters were formed by explosions when meteoroids struck the Moon. (The name meteoroid applies to solid inter-planetary particles that range in size from asteroids, which can be hundreds of kilometers in diameter, to dust particles. When these projectiles strike the surface of a planet or satellite, they are called meteorites.)

The rays that radiate from some of the lunar craters are further evidence that the Moon has no atmosphere. Although they were once thought to be impact cracks in the Moon's surface made bright by escaping gases, they are now known to consist of strewn dust, sand-like particles, pebbles, and

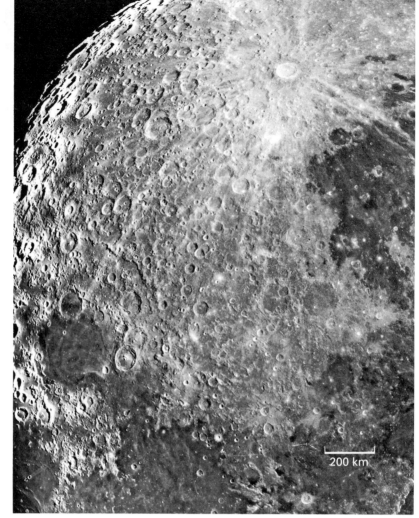

Figure 2.5. Lunar rays. White rays splash out across the Moon from crater Tycho at the upper right. Tycho is a large, young crater with a diameter of 85 kilometers and an age of 107 million years. Only relatively recent craters retain their white rays, for those of older craters are darkened and worn away by continued meteorite impact. The dark, flat circular feature in the lower left is Mare Nectaris (Sea of Nectar). This clear image was produced using the unsharp masking technique that permits high contrast and fine resolution. (Anglo-Australian Telescope © 1976. Photo prepared by David F. Malin.)

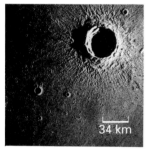

Figure 2.6. Timocharis. The medium-sized lunar crater Timocharis is a comparatively young crater that still exhibits the fresh details of impact. The deposits and ejecta have been thrown radially outwards by the impact that created the primary crater with its circular rim (34 kilometers in diameter). The smaller secondary craters become prominent beyond the radial ejecta (lower left). These small craters, which are often grouped in clusters or chains, were formed by rocks thrown out of the primary crater. (Courtesy of NASA.)

rocks ejected by the explosive formation of the central crater. Their long flights around the Moon would have been impossible in the presence of a resisting atmosphere. The smaller particles would have burned; the inter-mediate-sized particles would have fallen much closer to the source, and only the rare boulders would have flown up to about a thousand kilometers or so. A sizeable fraction of the smaller particles may even have flown off into space; particles moving with a speed greater than 2.38 kilometers per second can escape the Moon's gravity field.

Seen from above, the Moon's craters are round, despite the fact that the projectiles that produced them must have fallen in a variety of directions – some nearly vertically, others near a glancing angle. The large craters are shallow while the smaller craters have steep inner walls and bowl-shaped interiors. Further, the amount of material scooped out to make the interior approximately equals the amount in the rim, so if the rim were pushed back the crater floor would rise to the level of the neighboring surface. The simplest and most convincing explanation of these facts is that the craters were formed by explosions; they are not produced by bombs in the usual sense, but by the impact of meteoroids (Fig. 2.7). Let us try to imagine the process.

Focus 2A Lunar craters – volcanoes or bombs?

Early interpretations of the lunar craters suggested they were formed by volcanic activity. Two lines of evidence seemed to support this idea. First, almost all the craters are round, and in this respect they resemble volcanic craters on Earth. Second, for more than two centuries reputable astronomers had reported seeing activity resembling volcanic eruptions, especially near Aristarchus, the brightest feature on the Moon, and the crater Alphonsus, that is pictured here.

Gradually, the evidence began to favor the idea that the craters are formed by meteoritic impact. In the first place, it was found that the floors of most craters are slightly depressed below the surrounding level, in contrast to Earthly volcanic craters. Second, the craters were seen to resemble impact craters, because the amount of material piled in the rims is nearly equal to the material excavated from the interior; and because large impact craters are created by explosive forces and are quite round regardless of the direction of flight of the projectile. The overlapping of the rims of many craters was difficult to understand on the volcanic theory, and the pattern of lunar rays streaming out of the younger craters also seemed to suggest impacts.

But the debate was not ended until rocks were returned from the Moon. Samples from the highland craters and larger basins are conglomerates. That is, they consist of fragments of pre-existing rocks that have been welded together, as though by the enormous pressures that would be

119 km

produced by impact. The battered highland crust is a museum of impact scars created during an ancient bombardment.

Although the lunar maria were filled during ancient episodes of volcanism, the Moon has apparently been volcanically inactive for 3 billion years. The supposed outbursts reported by several astronomers were evidently optical illusions.

Yet the crater Alphonsus had one more surprise. On its floor are elongated craters with dark haloes. These are associated with fissures and might be volcanic, but they are more likely to be secondary impact craters.

Figure 2.7. Cross-sectional anatomy of a crater. An impacting meteorite excavates a circular crater that is almost 40 times the diameter of the meteorite. The depth of the crater is roughly one tenth its diameter, and the crater floor is depressed below the surrounding terrain. Although the meteorite vaporizes on impact, the explosion excavates material and hurls it outwards, creating a raised rim, radial ejecta and secondary craters. Large craters have central peaks created by the rebound of the underlying lunar surface.

Picture a meteoroid moving in an eccentric orbit about the Sun at a speed of, say, 30 to 40 kilometers per second. It is gravitationally attracted by the Earth and deflected from its orbit. But the Moon intercepts it and the rock crashes onto the lunar surface at a speed of 2 to 70 kilometers per second, depending on its direction of approach. (Projectiles from behind must catch up with the Moon, which is moving at about 30 kilometers per second about the Sun; so they will strike more gently.) The impacting rock is stopped; its energy of motion is suddenly transformed to heat, creating intense pressure and high temperature, generating shock waves that compress and shatter the nearby rock. The result is an explosion that excavates a crater. The impacting rock's energy of motion is sufficient to vaporize the rock and a portion of the Moon that is many times as massive as the meteoroid itself. A typical meteorite travelling with a speed of 25 kilometers per second has enough energy to melt nearly 1000 times its own mass or to vaporize 100 times its own mass. Moreover, common experience tells us that melting or vaporizing are not the only important effects of a colliding meteorite. If we strike a pebble or an ice cube with a hammer we melt or vaporize only a small portion of it; we crush most of it into small particles. Instead of tearing off individual molecules, the energy of motion is used to pull off clusters of molecules in the form of grains. These grains were initially surrounded by defects and weak layers in the crystal lattice, so their separation does not require much energy and they can remain whole. In this way the pulverizing process can liberate an even larger mass than melting or vaporizing.

It is difficult to say how much of the Moon will be pulverized or vaporized by the impacting meteorite, but experiments suggest that it can excavate a volume up to 10 000 times its own mass. The vapor will carry particles with it, and the force is outward from the point of impact; hence the circular craters.

The formation of large lunar craters can be visualized by imagining the effect of throwing a stone into gelatin. A large hole would be formed, but the strength of the gelatin would not be sufficient to maintain it as a hole; the floor would rebound, and a central peak might be formed.

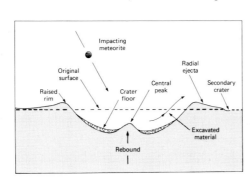

Table 2.1. *Large impact basins and the lunar maria*

Maria (Latin)	Seas (English)	Basin diameter (kilometers)
Oceanus Procellarum	Ocean of Storms	3200
Mare Imbrium	Sea of Rains	1500
Mare Crisium	Sea of Crises	1060
Mare Orientale	Eastern Sea	930
Mare Serenitatis	Sea of Serenity	880
Mare Nectaris	Sea of Nectar	860
Mare Smythii	Smyth's Sea	840
Mare Humorum	Sea of Moisture	820
Mare Tranquillitatis	Sea of Tranquility	775
Mare Nubium	Sea of Clouds	690
Mare Fecunditatis	Sea of Fertility	690

How large the crater will be is illustrated in Figure 2.8. This shows that the depths of lunar craters are about a tenth of their diameters and, for simplicity, we assume that the crater has the shape of a shallow cylinder. The radius of the crater will then be about 40 times that of the meteoroid. For instance, a crater with a radius of 10 kilometers will be formed by a meteoroid with a radius of 0.25 kilometers.

The largest craters are impact basins (see Table 2.1), and a typical one is the Imbrium Basin, which is prominent on the near side of the Moon. Its outline can be seen with the naked eye, forming an "eyesocket" of the face of the "Man in the Moon," and it has a diameter of 1500 kilometers. This region may have been blown out by an impacting meteorite about 40 kilometers in diameter. Basins this size were created early in the Moon's history and they have since been flooded and nearly filled with molten lava from the interior.

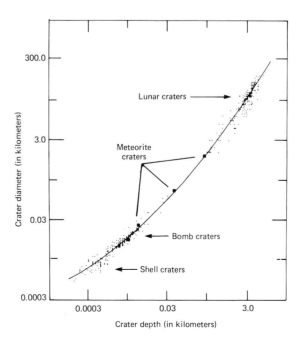

Figure 2.8. Crater depth and diameter. The continuous sequence in the relationship between the depths and diameters of man-made explosions and lunar craters. This plot provides evidence for the explosive origin of lunar craters. Craters formed by explosion upon impact will have a circular form regardless of the angle of impact. The gap between man-made explosion craters and the lunar craters is filled by terrestrial meteorite craters. (Adapted from Ralph B. Baldwin's "The Face of the Moon", University of Chicago Press, 1949 – page 132.)

The projectile that created the Imbrium Basin was almost powerful enough to break the Moon into pieces. It gouged out radial ridges and valleys that went a quarter of the way around the Moon, and it scattered a thick blanket of ejecta, now called the Fra Mauro Formation, over most of the near side of the Moon.

For all craters larger than about 120 kilometers, the energy of impact was so great that the floor of the crater rebounded, surging up and down, creating multiple rings as the excavated material that formed the rim was repeatedly jostled down into the crater. After being struck, the lunar surface vibrates like the head of a drum. An example is the Imbrium Basin. Its outer rim is defined by prominent mountain ranges including the Apennines to the east (right), the Alps above and the Carpathians below (see Fig. 2.9). Two inner rings are outlined by circular mare ridges and isolated mountain peaks, indicating that the mountainous rings formed by the explosion have been submerged in the lava that filled the Imbrium Basin, thus forming Mare Imbrium. In much the same manner, many of the conspicuous mountain ranges were formed long ago as rims, or inner rings, of impact basins that later filled with lava. There is no evidence for recent

Figure 2.9. The Apennine Mountains. Mare Imbrium (upper left) is rimmed by the Apennine Mountains (lower right). The radial structure and steep inner slopes of these mountains provided the first clue to the impact origin of the Imbrium Basin. The largest crater in this photograph is named Archimedes; it has a diameter of 83 kilometers. (Lick Observatory Photograph.)

83 km

mountain building, in contrast with the Earth. The fact that the lunar mountains were nearly all formed from impact debris probably accounts for their relatively low elevations, rarely more than 2.5 kilometers, despite the low gravity on the Moon which should make it possible to support much higher mountains if they had formed.

2.2 The Moon close up

(a) Voyage to the Moon

The space-age began October 4, 1957, when the Soviet Union launched its first artificial Earth satellite, named Sputnik 1. Two years later, the Russians sent their Luna 3 spacecraft around the Moon to take the first pictures of the normally invisible far side. (See Focus 2B, The far side of the Moon.) When the pictures were returned to Earth they showed that the far side is quite different from the near side and has few maria.

Before manned landings were accomplished by the United States, three types of robot spacecraft were sent to reconnoiter and answer two main questions for the proposed lunar landing. The first concerned the danger of encountering rocky terrain, where it would be impossible to land without capsizing. The second was the prediction, by some astronomers, that the lunar surface was covered with a layer of dry dust, perhaps a kilometer thick, that would make travel impossible. In fact, the astronauts might sink into the dust and vanish! The lunar surface was known to have been battered, churned and worn down by a hail of meteorites over the past 4 billion years. This bombardment had presumably smashed the surface into a loose debris of rocks, pebbles, grains, soil and dust, called the lunar regolith, that overlies the solid bedrock and forms the visible surface of the Moon. Tiny meteoroids, which are much more numerous than the larger ones that produced the visible craters, might have "sandblasted" the rocks and produced thick dust layers. (See Fig. 2.10 and Fig. 2.11.)

To start resolving these uncertainties, three Ranger spacecraft crashed onto the Moon, transmitting television pictures back to Earth as they rapidly approached the lunar surface. (Watching these pictures was a dizzying experience and the transmission of the final frames was interrupted by the crash itself.) These were followed by five Lunar Orbiters that

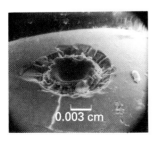

Figure 2.10. Micrometeorite crater. A high-velocity impact pit, or microcrater, on a glass sphere returned from the Moon. The central pit is 30 microns (0.003 centimeters) in diameter. These microcraters are formed by the impact of dust-sized particles travelling with a speed of several kilometers a second. (Courtesy of NASA.)

photographed most of the Moon's surface, missing only the polar regions, from a distance of several hundred kilometers, in search of suitable landing sites. In the process, they discovered large concentrations of mass just below the lunar surface. (See Focus 2C, Mascons.) The final stage of preparation involved six soft landings by the Soviet Luna 9 and five American Surveyors. While the ground-control crews watched anxiously, the feet of the spidery three-legged robots sank only a few centimeters into the lunar soil, showing that there was no thick dust layer and people could, indeed, walk on the Moon without sinking in over their head.

The Apollo astronauts first orbited the Moon in December 1968 (Apollo 8) and landed in July 1969 (Apollo 11). The actual landing was performed in the bug-like Lunar Module that separated from the main spacecraft while in orbit about the Moon. Michael Collins remained in orbit, while Neil Armstrong and Edwin Aldrin made the descent and landing. Then, while an estimated half billion people watched, Armstrong groped cautiously down the ladder to the lunar surface. He stood firmly on the fine-grained surface and an ancient dream had come true – man had set foot on another world. (See Focus 2D, Man on the Moon.) As Armstrong put it: "That's one small step for a man, one giant leap for mankind." The next day, the Italian newspapers put it more succinctly: "Fantastico!"

The astronauts' cameras recorded an eerie wasteland, scarred with

Focus 2B The far side of the Moon

In October, 1959, two years after launching the first artificial Earth satellite, the Russians launched the Luna 3 spaceship to the vicinity of the Moon. This opened a new era of the space-age, an era of remote sensing and direct exploration of the Moon.

Luna 3 swung once around the far side of the Moon and photographed its surface. When it returned to the vicinity of the Earth, a small television camera scanned the photographs and transmitted them to a receiver on the ground. The pictures (left) lacked fine detail but they showed that the far side of the Moon is strikingly deficient in maria. One mare, 445 kilometers in diameter (labelled I in the picture) was found, and it has been named the Sea of Moscow. The other dark maria that appear in the lower left of the picture are maria that appear on the Moon's eastern limb as seen from Earth, including Mare Crisium (II), Mare Marginis (III), and Mare Smythii (IV).

American astronauts surveyed the far side in more detail (right) using cameras aboard the Apollo 16 spacecraft. They confirmed that the far side is made up almost exclusively of light-colored and heavily cratered highlands. There is a giant impact basin that rivals the near-side Imbrium Basin, but curiously the far-side basins contain almost none of the dark lava that flooded the near-side basins.

The reason for this striking contrast between the near side and far side of the Moon is reasonably well understood. It is related to the thicker crust on the far side, which might prevent the emergence of lava from the interior. And, if the Moon has been in synchronous rotation for 4 billion years, keeping the same face toward the Earth, the gravitational action of the Earth may have focused meteorites toward the near side of the Moon.

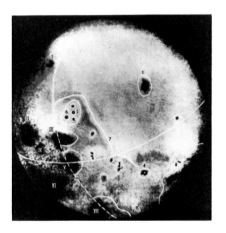

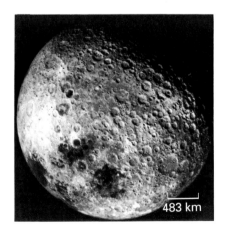

483 km

Focus 2C Mascons

Precise radio tracking of spacecraft on the near side of the Moon showed that their orbits are gravitationally deflected toward the circular maria. The spacecraft behaved as though the maria contained mass concentrations, hence the name mascon. Virtually all the circular maria on the near side showed this unexpected feature, and the excess mass in each is about 10^{21} grams, or $1/100\,000$ the total mass of the Moon.

What are the mascons? The most likely explanation is that they are regions where the low-density crust is thinner, leaving more room for the denser basalt of the mare and the upwelling of the high-density mantle during previous impacts. If the Moon had been entirely molten when the mascons were formed, they would have tended to sink beneath the surface because of their high density. This suggests the mascons were formed when the Moon had become fairly rigid between 4 billion years ago, when the impact basins were formed, and 3 billion years ago when the great lava flows stopped.

Because radio tracking of the orbiting spacecraft was not possible when they passed to the far side, it is not known whether there are mascons on the far side of the Moon.

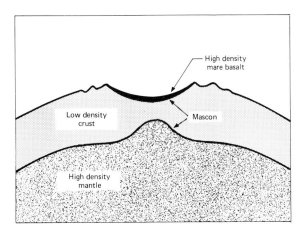

craters of all sizes and covered with dust. It clung to the astronauts' clothing and equipment. It showed the sharp outline of their bootprints, but there were no clouds of dust above the airless surface. Walking on the lunar soil was like walking on plowed earth or wet sand, and most of the finer dust had evidently been carried down into the regolith by the churning of the meteorites.

In all, there were six manned landings on the Moon, ending with Apollo 17 in December 1972. (See Fig. 2.12 and Fig. 2.13.) All the landing sites were on the near side (Fig. 2.14). The sites were chosen to provide samples of a wide variety of terrain. At first the astronauts travelled on foot (Fig. 2.15), then in roving vehicles with large wire-mesh tires (Fig. 2.16). Spacesuits shielded them from the brilliant Sun, and provided self-contained atmospheres to breathe, so the astronauts often stayed several days. Each time they blasted off, they left behind instruments to measure the flow of heat from the Moon's interior, as well as seismometers to monitor the

Figure 2.11. Light and dark soil. Aeons of weathering by micrometeorite impacts have darkened the lunar soil. The material that was excavated during the creation of these relatively young craters is lighter in color than the surrounding surface. The light-colored soil has been exposed to relatively little weathering, but it will eventually darken after billions of years of micrometeorite impacts. (Courtesy of NASA.)

vibrations of moonquakes and meteorite impacts. Mirrors, in the shape of the three-sided corners of a box, were also left behind to reflect laser light from Earth, permitting astronomers to bounce light off the Moon and measure its distance time and again, tracking its complicated motion with an accuracy of a few centimeters!

Returning to the orbiting craft, the astronauts jettisoned the landing module and headed for Earth, arriving home about three days later. Biologists felt there was a chance that the astronauts, or the returned samples, might infect the human race with some deadly lunar virus. The astronauts from the first three lunar landing missions were therefore placed in quarantine for three weeks after their return. They remained in fine health and the quarantine was not imposed on subsequent crews.

Focus 2D Man on the Moon

The stunning Russian accomplishments galvanized the American space program, and on May 27, 1961, President John F. Kennedy delivered his now-famous address to Congress, including the declaration: 'I believe that this nation should commit itself to achieving the goal, before the decade is out, of landing a man on the Moon and returning him safely to Earth."

Just over eight years later this goal was achieved for the first time. On June 24, 1969, Neil Armstrong became the first man to walk on the Moon. His footprint (shown here) reveals a thin layer of Moon dust, about a centimeter thick. Because there is no atmosphere, water or weather on the Moon, the footprint will probably remain visible for about 10 million years. By that time, micrometeorites will have erased it. Altogether, 12 astronauts have walked on the Moon.

14 cm

Figure 2.12. Light-painted Moonscape. A flash of sunlight accidentally leaked onto exposed film, painting the normally gray lunar landscape with red and gold. The rolling slopes in the background are the foothills of the Taurus Mountains. (Courtesy of NASA.)

Figure 2.13. Moon and Earth from space. During their final orbit of the Moon, the Apollo 17 astronauts watched the crescent Earth rise over the rolling terrain on the far side of the Moon. (Courtesy of NASA.)

(b) Rocks from the Moon

The astronauts brought back 382 kilograms, or almost half a ton, of rocks – and not an ounce of cheese (Fig. 2.17). All the rocks were formed by the cooling of molten lava. The maria rocks are similar to basalt, a dark, almost black rock that is fine grained and, on Earth, associated with lava flows that have solidified in the form of tall columns. The highland rocks are relatively rich in calcium and aluminum, while those of the maria are richer in magnesium, iron, and titanium.

After the lunar rocks solidified from molten lava, they were broken and flung about by meteorite impacts. Subsequent impacts then compacted and welded them into aggregates called breccias. These welded fragments are the most common type of rock on the Moon.

In their chemistry, the lunar rocks are quite similar to the rocks of Earth, but with an important difference. Compared with the Earth, the Moon is depleted in the volatile elements, those that vaporize easily, such as sodium and potassium. The Moon is also depleted in iron and elements that dissolve in molten iron, such as gold and nickel. This depletion suggests that the material of the Moon was at one time exposed to higher temperatures than the Earth. The Moon is relatively enriched in refractory elements that could have withstood higher temperatures.

There is no water on the Moon and, apparently, there never was any.

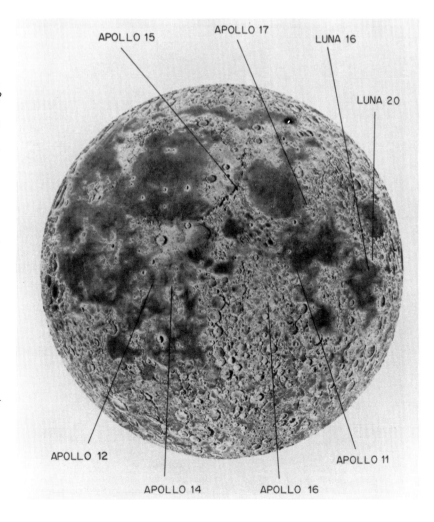

Figure 2.14. Landing sites. The Apollo landing sites were designed to obtain samples from a wide variety of terrain. Apollo 11 and 12 respectively landed on Mare Tranquillitatis and Oceanus Procellarum. The rocks returned from these missions showed that the maria were filled with lava between 3.0 and 3.4 billion years ago. The spot chosen for Apollo 14 was the Fra Mauro Formation. It is covered with material ejected during the impact that created the Imbrium Basin 4.0 billion years ago. By landing at a point just inside of the Apennine Mountains, the Apollo 15 astronauts could sample highlands, maria and the Hadley Rille. The Apollo 16 mission sampled the highlands near crater Descartes, while Apollo 17 landed in the highlands near Mare Serenitatis. The unmanned Luna 16 and 20 spacecraft returned small samples of highland rock and mare basalt from sites near or on Mare Fecunditatis.

The Moon rocks show no signs of ever having been exposed to water, and the rocks contain no moisture or hydrated minerals. The oxygen that on Earth forms rocks and water, forms only rocks on the Moon. There is also a scarcity of minerals on the Moon – only about 100 compared to the Earth's 2000. This is probably due to the lack of water and air, a lack which would also have prevented the formation of ore deposits of metals such as copper.

Rocks on the Moon have never been exposed to water or free oxygen, so contact with the Earth's atmosphere would alter their composition. They are kept in cabinets filled with a dry, oxygen-free atmosphere of nitrogen and are manipulated with long gloves sealed to the walls of the cabinets. When not under investigation, the rocks are kept in a massive steel-lined vault at the Lunar Receiving Laboratory of NASA's Johnson Space Center at Houston, Texas.

The samples appear to have contained no fossil life, no living organisms, and no organic material. The Moon is a desolate place barren of life.

(c) Of time and the Moon

In laboratories on Earth, the lunar rocks have not only revealed the chemistry of the Moon's surface, but its age as well. The method is known as radioactive dating, and it works this way. Certain types of nuclei, known as unstable parent isotopes, decay at a constant rate into stable lighter isotopes known as daughters. By measuring the amount of daughter

Figure 2.15. Moonwalk. An astronaut strolls past a rimless crater formed by the redistribution of material in the fine-grained regolith. Small impacting particles have sandblasted the lunar surface, producing smooth, undulating layers of fine dust and rounding the surfaces of lunar rocks. Larger meteorites pound and churn the surface, producing a layer of ground-up rocky debris. (Courtesy of NASA.)

Figure 2.16. Lunar rover. The limited moonwalks on the flat maria during the Apollo 11 and 12 missions were succeeded during subsequent missions by long drives in a lunar rover across more rugged terrain. This lunar rover stands at the foot of the Apennine Mountains. Mount Hadley, the highest peak of this chain of mountains, is in the background. It is almost five kilometers high. The rover remained on the Moon. Free from wind, rain and rust, it might last for millions of years, but micrometeorites will continue to pit its surface. (Courtesy of NASA.)

Table 2.2. *Radioactive isotopes used for dating*

Parent isotope	Daughter isotope	Half-life (years)
Rubidium 87	Strontium 87	49 billion
Rhenium 187	Osmium 187	43 billion
Lutetium 176	Halfnium 176	35 billion
Thorium 232	Lead 208	14 billion
Samarium 147	Neodymium 143	11 billion
Uranium 238	Lead 206	4.5 billion
Potassium 40	Argon 40	1.25 billion
Uranium 235	Lead 207	704 million
Plutonium 244	Xenon 132, 134, 136	82 million
Iodine 129	Xenon 129	17 million

material and knowing the rate of decay, the age of the rock can be estimated. (See Table 2.2)

This method is something like determining how long a log has been burning by measuring the amount of ash and watching a while to determine how rapidly the ash is being produced. Of course, this method assumes that the ash is not being blown away by the wind, otherwise the estimated age would be too small. And similar restrictions apply to the dating of rocks. The daughter isotopes must be trapped in the rock and not escape or the estimated age will be too short. In fact, the daughters, being relatively small atoms, can escape quite easily when the rock is molten; only when it cools and solidifies do the daughters start to accumulate. For this reason, the ages

Figure 2.17. Sampling rock. An Apollo 17 astronaut collects samples from a boulder believed to have originated high up on the North Massif of this landing site. The huge rock rolled down about a billion years ago, splitting into five pieces during the fall. The total length of the boulder, when reassembled, is about 20 meters. The South Massif can be seen on the other side of the valley. (Courtesy of NASA.)

2 meters

determined for the rocks are really the times since the rock became solid. And if the rock is remelted, say by the impact of a meteorite, the radioactive clock is reset, and the age will measure the time since the last solidification.

When these techniques are applied to meteorites that have fallen to Earth, an age of about 4.6 billion years is obtained, and this is generally taken to be the time of formation of the planetary system. Almost all the highland rocks returned from the Moon give ages that range around 4.0 billion years; a very few of the rocks record the Moon's actual formation 4.4 billion years ago.

So the chronology of the Moon probably began about 4.4 billion years ago, when the solar nebula gathered into the objects of the planetary system. The Moon may have accumulated from smaller objects, the planetesimals, but its origin and very early history are still lost in the mists of time. We will step around this gap for a moment, returning at the end of the chapter after discussing new evidence on the Moon's interior and the effect of tides.

2.3 Inside the Moon

(a) Moonquakes

As the Earth has earthquakes, so the Moon has moonquakes which were first detected by the sensitive seismometers placed by the Apollo astronauts at four widely spaced locations on the lunar surface. Because the Moon is not shaken by winds, ocean tides and road traffic, the lunar seismometers can show moonquakes that are much smaller than any known earthquakes. The moonquakes never exceed a magnitude of 2 on the Richter scale. If you stood directly over the strongest moonquake you would not even feel your feet shake. (The magnitude of an earthquake is measured on a logarithmic Richter scale, which is based on the amplitude of seismic waves recorded by seismometers. Earthquakes with Richter magnitude 1 to 3 are recorded but not felt by humans. Damage to buildings begins at magnitude 5 and increases to total destruction in the largest earthquakes, of magnitude 8.)

The moonquakes are not only gentler than earthquakes, they also have distinctly different behavior. While tremors on the Earth start suddenly and persist for only a few minutes, the moonquake waves build up gradually and continue for more than an hour. Evidently the body of the Moon is an almost perfect medium for the propagation of seismic waves.

Some moonquakes are produced by meteoroids striking the surface, and their timing is quite irregular, but many of the deeper moonquakes repeat themselves at semi-monthly (14 day) intervals. These recur at the time of month when the Earth–Moon distance is smallest (perigee) or greatest (apogee). This pattern suggests that they are triggered by tidal stresses as the Moon swoops toward and away from the Earth in its elliptical orbit. Probably the body of the Moon is adjusted to the average tidal force, and when the force is greater or smaller, the internal stress of the Moon causes its rocks to slip against each other, producing moonquakes.

Not only are their schedules fixed, some of their locations repeat as well. The origins of the deep moonquakes lie 800 to 1000 kilometers below the surface, much deeper than earthquakes (usually about 100 kilometers), and there are certain locations that reappear again and again in the seismometer records.

(b) The internal structure of the Moon

Such records have revealed the travel-speed of seismic waves inside the Moon and permitted the building of a model for the lunar interior, in much the way that geologists have modelled the Earth. (See Chapter 5.) There are two types of seismic tremors: longitudinal waves, that resemble ordinary sound waves and can be propagated in both solid and liquid rock; and transverse waves, that correspond to the sideways jiggling that can be seen in a bowl of gelatin. Transverse waves require stiffness and cannot propagate through a liquid. These two types can be distinguished on the Moon because they move the seismometers in different ways, and the comparative behavior of the longitudinal and transverse waves reveals regions of molten rock.

According to the current model, the Moon has a crust of light material some 60 to 100 kilometers thick, and under this is a 1000 kilometer thick lithosphere of denser rock that is cold and solid. (See Fig. 2.18.) Beneath this, the Moon becomes warmer and to a depth of about 1400 kilometers there may be a zone of molten rock. Curiously, the moonquakes appear to be generated near the boundary between the solid and the molten rock. The Moon may also have a metallic core, like the Earth, but there is not a sizeable core of dense iron comparable to the Earth's. This lack of core is consistent with the fact that the Moon has no global magnetic field. (See Focus 2E, The magnetized Moon.)

(c) Formation of the highlands and maria

The composition and great age of the highland rocks suggests that they formed when the Moon was very young and its outer layers were molten – forming a magma ocean. Less dense material floated toward the surface, and denser material sank to the interior. The differentiated magma ocean then began to cool about 4.4 billion years ago, forming a thin, low-density crust. Portions of this crust are today's highlands, rich in light elements such as calcium and aluminum.

During the next phase (lasting until 3.9 billion years ago), the crust was bombarded with left-over planetesimals and rocks. An intense bombardment created the large impact basins and most of the highland craters by about 4.0 billion years ago.

Gradually the hail of planetesimals decreased, and the outer layers continued to cool. (See Fig. 2.19.) The radioactive decay of long-lived unstable elements, such as uranium and thorium, gently produced heat that gradually warmed up the interior. There followed an era of volcanism, from 4.2 to 3.1 billion years ago, as molten basaltic lava welled up from the interior, flooding the great impact basins and producing the dark circular maria that can be seen today. (See Fig. 2.20.) The liquid lava spread quickly away from its vents, flowing out in thin sheets and covering its sources rather than piling up, so no volcanic mountains were created (Fig. 2.21). Successive lava flows left their marks in some of the maria,

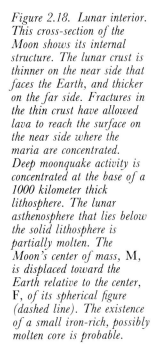

Figure 2.18. Lunar interior. This cross-section of the Moon shows its internal structure. The lunar crust is thinner on the near side that faces the Earth, and thicker on the far side. Fractures in the thin crust have allowed lava to reach the surface on the near side where the maria are concentrated. Deep moonquake activity is concentrated at the base of a 1000 kilometer thick lithosphere. The lunar asthenosphere that lies below the solid lithosphere is partially molten. The Moon's center of mass, M, is displaced toward the Earth relative to the center, F, of its spherical figure (dashed line). The existence of a small iron-rich, possibly molten core is probable.

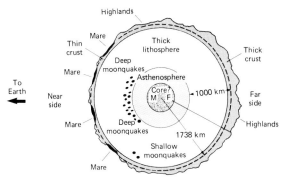

showing that the maria were not formed in a single quick pulse of volcanism. We know this from the great spread of ages for the lunar basalts – over one billion years.

Focus 2E The magnetized Moon

Measurements from spacecraft orbiting 100 kilometers above the surface of the Moon indicate that large blocks of lunar crust are magnetized. These localized magnetic fields are as broad as 100 kilometers, but they do not combine into an overall global pattern. The Moon has no north and south magnetic pole, and the Moon's global magnetic field is less than a millionth (10^{-6}) that of the Earth.

The lunar rocks returned to Earth also show signs of remnant magnetism that has been preserved since they solidified from molten material more than 3 billion years ago. (Their magnetic fields are about 0.003 gauss. The Earth's field is 100 times stronger, 0.35 gauss, at the surface of the equator.) This lunar magnetism is a fossil of an ancient magnetic field that may have been as strong as the present-day field of the Earth.

The source of this ancient field is not well understood, but there are several possible explanations. The Moon may have had its own field, generated in a molten core, as the Earth generates its own field. As the Moon's core cooled and the rotation of the Moon decelerated, the magnetic field would have gradually become weaker, eventually reaching its present low level. On the other hand, the ancient field may have been swept in from the solar nebula during the early stages of formation. Finally, the Moon may have been magnetized by the Earth, at a time when the two bodies were much closer together.

This is one of the remaining mysteries yet to be solved by the space-age exploration of the Moon.

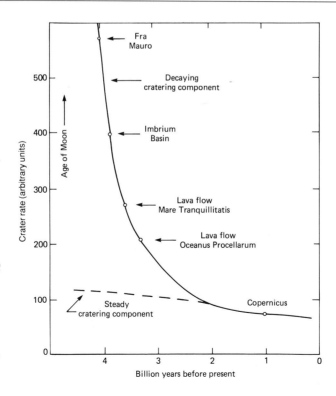

Figure 2.19. Varying crater rate. The cratering rate on the lunar surface is plotted against time. The arrows point to the crater rate and rock ages at various Apollo landing sites. The cratering rate was very high during an intense bombardment that occurred 4.0 billion years ago. The rate of cratering dropped rapidly during the subsequent billion years, giving way to the lower steady rate of crater production that has persisted for the last 3 billion years.

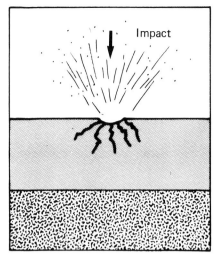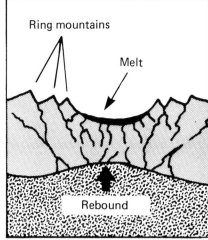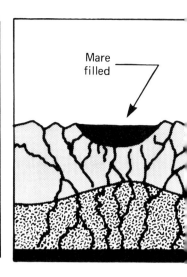

Figure 2.20. Mare formation. Disintegrating and vaporizing as it strikes, a meteorite blasts an impact basin out of the lunar surface (left), while the associated shock wave creates fractures in the rock beneath the basin. The blast hurls up mountain rings around the basin (middle), and the underlying rock adjusts to the loss of mass above it by rebounding upward. The uplifted mantle causes additional fractures in the rock, while a pool of shock-melted rock solidifies in the basin. All of the major impact basins on the Moon were created in this way between 3.9 and 4.3 billion years ago. Later, interior heat from radioactivity caused partial melting inside the Moon, and lava rose along the fractures, filling the basin layer by layer to form a mare (right). The lunar maria were filled in this way between 3.1 and 3.9 billion years ago.

know this from the great spread of ages for the lunar basalts – over one billion years.

The lava inundated all craters in its path, wiping the slate clean of previous impacts and preparing a fresh surface to record new impacts, which, by this time, had greatly diminished in intensity (Fig. 2.22). Thus the maria are relatively unscarred and most of their craters are small and relatively young.

Lunar volcanism continued for over 800 million years, as the outer zone of solid rock gradually became thicker and the lava worked its way from deeper and deeper in the Moon, eventually stopping altogether.

Finally the Moon became a desolate, quiet place altered only by the continued bombardment from space. (See Table 2.3). Except for the occasional large impacts that produced the rayed craters, and churning by small meteoroids, the Moon has remained essentially unchanged for the past 3 billion years.

In fact, almost all the rocks and soil currently at the surface of the Moon have remained within a few meters of the surface for hundreds of millions of years. During that time, the Alps and the Atlantic Ocean formed on Earth, but the lunar rocks remained unmoved.

2.4 Tides and the once and future Moon

Walking along the ocean beach some morning, we might notice that the waves seem to be reaching farther and farther up the sand. The tide is flooding the beach. A few hours later, it hesitates and then begins to ebb, retreating onto the flats where the clams may often be found. Every 12.5 hours, it returns, although not precisely to the same height.

The Sun and Moon both contribute to the formation of the tides, but the major portion of this rhythmic ebb and flood is driven by the Moon, whose tide is 2.2 times as high as the Sun's. For the moment, we will ignore the Sun, because its effect is similar to the Moon's, but smaller.

The Moon creates two high tides each day because the gravitational force of the Moon draws the ocean out into an ellipsoid, the shape of an egg that is the same size at both ends. As the Earth's rotation carries the continents past these tidal humps, we experience the rise and fall of water. In mid-ocean the tide is only 10 to 30 centimeters in height and goes unnoticed. But, where it is blocked by a shore it often runs 2 or 3 meters, and in some

Table 2.3. *Key events in the history of the Moon*

Feature created	Time (billions of years ago)	Process
Moon	About 4.5	Accretion
Magma oceans	4.4	Accretion and melting
Crust	4.4 to 4.2	Differentiation and cooling
Highlands, impact basins	4.2 to 3.9	Intense bombardment
Maria	4.2 to 3.1	Volcanism
Regolith, smooth surface	3.1 to present	Weak bombardment

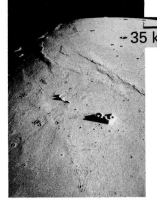

~ 35 km

Figure 2.21. Imbrium lava flows. Lunar volcanism is seen frozen in place on Mare Imbrium when it is illuminated under low Sun angle (sunlight near the horizon). The highly viscous lava flowed in the northeast direction (upper right) for distances of up to 600 kilometers, but the flow fronts only average 30 meters in height. The great length of these flows suggests a high rate of lava eruption. Yet, there is no sign of explosive activity and no volcanic cone or vent can be found at the place from which the lava issued. The lava was so fluid that it flowed rapidly out from the lunar interior, moving away from the fissures through which it was vented and spreading out into thin extensive sheets rather than piling up to form volcanoes. (Courtesy of NASA.)

large bays it can be 10 to 20 meters high, as in the Bay of Fundy in Nova Scotia.

Earthquakes and volcanoes can produce enormous waves that spread across an ocean and pile up on the shore; these waves are often called "tidal waves" although they have nothing to do with tides. Their proper name is *tsunami*, a Japanese word for large harbor wave.

(a) The pattern of the tides

The gravitational force of the Moon decreases with distance, so the Moon pulls hardest on the ocean facing it, and least on the opposite ocean; the Earth between is pulled with an intermediate force. As a result, the water directly beneath the Moon is pulled up away from the Earth's center, and the Earth's center, in turn, is pulled away from the water on the opposite side, causing another high tide. Thus the differences in the gravitational attraction of the Moon on opposite sides of the Earth produce two tidal bulges – one facing the Moon and one facing away. (See Fig. 2.23.)

In the course of a month, the changing alignment of the Sun and Moon causes the tides produced by these two bodies to alternately reinforce and interfere, leading to the cycle of spring tides and neap tides. The spring tide occurs near new and full moons, when the Sun and Moon reinforce each other's tides, and the neap tide occurs near first and third quarter, when they interfere with each other. The spring tides are usually 2 or 3 times as high as the neap tides (Fig. 2.24).

The time of high tide varies from place to place. When the flood tide moves in from the ocean, it may have to work its way among islands and peninsulas and along channels; this twisted path will delay the arrival by amounts that vary with location and with the time of the month. This time delay is called the "establishment" of the port, and the result is that the high tide rarely occurs when the Moon is overhead, usually an hour or two later and occasionally more. Similar delays can be noted in tidal pools. They begin filling when the tide reaches them, are filling at their greatest rate at the time of high tide, and do not begin emptying until the tide has pulled back. They are lowest long after low tide.

On a slowly rotating planet without continents, the tide would be highest along the line joining the centers of the Earth and Moon, that is, when the Moon is overhead. This is not the case for the Earth. The friction of the continents and the rapid rotation of the Earth carry the ocean's tidal bulge forward so it precedes the Earth–Moon line by about 3 degrees. This means that in the open ocean high tide actually occurs about 12 minutes after the Moon is overhead.

(b) The days are getting longer

Our planet meets resistance in its daily rotation. As the tides flood and ebb, they create eddies in the water, producing friction and dissipating energy at the expense of the Earth's rotation. The ocean water is heated ever so slightly by the motion of the tides and the Earth's rotation is slowed. The tides therefore act as brakes on the spinning Earth, slowing it by friction in much the way that the brakes of a car slow its wheels and become warm. The friction of tides dissipates energy at the rate of 5 billion horse-power

Figure 2.22. Ghost crater. An old ghost crater (bottom) has been filled and nearly obliterated by the lava that created Mare Imbrium. The younger crater Lambert (top) still exhibits the radial deposits and ejecta that were thrown out by the impact that created this crater. The photograph indicates that large, crater-forming impacts occurred before and after the lava flows that filled Mare Imbrium. (Courtesy of NASA.)

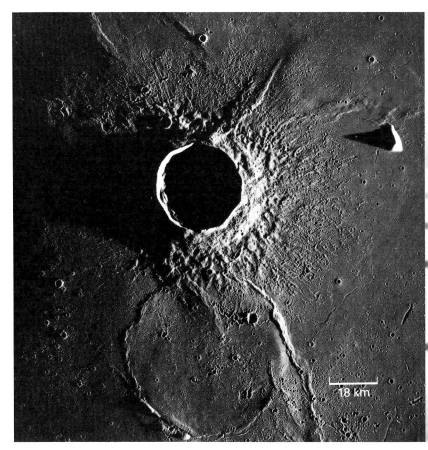

Figure 2.23. Cause of the tides. The Moon's gravitational attraction causes two tidal bulges in the Earth's ocean water, one on the closest side to the Moon and one on the farthest side. The closest bulge twists ahead of the Moon as the Earth rotates. This produces a lag in time between meridian transit of the Moon (when the Moon is directly overhead) and the highest tide. Because the tidal bulge nearest the Moon is twisted off the Earth–Moon line, it produces a force that tends to pull the Moon ahead in its orbit, causing the Moon to spiral slowly outward.

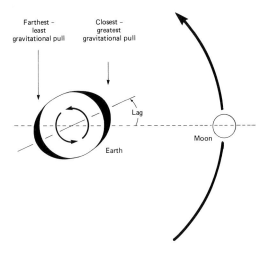

$(4 \times 10^{19}$ ergs per second). Tidal friction is slowing the rotation of the Earth, and the day is becoming longer at a rate of 2 milliseconds (0.002 seconds) per century, or one second every 50 000 years. Tomorrow will be 60 billionths of a second longer than today.

This lengthening of the day is consistent with the displaced paths of total eclipses observed in ancient times. If the current (slow) rotation rate is used to wind the Earth backward 2500 years to the time of an eclipse, the Earth would rotate about a quarter-turn too little, putting those predicted eclipse paths several thousand miles west of their actual locations. These paths would then conflict with the reported occurrences. (The full story is a bit more complicated, because the length of the month is also increasing and this reduces the discrepancy considerably.)

In this way, an understanding of the effect of the tides can help us understand the ancient astronomical records, and conversely those records also help us understand the effect of the tides. Aside from such historical determinations, this change of the Earth's rotation is imperceptible to humans, and it has not yet been measured directly. It is also mixed in with an erratic rate produced by the vagaries of the weather and the seasons, so all in all the Earth's rotation is no longer the best choice of clock. Astronomers now prefer to rely on atomic clocks due to their stability and continued accuracy. (See Focus 2F, Earth's erratic clock.)

Indirect historical measures of the Earth's rotation have been made by paleontologists through studies of fossil corals. The growth patterns of these corals consist of annual bands and fine daily ridges, produced by the effect of seasonal and daily changes of water temperature on the growth rate. The days were shorter in the past, but the year was the same, so the number of days per year increases as we go back in time. Ancient corals confirm this, and they show a greater number of daily ridges per annual band than modern corals. Careful counting reveals that the day was only 22 hours long when we look back 400 million years.

Studies of daily growth increments have recently been extended to the fossilized algae called stromatolites, which indicate that the day may have been only 10 hours long 2 billion years ago. However, any attempt to reconstruct the variations of the length of the day from geological evidence are made uncertain by the drastic changes that have occurred in the relative positions and sizes of the continents, their shelves, and the oceans.

(c) Earth's tidal influence on the Moon
The Moon pulls the Earth's oceans and the oceans pull back, in accord with Newton's third law that every action has an equal and opposite reaction.

Figure 2.24. Spring and neap tides. The height of the tides and the phase of the Moon depend on the relative positions of the Earth, Moon and Sun. When the tide-raising forces of the Sun and Moon are in the same direction, they reinforce each other, making the highest high tides and the lowest low tides. These spring tides (left) occur at new or full Moon. The range of tides is least when the Moon is at first or third quarter, and the tide-raising forces of the Sun and Moon are at right-angles to each other. The tidal forces are then in opposition, producing the lowest high tides and the highest low tides, or the neap tides (right). Of course, the total volume of ocean water remains the same during spring and neap tides. As illustrated in the drawing, the Moon's tide-raising force is about twice the Sun's. The heights of the tides have been greatly exaggerated in comparison to the size of the Earth.

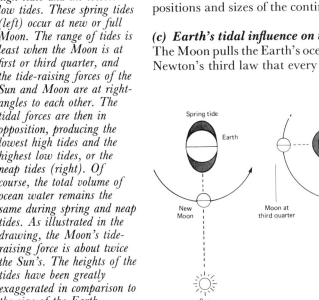

Focus 2F Earth's erratic clock

The Sun's and Moon's tidal action on the Earth causes the Earth's rotation to slow down. As a result, the day is steadily lengthening by 2 milliseconds (0.002 seconds) per century.

In addition, the rotation also runs fast and slow over intervals of months and decades. Until recently, astronomers used the rotation of the Earth as a clock, and this erratic behavior appeared as an irregular discrepancy between the observed and predicted positions of the Moon. This was not so surprising, but when these discrepancies also appeared in the behavior of the planets, astronomers knew the trouble was in their clocks. They needed to replace the old clock, and the first step was to redefine time in terms of the orbital motion of the Earth about the Sun. This was called "ephemeris time," but even that proved unsatisfactory for studying the motion of the Earth, so astronomers looked for a third method to keep time. The development of atomic clocks, based on the regular vibrations of atoms, was the answer, and today "atomic time" is used to study the motions of the Earth, Moon, and planets. The figure here compares the time kept by the Earth's rotation with atomic time.

The irregularity of the Earth's rotation has several causes. There are two types of seasonal fluctuations in the length of the day amounting to about one millisecond (0.001 second). The first type is a yearly variation, fast in the autumn and slow in the spring, that is produced by seasonal changes in the hemispheric wind patterns. The second type is a half-year variation in the length of the day probably produced by tidal distortion of the Earth. In addition, there is a slower variation, over an interval of decades, that may be generated by changes in the Earth's liquid core. This core rotates more slowly than the solid mantle, producing a westward drift of the geomagnetic field correlated with the decade variations in the length of the day.

So the Earth's clock has become more interesting for its irregularity than for its time-keeping ability.

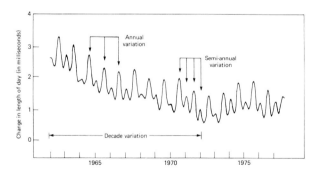

The net effect is to swing the Moon outward into a more leisurely orbit. Referring back to Figure 2.23, notice that the tidal bulge on the side facing the Moon is displaced ahead of the Moon. Ths bulge pulls the Moon forward; conversely, the bulge on the far side pulls the Moon backward. The forward pull is stronger because that bulge is closer to the Moon, so the Moon gains angular momentum from the Earth.

It is not hard to see that this will swing the Moon away from the Earth, if we look at the key equations. When we do the arithmetic, we find that the change of 0.002 seconds per century in the length of the day implies an outward motion of the Moon amounting to 4 centimeters per year. Small as it is, this value is just measureable with the laser reflectors planted on the Moon by the Apollo astronauts.

Because more angular momentum is transferred when the Moon is closest to the Earth (perigee) than when it is farthest (apogee), the orbit is also becoming more elongated; its eccentricity is increasing. The Moon's orbit was rounder in the past, and it was probably closer to the Earth.

Will the Moon's outward motion carry it away from the Earth altogether? Probably not, because there is not enough energy in the Earth–Moon system for these bodies to overcome their binding energy and go their separate ways. Only the intrusion of a massive third body could achieve that (or some fantastic project to attach enormous rockets to the Moon and launch it into space!). What will ultimately happen is the following. The Earth's day will become longer and will eventually catch up with the length of the month. When the day and the month are equal, the Moon-induced tides will cease moving; the oceans will rise and fall much more gently, under the influence of the Sun. The Moon will hang motionless in the sky and will be visible from only one hemisphere. (Perhaps travel agents will offer special tours to see the Moon from the other side of the Earth.) At that stage, the recession of the Moon will stop.

Then, billions of years from now, the Sun's tidal action will take over, slowing the Earth's rotation even further, until the day becomes longer than the month. At this point, angular momentum will be drawn from the Moon, and it will begin approaching the Earth, heading on a course of self-destruction until it is finally torn apart by the tidal action of the Earth. Perhaps it will form a ring around our planet. In any case, it will probably end its years where it apparently began – close to the Earth. By this time, however, the Sun will have expanded into a giant star, engulfing the Earth and Moon.

2.5 The Moon's uncertain parentage

The Apollo voyages permitted a more precise statement of what the theories for the Moon's origin must explain, and although the answer is still uncertain, a reasonably satisfactory theory has emerged.

(a) Constraints on theories of the Moon's origin

It has been known for a long time that the Moon is a relatively large companion compared to the satellites of other planets. Its orbit is peculiar in that it lies neither in the plane of the Earth's orbit nor the Earth's equator. The Earth–Moon pair also carries an anomalously large rotational momentum. Perhaps even more important is the low density of the Moon, which matches the outer layers of the Earth. The Moon cannot have the same overall composition as the Earth; for example, it cannot have a sizable core of iron.

Space-age explorations have led to additional constraints. (See Table 2.4) For example, radioactive dating indicates that some of the Moon's rocks were formed 4.4 billion years ago, and that it has been a chemically differentiated object for at least that length of time. The Moon must have formed as a discrete object at about the same time as the solar system.

There are also significant differences in the chemical composition of the Earth and Moon that provide further constraints, and perhaps some clues to the origin. The Moon is relatively depleted in volatile elements that would have boiled out into space if the lunar material were hotter at an early stage, leading to its relatively low density and enrichment in elements that could have withstood higher temperatures.

Table 2.4. *Constraints on models for the origin of the Moon*

Constraint	Implication
Mean density = 3.344 g/cm³	No large iron core
Secular acceleration of orbit	Moon once closer to Earth
Moon rocks 4.4 billion years old	Moon as old as Earth
Depletion of metals	Removal of iron prior to formation
Depletion of volatiles	Formation at high temperature
Enrichment of refractories	Condensation of protolunar cloud at high temperature
Oxygen isotope ratios	Earth and Moon formed in same region of solar system
Large rotational momentum	One massive glancing impact

Figure 2.25. Classical origin theories. According to the fission theory (top left), the rotational speed of the young Earth was great enough for its equatorial bulge to separate from the Earth and become the Moon. In the capture theory (top right), a vagabond moon-sized object once passed close enough to be captured by the Earth's gravitational embrace. We have pictured disruptive capture, with subsequent accretion, but the Moon could have been captured intact. The accretion theory (bottom) asserts that the Moon formed from a disk near the Earth.

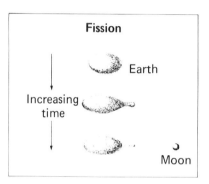

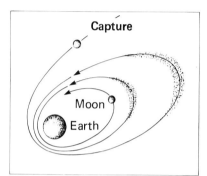

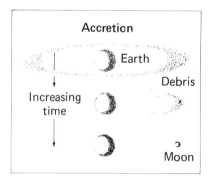

These are the broad facts that must be explained. Thus far, there are three classical theories (see Fig. 2.25). But, as Sherlock Holmes said in *The Adventure of Silver Blaze*, "I am afraid that whatever theory we state has very grave objections to it." A more recent theory, the giant impact theory, does not seem to be in a such a grave state, however.

(b) The fission theory
This theory is suggested by the fact that the Moon is now moving away from the Earth and the two must have been much closer together in the distant past when the Earth was spinning much faster.

The fission theory supposes that as the proto-Earth contracted, it spun up so fast that the centrifugal force pushed out the equator, distorting it

Figure 2.26. Modern fission theories. Simulations on a supercomputer (right) indicate that a molten, rapidly-spinning Earth would become unstable, developing long spiral arms of matter that would eventually wrap around the planet in a ring. The Moon might have formed from the densest part of the ring, while the rest of the ring scattered into space and carried away angular momentum. If the proto-Earth was nearly totally molten, its viscosity would have been negligible and fission might have occurred this way. However, when the dissipative effects of viscosity are taken into account (left), an initially distorted, rapidly-rotating Earth (a) evolves through (b) and (c), arriving at (d) one rotation period later. Because of the presence of dissipation, rotational instability is avoided and fission in the classical sense does not occur. (Courtesy of Richard H. Durisen, Indiana University (right) and Alan P. Boss, Carnegie Institution of Washington (left).)

(a)

(b)

(c)

(d)

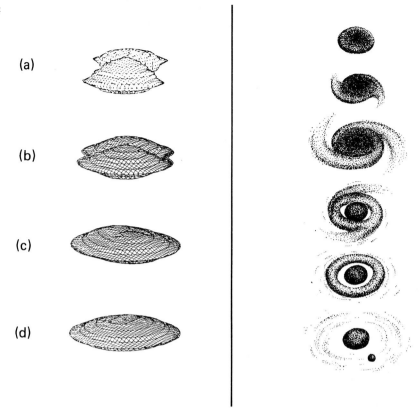

from a sphere to an elongated pear. Further spin up would elongate the "neck" and pinch off a great glob, forming the Moon. If this occurred after the Earth's iron had settled to the center, the Moon would naturally be depleted in iron and would chemically resemble the outer layers of the Earth. (It is conjectured that the settling of iron toward the center of the Earth might have caused an additional spin up, accelerating the formation of the Moon.) Once the Moon had separated, tidal friction caused it to move slowly away toward its present orbit.

This theory is particularly attractive because it seems to explain the low density of the Moon as well as the Moon's rather large mass. But it does this almost too well, because there are differences in chemical composition that are difficult to understand if the two bodies were taken from the same matter.

And there are two dynamical difficulties that seem almost fatal to the theory. First, the primordial Earth would have had to spin with a period less than 2.65 hours if it were to throw off the Moon. This seems unlikely to have been the case, because the angular momentum of the proto-Earth would have been 4 times the present total angular momentum of the Earth–Moon system. And, as angular momentum is very difficult to shed, this picture requires conjuring up a process to distribute this angular momentum to the rest of the solar system. With ingenuity, this can be done, but the fission theory loses its attractive simplicity when it is doctored this way. (See Fig. 2.26.) Second, there is the Moon's orbit plane, which is tilted from the equatorial plane of the Earth. If the Moon spun off the Earth's equator, the two planes ought to be coincident, but a later impact might have knocked the two planes out of alignment.

(c) The capture theory

If the Moon was not plucked out of the Earth, perhaps it formed elsewhere and was captured during a close approach to the Earth. The primary advantage of this capture theory is that it easily permits chemical differences between the Earth and Moon; the farther away the early Moon was, the more different it might have been. But even this is not sufficient as a theory, because the Moon appears to be chemically unique, unlike any other known satellite or planet.

Another severe obstacle to accepting the capture theory is the difficulty of understanding how the capture could have taken place at all. In order to produce an orbit about the Earth, some energy must be dissipated so the system will not exceed its binding energy. For this to occur, the Moon would have to be slowed down as it approached, and gravitational forces alone do not behave this way. Objects passing near the Earth without colliding will be thrown into new orbits about the Sun and will soon leave the Earth's neighborhood.

That is, an object hurtling near the Earth would be accelerated by the Earth's gravity, and it would gain enough speed to fly away again – provided it did not collide. It would behave like a toy car that gains enough speed rolling down one hill to carry it the same height up the next hill. As an example, except for an occasional direct hit, comets speed past the Sun and return to the outer solar system without being captured by the Sun's gravity into a planet-like orbit.

Advocates of the capture theory suggest that the Moon might have approached the Earth rather slowly, from a neighboring orbit, and the excess energy might have been dissipated by collisions with planetesimals around the Earth or by tidal friction. Others say that this is extremely unlikely.

(d) The accretion theory

The third classical theory suggests that the Moon was formed in the neighborhood of the Earth through a process not unlike the probable formation of the planets around the Sun. Like the planets, the Moon is imagined to have accumulated from a cloud of planetesimals, either at the same time as the formation of the Earth or immediately after, from the Earth's debris.

Such a theory seems to apply nicely to planets such as Jupiter that have families of satellites resembling the solar system. But where does that leave the planets that have no known moons, Mercury and Venus? And what about Mars, with only two minuscule companions? If the process that formed our massive Moon is the natural way of things, we have difficulty understanding these other systems. And why should the chemistry of the Earth and Moon be so different? How were the volatile elements driven out of the Moon if it always orbited near the Earth?

Special assumptions can help extricate the accretion theory from its difficulties. Volatile elements may have been distilled out of the Moon early in its history, and perhaps Mercury and Venus once had moons that have crashed into their planets under the influence of the solar tides. Perhaps a spillover from the belt of asteroids robbed Mars of its moons and left it with two calling cards. Even so, the lack of iron in the Moon is unexplained, and the accretion theory loses its appeal when such special assumptions are introduced.

(e) The giant impact theory

Many astronomers have recently embraced a giant impact theory that differs from the three classical theories (Fig. 2.27). According to this new theory, a Mars–sized body collided with the young proto-Earth in a glancing blow. The collision may have occurred 4.5 billion years ago, when a rocky crust was beginning to congeal around the partially molten, differentiated Earth. The energy of the collision crushed and vaporized the surfaces of the colliding objects. Surface material from both bodies squirted out, and some of it remained in orbit about the Earth forming a disc that collected into moonlets and finally formed the Moon we know.

If the Moon formed from material in the mantle of the impactor, this would explain the chemical difference between the Moon and Earth. For example, the Moon has little or no iron because the iron of the impactor had sunk to its core, out of reach of the collision. And the Moon lacks volatile elements because they would have boiled away and vaporized during the collision. Thus, the new theory seems to explain the facts with a minimum of assumptions. As one astronomer stated, "it requires no magic, no special pleading, no extra twiddling and no deus ex machina." One special assumption is required – that a Mars-sized object found its way to a collision with Earth. But, of course, the unusual Earth–Moon system calls for unusual explanations, and there is abundant evidence for massive collisions in the early history of the solar system.

All current theories have their strengths and weaknesses, although there seems to be a consensus that the impact theory is viable.

Exploration of the Moon has therefore pointed toward a solution of the

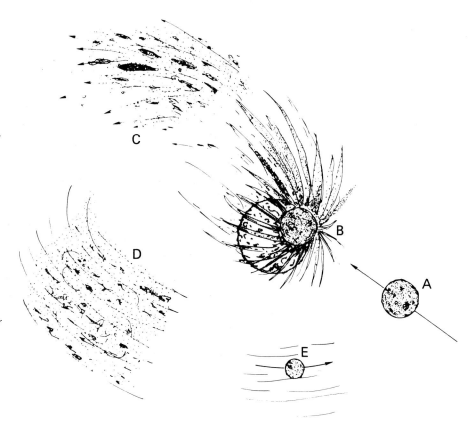

Figure 2.27. Giant impact theory. According to the giant impact hypothesis, a Mars-sized planetesimal (A) impacts with the newly formed proto-Earth (B) tangentially, resulting in a gigantic explosion and the jetting outward of both planetesimal and proto-Earth mass. Some fraction of this mass remains in Earth orbit (C), while the rest escapes Earth or impacts again on Earth's surface. A proto-Moon begins to form from the orbiting material (D), accreting neighboring matter and finally becomes the Moon (E). It may be mostly derived from the mantle of the impactor. (Courtesy of Alan P. Boss, Carnegie Institution of Washington.)

Focus 2G The Moon – summary

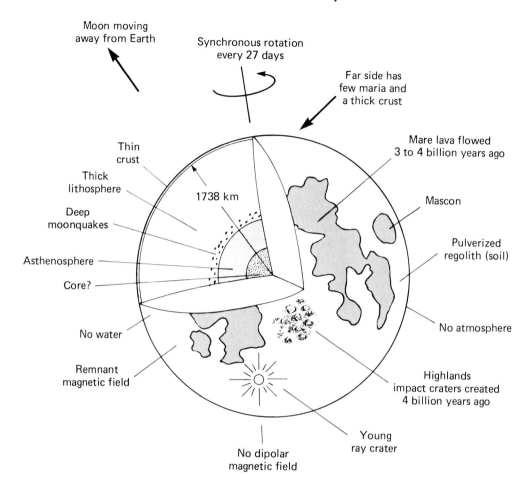

Mass: 7.353×10^{25} grams = 0.0123 M_{E} (Earth = 1)
Radius: 1738 kilometers = 0.2725 R_{E} (Earth = 1)
Mean density: 3.344 g/cm^3
Rotational period: 27.322 Earth days
Orbital period: 27.322 Earth days
Mean distance from Earth: 3.844×10^5 kilometers

mystery, although the record of the Moon's first 600 million years was wiped out by an intense bombardment 4.0 billion years ago. The voyage to the Moon also opened the path to explorations of the rest of the solar system. The Moon was the first port of call in our voyage to the planets and, perhaps, to the stars.

Mercury's southern hemisphere. The heavily cratered surface of Mercury closely resembles the lunar highlands. Both the Moon and Mercury lack any substantial atmosphere, so they have no surface erosion from winds and flowing water. They therefore preserve the record of an intense bombardment early in the history of the solar system.

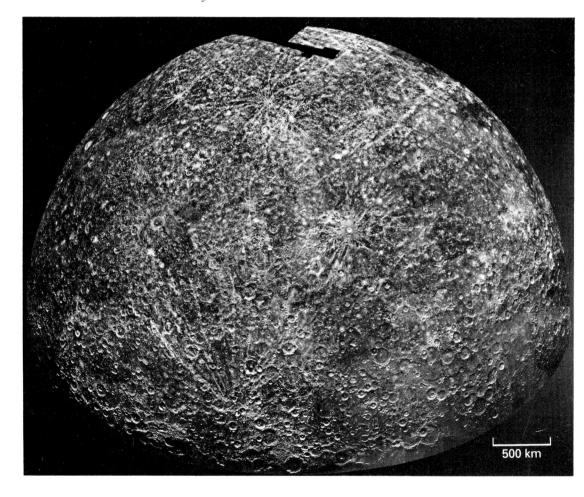

500 km

3 Mercury: a battered world

Mercury looks like the Moon on the outside, but it resembles the Earth on the inside.

This planet may have been blown apart by an ancient collision with a Moon-sized object.

3.1 Introduction

The planet Mercury, named for the wing-footed messenger of the gods in Roman mythology, is a tiny world that revolves closer to the Sun than any other known planet. It therefore has the shortest year – 88 days – and the highest orbital speed. Like a moth about a flame, Mercury races around the Sun at an average speed of 48 kilometers per second, pulled by the powerful solar gravitational field.

Little was known about this planet until it was explored by an unmanned spacecraft on an interplanetary fly-by mission, and by radar. These observations disproved several old ideas about Mercury, and in 1985, Earth-based observations revealed another surprise. These topics are the subject of this chapter.

3.2 A tiny world in the glare of sunlight

Mercury's small apparent size and its proximity to the Sun make it difficult to see from Earth. The planet is surpassed in brightness by only a few stars, but its orbit never reaches into the dark night sky. Hence, it is visible to the naked eye only in twilight when it is low in the sky and must be seen through a thick layer of air (Fig. 3.1). Astronomers have therefore taken to observing Mercury near mid-day, when it is far from the horizon and can be seen through a relatively thin layer of air. At such times, Mercury can only be

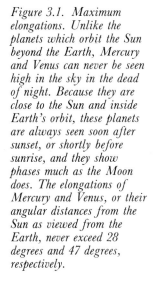

Figure 3.1. Maximum elongations. Unlike the planets which orbit the Sun beyond the Earth, Mercury and Venus can never be seen high in the sky in the dead of night. Because they are close to the Sun and inside Earth's orbit, these planets are always seen soon after sunset, or shortly before sunrise, and they show phases much as the Moon does. The elongations of Mercury and Venus, or their angular distances from the Sun as viewed from the Earth, never exceed 28 degrees and 47 degrees, respectively.

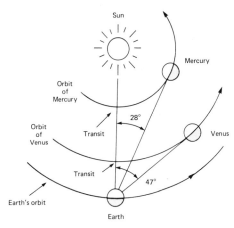

seen with telescopes, and most astronomers have not seen Mercury with the naked eye. Needless to say, these observations are difficult to perform.

With the exception of distant Pluto, Mercury has the most inclined (7 degrees to the ecliptic) and eccentric orbit of any planet in the solar system. Only the comets and an occasional asteroid (see Chapters 7 and 11) can outdo Mercury and Pluto in this regard, and it is probably no accident that these two are at the extreme edges of the system of planets. Their early history as outliers of the solar nebula may have left them room for eccentric behavior.

Sunlight is ferocious on the surface of Mercury. It has ten times the intensity found on the Moon's surface, and noon-time temperatures soar to more than 700 degrees Kelvin (800 degrees Fahrenheit). This is hot enough to melt tin, lead, and even zinc. But around midnight, exposed to the cold of interplanetary space, the far side of Mercury cools to 100 degrees Kelvin (−279 degrees Fahrenheit), so the temperature range of the surface of Mercury is greater than that on any other moon or planet in the solar system.

Sunlight reflected from Mercury's dark surface has a peculiar property that imitates light reflected from the Moon; it is slightly polarized. This effect is invisible to the human eye, but it can be detected by using polarizing filters; this is a sign that the planet's surface is covered with a blanket of dust that has been pulverized by meteoritic impacts.

In addition to affecting the quality of reflected sunlight, Mercury's hot surface gives the atoms of its atmosphere an opportunity to escape into space. Sped up by their contact with the surface, and held only loosely by Mercury's weak gravity, the atoms of its atmosphere have little difficulty escaping into interplanetary space.

As planets go, Mercury is a tiny world, not much larger than our Moon and slightly smaller than Jupiter's satellite Ganymede and Saturn's Titan. The linear radius is relatively easy to measure from its angular radius and distance; the most recent value is 2439 kilometers, which is about 40 percent larger than the Moon's radius. Mercury's mass (like that of Venus) has been more elusive because the planet has no moons, and its mass was originally determined by its gravitational influence on Venus and Earth. The estimate was improved when a space probe, Mariner 10, flew to within 5800 kilometers of Mercury's surface. From the deflection of its trajectory, the mass of Mercury has been determined as 3.302×10^{26} grams.

Mercury is surprisingly massive for its size. Its volume is only slightly larger than the Moon's, and yet it has four times the Moon's mass. This implies an average density of about 5.43 g/cm^3, which is nearly as high as that of the Earth, 5.52 g/cm^3. (Both of these densities are slightly increased by the high pressures towards the planets' interiors, and the effect is greater for the Earth. In uncompressed states the material of Mercury would actually be slightly denser than the material of the Earth.) The most natural explanation is that Mercury's high density is due to an unusually high fraction of iron, which is cosmically the most abundant of the denser elements. If this iron is concentrated in the core of the planet, then the core takes up three-quarters the radius of the whole planet.

3.3 Mercury's thin atmosphere

Atmospheres of the other planets have been detected in three ways: (1) by sightings of clouds and haze layers floating above the surface; (2) by spectroscopic studies of the daytime side that reveal absorption of sunlight produced by molecules in the atmosphere; (3) by refraction of starlight passing near the planet. All of these techniques require a moderately thick layer of gas, and none of them had given any signs of an atmosphere around Mercury.

A fourth method is to look at the night-time side for the faint light emitted by the atmosphere, similar to aurorae on the Earth. Many of these emissions are in the ultraviolet spectrum, and when the Mariner 10 mission flew by Mercury it scanned the planet with an ultraviolet spectrometer. Oddly, it found none of the expected emission – only small quantities of helium, atomic hydrogen and a trace of atomic oxygen. The composition of this gas implied that it was a transient wisp of the solar atmosphere carried outward and temporarily captured by the planet. Such gas is called the solar wind and it had been known to account for the behavior of comet tails (see Chapter 11) and the Earth's aurorae. Mercury's thin atmosphere is probably replenished continuously by particles captured from the solar wind.

From the strength of the atmospheric emission, the number density of emitting atoms was calculated to be about 4500 helium atoms and 8 hydrogen atoms per cubic centimeter. From these numbers and the known temperature and gravity, an upper limit on the atmospheric pressure at the surface of Mercury was estimated. The result, 2×10^{-10} millibars, is only one-fifth of a million-millionth of the pressure at the Earth's surface. (On a typical day on Earth, the ground-level pressure is 1000 millibars or 1 bar.) This thin atmosphere is a far better vacuum than can easily be produced in a laboratory on Earth.

There the matter seemed to rest: the atmosphere of Mercury had been "borrowed" from the Sun, and in the process Mercury had lost its own original gases. Then, in 1985, a fifth method for detecting an atmosphere was adopted and another species of atom was discovered in Mercury's atmosphere. This method involved an ingenious observation with an Earth-based telescope equipped with a spectrometer. The telescope was turned on Mercury during the day when the planet was high above the horizon. Due to the brightness of the sky, the observed spectrum was a composite of Mercury's emissions and the interfering emission of the sky, so the telescope was alternately pointed at the sunlit side of Mercury and at the sky. By using a computer to subtract out the spectrum of the sky, a relatively pure spectrum of the planet was obtained. The yellow portion of the spectrum revealed strong absorption features produced by sodium in the Sun's atmosphere and superposed were two narrow spikes of emission. These spikes are not normally seen in the Sun's spectrum, so they must be produced by the atmosphere of Mercury. But they were not seen on Mercury's dark side, so they must be produced by the action of intense sunlight.

The strength of this sodium light from Mercury is so great that there must be about 150 000 atoms of sodium in each cubic centimeter of the atmosphere (giving a pressure of 1.2×10^{-11} millibars). This makes sodium one of the most abundant elements in Mercury's atmosphere and raises the question: where did it all come from? The most likely source is the rocky material of the surface of Mercury. Perhaps the sodium atoms are "chipped" (or sputtered) out of the material on the surface by high-speed atoms in the solar wind. This would also be consistent with the subsequent detection of potassium in Mercury's atmosphere.

3.4 The halting spin of old age
There is another story about Mercury that has an ironic twist. This one was caused by telescopic observations that were interpreted incorrectly until radar signals gave a new view of the planet.

Astronomers once supposed that solar tides in the body of Mercury would cause the planet to rotate on its axis once every 88 days, in step with its orbital period. Just as the Moon always presents the same face to the Earth, it was thought that one side of Mercury always faced the Sun. To test this idea, the Italian astronomer Giovanni Schiaparelli (1835–1910) watched the surface markings seen through his 18-inch (46 centimeter) telescope and he concluded that the same side of Mercury did, indeed, always face the Sun. For three-quarters of a century, telescopic observers agreed with his conclusion and in the middle of the twentieth century one astronomer claimed that the rotational period equalled the orbital period with a precision greater than one part in a thousand. All of these astronomers were dead wrong!

In 1965, Mercury's true rotational period was determined with a radar signal that rebounded from the planet (Fig. 3.2). This signal was sent by an immense mega-watt transmitter and it was formed into a finely-tuned pulse with a narrow range of wavelengths. The pulse spread out like the ripples on a pond, sweeping past Mercury which reflected a small portion. The faint echo, in turn, spread out from Mercury and was received about ten minutes later by the same antenna, which was then acting as a receiver. As illustrated in Figure 3.3, the planet's rotation detuned the pulse, slightly spreading its range of wavelengths. One side of the globe was rotating away from the Earth, while the other side was rotating toward the Earth. These motions produced slight changes in the wavelength of the echo and from

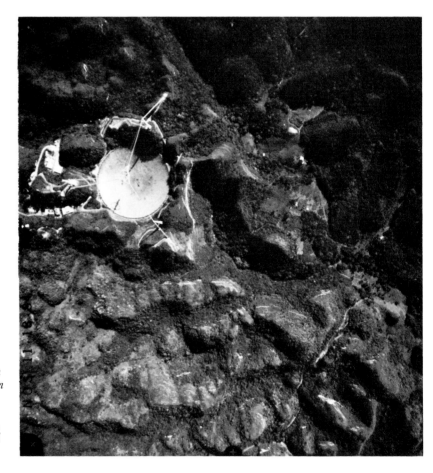

Figure 3.2. Arecibo Observatory. The world's largest radar telescope is nestled into the hills near Arecibo, Puerto Rico. It is 305 meters (1000 feet) in diameter. Powerful radar pulses sent from, and received by, this giant telescope first measured Mercury's rotation period in 1965. Until then it had been wrongly thought that Mercury kept one side permanently facing the Sun, with a rotation period equal to its orbital period.

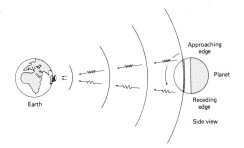

Figure 3.3. Radar probes of Mercury. The radar signal spreads out as a spherical wave, and only a small fraction is intercepted by the planet. As the wave sweeps by the planet, it is reflected in spherical wavelets whose wavelengths are Doppler-shifted by the rotational motion of the planetary surface. The waves from the receding side are red-shifted towards longer wavelengths and those from the approaching side are blue-shifted to shorter wavelengths. The amount of shift reveals the speed of the rotation. Knowing this speed, S, and the circumference of the planet, C, the period of rotation, P, follows from P = C/S.

these changes, the speed of the surface and the rotational period were calculated. The result came as a surprise. The rotation period was 58.6 days, or exactly two-thirds of the 88-day period that had been accepted for so long. Thus, with respect to the star background, Mercury spins on its axis three times during two full revolutions about the Sun. (This relationship follows from $3 \times 58.6 = 2 \times 88$.)

How could this observational error have occurred? There is no single explanation, but several things conspired to hide the true period. In the first place, the markings on Mercury's face are complex and difficult to see. If astronomers looking at the markings were already convinced that the predicted 88-day period was probably correct, they may have been prejudiced in their interpretation. And another factor was at work. Because three times the true period was equal to twice the orbital period, the markings would have returned to the sunlit side after two orbital revolutions. Thus, astronomers could have been fooled, because looking at Mercury after two of its orbital periods they would see the same markings on the sunlit side and would find no disagreement with the 88-day period. Only by observing it carefully at every 88-day interval would they have detected the discrepancy. Thus, half the telescopic observations that led to the 88-day period were actually correct, but many of the conflicting observations were ignored or missed. This is a striking example of curious observational circumstances and theoretical expectations that misled nearly everyone.

During one sidereal rotation on its axis, Mercury completes two-thirds of its orbit about the Sun, but the Sun appears to move only one-third of the way around the planet, so only one-third of a Mercurian day has passed. (See Fig. 3.4.) This means that a complete day will require three sidereal rotation periods, or two orbit periods. The time from one noon to the next, or the solar day, on Mercury is two Mercurian years.

But why does Mercury rotate with a period that is two-thirds of its orbital period? The answer is found in tidal forces on its elongated body and the

Figure 3.4. Rotation of Mercury and the Moon. The Earth's tidal force has drawn the Moon into synchronous rotation, and the bulge in the Moon's shape prevents it from rotating freely. As a result, the Moon always keeps the same hemisphere toward the Earth, and the lunar rotation period is equal to its monthly period of revolution about the Earth. Mercury's rotation was slowed by the Sun's tidal force, and because Mercury is elongated it has become locked into a rotation that spins it one and a half times during each orbital revolution. Mercury therefore rotates with respect to the stars in 58.6 Earth days, which is two-thirds of its orbital period of 88 Earth days. A Mercurian day lasts for two Mercurian years, as can be seen by tracing a full rotation of the planet.

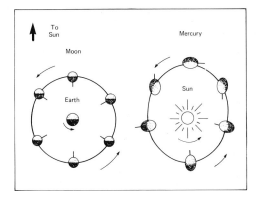

eccentric shape of its orbit. If Mercury had a nearly circular orbit, its rotation would have slowed to synchronism with its 88-day orbit. Like the Moon, it would have rotated with one face toward the parent body. But the tidal force of the Sun increases when Mercury is closest to the Sun, and this force gives an abrupt twist to its elongated shape. These twists tend to speed up the rotation, and they may have knocked the planet into the shorter, 58.6-day period by "spin–orbit coupling." Once this period has been reached, it is maintained. If the planet tends to rotate a little too fast, the timing of the tidal twist is altered and the planet is slowed down; if it rotates too slowly the tidal twist speeds it up slightly, and the synchronism is re-established. Thus, the 3 to 2 resonance is likely to persist during the remaining history of Mercury.

The fact that Mercury now rotates so slowly may account for its lack of satellites. If a planet rotates more slowly than its satellite's orbit, tidal interaction between the planet and its satellite can pull them together. Venus also rotates slowly and has no known satellites. Some astronomers have even argued that Mercury is a former moon of Venus that might have escaped by tidal interaction into its own solar orbit.

3.5 A Moon-like surface

On March 29, 1974, the glare surrounding Mercury was penetrated by the Mariner 10 spacecraft as it sped past the planet. The spacecraft had been flung toward Mercury during a close encounter with Venus seven weeks earlier. After passing Mercury's night side, Mariner 10 went into orbit around the Sun, mimicking a tiny planet, and as the spacecraft and Mercury looped around the Sun, they each returned to nearly the same place every six months, allowing repeated observations of the planet. Closeup photographs were taken during three such encounters, but the spacecraft's supply of maneuvering gas was finally depleted, so the radio transmitters were turned off.

Mariner 10's three encounters provided photographs of about half of Mercury's surface and showed features 1/5000 the size of features previously seen on the best photographs from Earth. These closeups revealed a Moon-like landscape that had never been seen before, and they gave us a glimpse into the planet's past. But the space-age exploration of Mercury is still at an early stage because Mariner 10's photographs are little better than telescopic observations of the Moon from Earth. Much more detail remains to be seen.

The world revealed by Mariner 10 resembles our Moon. Like the lunar surface, the surface of Mercury is pockmarked with craters ranging in diameter from impact basins 1000 kilometers across to craters only 100 meters in diameter, the limit of Mariner 10's photographic resolution (Fig. 3.5). The ubiquitous craters of Mercury strongly resemble their lunar counterparts, indicating that they were formed, and then eroded, by meteoritic impacts. As on the Moon there are small bowl-shaped craters on Mercury and larger ones with central peaks, flat floors, and terraces. Both worlds also contain fresh young craters with bright rays as well as older craters without rays.

Unlike those on the Moon, the craters on Mercury have been named after distinguished artists, composers, and writers. The largest crater is named Beethoven, followed by Tolstoy, Raphael, Goethe, and Homer in order of decreasing size. Mercurian craters have also been named for Mozart, Matisse, and Mark Twain.

There are also subtle differences caused by the planet's higher gravity. Because the force of gravity on the surface of Mercury is twice that of the Moon, material ejected from a crater on Mercury is not thrown as far. Also the transition from simple to complex craters – produced by gravitational slumping of material along the walls – occurs at a smaller diameter among Mercury's craters, and the mountains are not as high as those on the Moon. This comparative lack of relief is probably a result of the greater strength of gravity on Mercury that makes the rocks heavier in comparison to their internal strength and inhibits mountain-building. In effect, the greater gravity pulls the mountains down.

Mercury's surface also contains numerous multi-ringed impact basins, many of them more than 200 kilometers in diameter. The largest of these has been named Caloris, the Latin name for heat, because it is located at one of the two spots on Mercury that face the Sun at perihelion (Fig. 3.6). During all three Mariner 10 encounters, half of the Caloris Basin was hidden in shadow beyond the terminator. However, it appears to be surrounded by a ring of irregular mountains 1300 kilometers in diameter and 2 kilometers high. It vaguely resembles the lunar impact basins, Orientale and Imbrium, but unlike them it appears to have been flooded by lava almost immediately after formation, because the inner and the outer rims contain comparable numbers of craters.

The similarity of the surfaces of the Moon and Mercury, despite their differing masses and locations in the solar system, suggests that impacting objects were spread throughout the inner solar system during its early days. Mercury could have been bombarded at about the same time as the Moon, but we cannot be sure, for we do not know the ages of the rocks of Mercury, nor do we have an absolute chronology for its surface features.

Mercury's surface, like that of the Moon, exhibits a striking two-fold division into heavily cratered highlands and lightly cratered lowlands. All in all, Mercury seems to have undergone a series of events similar to the history of the Moon. Both objects, despite their different locations in the solar system, appear to have been subjected to an intense bombardment followed by the rise of magma to the surface. But the volcanic vents on Mercury were probably then squeezed shut during a global compression of the crust, while those on the Moon remained open.

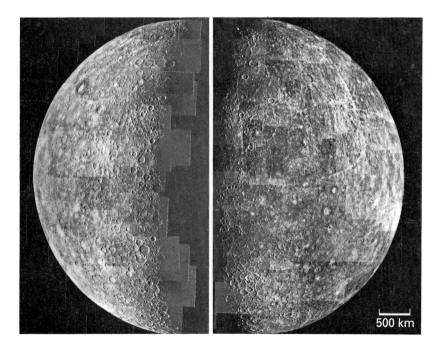

Figure 3.5. Spaceshots of Mercury. Photomosaics of Mercury photographed by the approaching spacecraft (left) and the departing spacecraft (right). The heavily cratered surface of Mercury resembles the lunar highlands. Bright rayed craters are also present on Mercury, as they are on the Moon. (Courtesy of NASA.)

500 km

Mercury's unique winding cliffs, called scarps, are probably due to shrinking as the interior cooled. They look like the skin of a shrivelled apple. They snake their way across craters and plains, attaining lengths greater than 500 kilometers and rising to 3 kilometers, as high as the Pyrenees. The scarps are thought to be the joints between great blocks of Mercury's crust that have shifted up on one side and down on the other, creating cracks that split mountains and craters as they moved. One of these, named Discovery Rupes, can be seen in Figure 3.7. The total shrinkage implied by the scarps is about 3 to 4 kilometers in the radius of the planet, about what is to be expected if the rocky mantle had cooled from a liquid state.

Close examination shows that the scarps deform the older craters but are

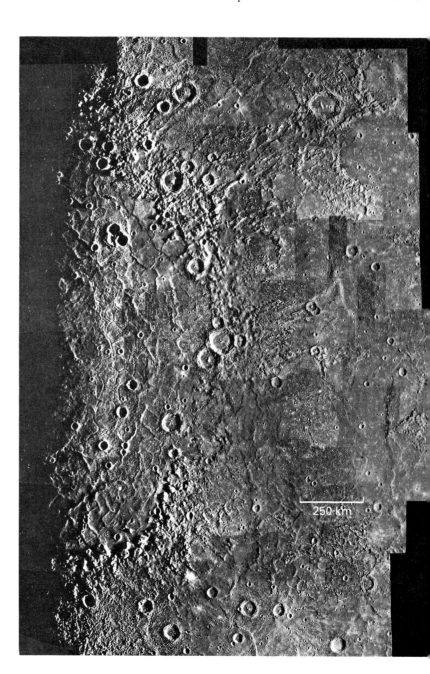

Figure 3.6. Caloris Basin. The multi-ringed Caloris Basin (left center) is 1300 kilometers in diameter. It is bounded by mountains that rise 2 kilometers above the surrounding terrain, and its floor is strongly disrupted by fractures and ridges. The Caloris Basin is similar in size and appearance to the Orientale Basin on the Moon, and both originated by impact of cosmic projectiles tens of kilometers in diameter. (Courtesy of NASA.)

250 km

themselves deformed by the younger ones, so they must have formed as the intense early bombardment weakened, and not during the planet's subsequent history. Curiously, Mercury, in contrast to the other terrestrial planets, did not experience any extensional processes during that time.

3.6 An Earth-like interior

Although Mercury resembles the Moon on the outside, it probably resembles the Earth on the inside. Its density matches the Earth's and is 1.6 times the density of the Moon. This high density is attributed to a massive iron core that is fully three-fourths the diameter of the planet. The iron core is as big as the Moon, and Mercury is, in fact, mostly core surrounded by a relatively thin mantle. The striking similarity of this mantle to the Moon's suggests that it, too, is formed of silicates that separated from the iron of the core during the early days after accretion when the planet was largely liquid. The high weight of the iron atoms would have slowly carried them down into the interior, leaving the silicates to form a lighter mantle floating on the surface.

Iron is relatively dense, and when molten it can flow from the outer part of a forming planet to the core. As it descends, it releases gravitational energy, much the way water gives up gravitational energy when it flows downward in a hydroelectric plant. Inside a planet, the heat released this way is sufficient to melt a large portion of the planet. But Mercury would then have cooled quickly, thanks to its relatively small radius, which makes its interior relatively close to its surface. The crust would have formed a single thick plate, eliminating the possibility of plate tectonics, of the type found on Earth (see Chapter 5).

Why does Mercury have so much iron and so little rock? Some astronomers think that Mercury once had a thick rocky mantle that was blasted off by a catastrophic collision with a smaller object.

Although heat can be generated throughout the volume of a satellite or planet, it is radiated to space only from the surface. The material of small bodies is relatively close to the surface, so it can cool more rapidly than the material of large bodies. For this reason, astronomers expected Mercury's entire interior to be solidified, while the Earth retains a liquid core. On the other hand, Mercury should have shrunk by about 40 kilometers if it had completely solidified, and this would be inconsistent with the height of its scarps. So, the paradoxical possibility of a remnant liquid core cannot be ruled out, for Mercury has a magnetic field.

3.7 Mercury's mysterious magnetic field

Mariner 10 carried a sensitive device for detecting magnetic fields, a magnetometer. As the spacecraft moved toward Mercury, the magnetometer plotted the fluctuating field of the solar wind. However, when it reached the neighborhood of the planet, it abruptly entered a new environment – a magnetic field that emanated from the planet. The increasing strength of this field as the spacecraft approached the planet indicated that the field of the planetary surface would be 0.01 times the Earth's surface magnetic field. Such a field was strong enough to carve out of the solar wind an elongated magnetic cavity with a tail pointing away from the Sun (Fig. 3.8).

Near the planet, the dipole magnetic field has a shape similar to the field around a compass, with the north pole along the rotational axis of Mercury. The Mercurian field and magnetic cavity (magnetosphere) is a scaled-down version of the Earth's field, except that Mercury occupies a larger fraction of its magnetosphere. As a result, the charged particles trapped in Mercury's field quickly collide with the planetary surface, where they are

Figure 3.7. Discovery Rupes. Mercury is distinguished from the Moon by having enormous cliffs, or scarps, that cut across its surface. The dark scarp shown here crosses nearly 500 kilometers of Mercury's intercrater plains, slicing through two craters that are 30 and 45 kilometers in diameter. It is called Discovery Rupes after Discovery, the name of the ship that Captain Robert F. Scott used in his first expedition to Antarctica, and Rupes, the Latin word for steep cliff. These long cliffs, which are as much as 4 kilometers high, were probably created when the planet cooled and contracted. (Courtesy of NASA.)

30 km

Figure 3.8. Mercury's magnetic field. The magnetic field of Mercury is a miniature version of the Earth's magnetic field, complete with bow shock, magnetosphere and magnetotail. Mercury's magnetic axis is closely aligned with its rotation axis, and its polarity is the same as the Earth's, with magnetic north corresponding to geographic north. The electrified solar wind compresses the magnetic field into a bow shock on the sunward side and draws it out into a tail on the opposite side.

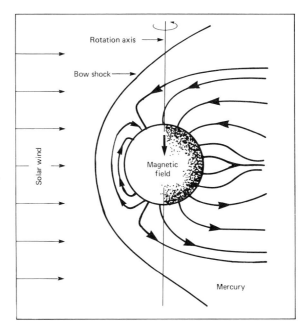

absorbed. Mercury does not have belts of trapped radiation comparable to the Earth's Van Allen belts (see Chapter 5).

The discovery of Mercury's magnetic field was completely unexpected, and its source is a mystery. Rapid rotation of a molten iron core, acting as a dynamo, was thought to be a prerequisite for generating a planetary magnetic field, but Mercury's core should have cooled and solidified long ago because Mercury is so small. And, even if the core were still molten, the planet's rotation is probably too slow to generate a field by the dynamo action that operates in the Earth. For instance, the planet Venus, which may have a liquid core, but is rotating very slowly with a period of 243 days, has no detectable magnetic field. The best guess is that Mercury's iron core remains liquified, perhaps as the result of heat generated from the decay of radioactive elements, and a magnetic field is generated in spite of the planet's slow rotation. Another possibility is that the Sun's wind generates Mercury's magnetic field.

As we shall see, every planet is unique in one way or another, and Mercury is certainly no exception to this rule.

Focus 3A Mercury – summary

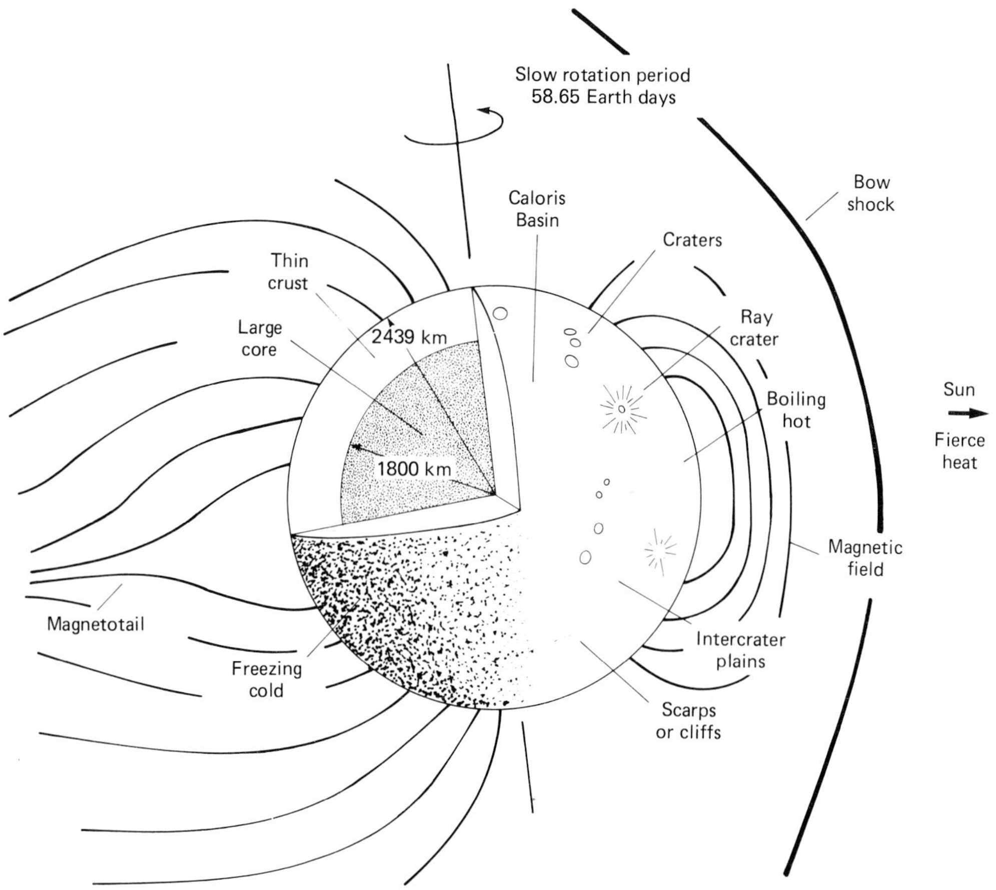

Slow rotation period
58.65 Earth days

Bow
shock

Caloris
Basin

Craters

Thin
crust

Ray
crater

Large
core

2439 km

Boiling
hot

Sun

Fierce
heat

1800 km

Magnetic
field

Magnetotail

Intercrater
plains

Freezing
cold

Scarps
or cliffs

Mass: 3.30×10^{26} grams = 0.055 M_E (Earth = 1)
Radius: 2439 kilometers = 0.382 R_E (Earth = 1)
Mean density: 5.43 g/cm^3
Rotational period: 58.6462 Earth days
Orbital period: 87.969 Earth days
Mean distance from Sun: 0.387 A.U.
Mercury has no satellites
Surface magnetic field strength = 0.0035 gauss

Yellow clouds of Venus. Bright clouds of sulfuric acid wrap around the polar regions of Venus. The creamy yellow veil of clouds circulates once around the planet in only four Earth days, moving at a speed of 360 kilometers per hour. The mottled features near the center may be due to convection caused by the Sun's heat. (Courtesy of NASA Headquarters.)

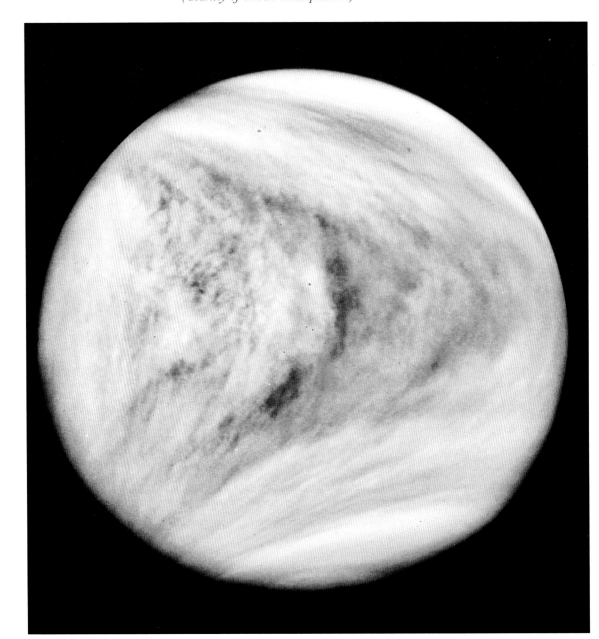

4 Venus: the veiled planet

There are no seasons on Venus.

The surface of Venus is hot enough to melt lead.

There are no daily or seasonal changes of temperature on Venus' surface.

If it were not for the greenhouse effect, Venus would be much colder.

Although Venus is now dried out, it may have once had vast oceans.

Volcanoes may have erupted on Venus.

Figure 4.1. The birth of Venus (detail). According to legend, Venus, the goddess of love, was daughter of the sky-god Uranus, and she entered the world by rising naked from the sea. Sandro Botticelli's beautiful painting depicts new-born Venus rising from the waves. (Courtesy of the Uffizi Gallery, Firenze.)

4.1 The goddess of love

(a) Goddess revealed

The planet Venus was named for the ancient Roman goddess of beauty and sensual love; her Greek counterpart was Aphrodite, the goddess of fertility. To the Babylonians, the planet was Ishtar, the mistress of the heavens, and to the Chinese it was "the beautiful white one." Thus, Venus has been the symbol of love and beauty since the beginning of civilization (Fig. 4.1). The Greeks worshipped Aphrodite on the island of Cythera, and therefore the

adjective "Cytherean" has often been applied to the planet. This is the adjective we will use, instead of the linguistic hybrid "Venusian."

In many ways, Venus is the Earth's twin sister. She has almost the same weight and waistline as the Earth (Fig. 4.2). Her radius is 95 percent that of the Earth, and her mass is 81 percent of the Earth's, so the feel of gravity at the planet's surface is similar to that on Earth.

The surface of Venus had never been seen before the 1970s because of its ubiquitous cloud cover, so scientists had been free to speculate, and some thought that its surface might be teeming with fascinating creatures. Because Venus is closer to the Sun, each square meter of its surface intercepts about twice as much sunlight as it would at the Earth's distance, and thus the climate beneath the clouds was thought to be warm and temperate. The dense clouds also suggested the presence of water. Therefore Cytherean creatures were thought to flourish in a warm, wet environment that included steamy swamps and jungles.

But this romantic vision has been drastically altered by space-age scientists, whose spacecraft have penetrated Venus' perpetual clouds and revealed a truly hellish and sterile surface. Beneath her pure, gleaming clouds, the planet of love is an inferno!

The main ingredients of Venus' atmosphere are the same as the Earth's, although the proportions are significantly different (see Table 4.1), and her atmosphere is mostly carbon dioxide (CO_2), with a whiff of nitrogen (N_2) and a trace of oxygen (O_2), which is only present at the minuscule level of 20 parts per million.

Venus has boiled dry, like a kettle left too long on a stove. And there are no seasons such as we know on Earth. Her terrain is gloomy; 98 per cent of the sunlight is captured at higher levels in the dense, cloudy atmosphere. As a result of the atmosphere's peculiar filtering action, the rocky surface of Venus is bathed in the dim light of an orange sky.

Figure 4.2. Comparison of Earth and Venus. As seen through the telescope, Venus is a featureless planet enshrouded by a thick, unbroken layer of clouds. Its radius is 95 percent of the Earth's and its mass is 81 percent of the Earth's mass. Venus orbits the Sun at a mean distance of 0.723 A.U. with an orbital period of 225 Earth days. Venus is, therefore, only a little closer to the Sun than the Earth. All of these similarities, combined with the fact that Venus is our closest planetary neighbor, gave rise to the idea that Venus is our sister planet.

(b) Comparative evolution of the Earth and Venus

The histories of the Earth and Venus have been quite different. Most of the carbon dioxide in the Earth's air was captured by life within the oceans and

Table 4.1. *Abundances of primary constituents of the atmospheres of Venus and Earth*

Constituent	Venus (lower atmosphere)	Earth (sea level)
Carbon dioxide (CO_2)	96.4%	0.03%
Nitrogen (N_2)	3.4%	78.08%
Water vapor (H_2O)	0.01% (variable)	5.0% (variable)
Oxygen (O_2)	Less than 20 parts per million by volume	20.95%

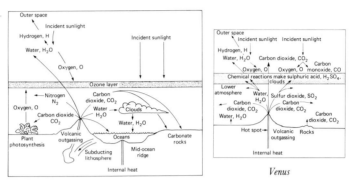

Figure 4.3. Fate of volcanic gas. Both water vapor and carbon dioxide were released by volcanic outgassing into the atmospheres of the Earth (left) and most likely Venus (right). On the Earth most of the water went into the oceans, and most of the carbon dioxide was locked up in carbonate rocks, dissolved in the oceans, or used in photosynthesis to make oxygen and carbohydrates. Venus eventually became so hot that any existing oceans evaporated, and any carbon dioxide in its rocks was released to the atmosphere. Because there is virtually no oxygen in Venus' atmosphere, there is no ozone layer, and ultraviolet sunlight gradually destroyed most of the water vapor, releasing its component hydrogen to outer space.

then locked into limestone and other carbonate rocks. Plants made the air breathable by supplying it with oxygen. In contrast, Venus evolved into a dry high-temperature world essentially devoid of atmospheric oxygen. The surface of Venus is now so hot that it would destroy any living thing, and any Cytherean oceans would boil away.

Venus and the Earth probably started out as nearly identical twins. We know that there is as much carbon dioxide in the Earth's carbonate rocks as in Venus' atmosphere. (See Fig. 4.3.) The critical difference may have been Venus' smaller orbit, keeping it closer to the Sun, coupled with the greenhouse effect, discussed later. The slightly higher initial temperature of Venus may have led to a runaway heating of the atmosphere which turned Venus into the dried-out lifeless sphere that we see today. Because the Earth is slightly further from the Sun, it evolved into a living world capable of sustaining a remarkable diversity of living things. The balance is a delicate one. If the Earth's distance from the Sun were slightly reduced, a runaway greenhouse effect might cause our oceans to boil away and life would be destroyed. Moreover, if we increase the amount of carbon dioxide in our atmosphere, by burning fossil fuels and destroying tropical forests, we might conceivably find ourselves in just such a runaway situation.

4.2 The atmosphere of Venus

(a) Oppressive heat and weight

Spacecraft that have descended through Venus' clouds have found that the lower atmosphere is uniformly and oppressively hot. As illustrated in Figure 4.4, there is an increase in temperature with decreasing height, culminating in a blistering 730 degrees Kelvin (855 degrees Fahrenheit) at the surface. The dense lower atmosphere stores so much heat and transports it so efficiently from one part of the globe to another that there are no daily, seasonal, or latitudinal changes in the temperature. The surface

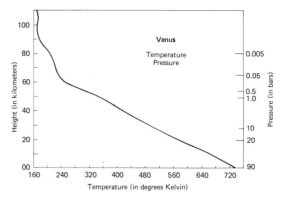

*Figure 4.4. A hot and heavy atmosphere. The temperature and pressure of Venus' lower atmosphere systematically increase with decreasing height. At about 60 kilometers above the surface, where the clouds reside, the meteorological conditions are similar to those in the clouds of our own planet, with a temperature of about 250 degrees Kelvin and an atmospheric pressure of 0.1 bars. But Venus' high-flying clouds are about 8 times higher than those on Earth, and conditions at the bottom of the thick atmosphere are very different. At the surface the temperature reaches a sizzling 730 degrees Kelvin, and the atmosphere weighs 90 times that of the Earth's atmosphere. (Adapted from Alvin Seiff et al., Science **205**, 47 (1979).)*

temperature varies by no more than a few degrees between the equator and poles, and the surface remains at an even temperature during the long, dark nights.

The surface of Venus also lies under the weight of an atmosphere that is 90 times more massive than the Earth's. The surface pressure on Venus is comparable to that experienced by a submarine 500 fathoms (3000 feet) below the surface of our terrestrial oceans.

Large amounts of carbon dioxide account for Venus' heavy atmosphere. If Cytherean temperatures existed on the Earth, our oceans would vaporize, and our rocks would liberate their carbon dioxide. Our atmosphere would then be about 300 times more massive than it is now. Even if the vaporized water somehow disappeared, the remaining carbon dioxide atmosphere would weigh about the same as Venus' present atmosphere.

(b) Acid rain and raging winds

Astronomers have known for centuries that Venus is perpetually covered with clouds (Fig. 4.5). A detailed study of the sunlight that is reflected from Venus' clouds indicates that the reflecting cloud particles have a spherical shape, implying that the particles are liquid droplets rather than ice crystals. Liquid water can be eliminated as the primary cloud ingredient because it would turn to ice at the low temperatures near the Cytherean cloud tops (250 degrees Kelvin). Water and most of the other plausible liquids could also be ruled out because they have the wrong optical properties. They have a refractive index that is quite different from that of Venus' clouds.

Baffled astronomers finally found the answer. The clouds of Venus are composed of sulfuric acid! Droplets of sulfuric acid produce the observed refractive index, 1.44. (Sulfuric acid is commonly used in car batteries and it contributes to the eye-stinging quality of smog in some industrial cities, especially near smelters.) Particles of sulfur at high altitudes also contribute to the haze.

The pale yellow clouds of Venus are distinctly different from our white terrestrial clouds. The clouds on the Earth result from the cooling of rising air. The decrease in temperature at higher altitudes causes the water vapor to condense out as liquid water droplets or to freeze into crystals of water ice. The clouds on Venus are more akin to terrestrial smog, for they are formed by a series of chemical reactions involving gaseous sulfur dioxide and water vapor. The chemical reactions above the cloud tops are photochemical reactions that derive their energy from sunlight, while those below the cloud tops are thermochemical reactions that are driven by the intense heat.

Gaseous sulfur and sulfur dioxide have probably been ejected into Venus' atmosphere by volcanic outbursts, as they now are on the Earth. The sulfurous gases eventually rise through the dry atmosphere and combine with water vapor to produce clouds of sulfuric acid. When the sulfuric acid droplets fall into the warmer atmosphere, they evaporate, and

Figure 4.5. Ultraviolet clouds. Although Venus' cloud tops are featureless when viewed in visible light, banded markings are detected at the ultraviolet wavelengths. This photograph of Venus was taken in ultraviolet light by the Mariner 10 spacecraft at a distance of 760 000 kilometers. Strong zonal winds combine with weaker poleward winds to carry the clouds in a slow spiral towards the poles. (Venus' atmosphere is actually yellow in color, and the blue color shown here is an artificial color used to enhance the ultraviolet markings.) (Courtesy of NASA.)

Figure 4.6. Raging winds. A 100 meter per second wind (360 kilometers per hour) whips the upper layer of Venus' cloud deck around the planet's equator once every four days. This is dramatically illustrated by these photographs taken at ultraviolet wavelengths on consecutive days by the Pioneer Venus spacecraft. The Y-shaped clouds move towards the west (left). The winds of Venus are dominated by a zonal (east-to-west) circulation that drives the high-flying clouds around the planet in only four Earth days. The planet also rotates westward, but with the much slower period of 243 days. (Courtesy of Larry Travis and NASA.)

the gas rises again to the cloud layers. Thus, the acid rain on Venus is probably evaporated in the hot, dry atmosphere before it reaches the surface. In contrast, the sulfur in the Earth's atmosphere dissolves in water clouds and falls to the surface, where it damages forests and lakes.

Sequential ultraviolet photographs have shown that raging winds are blowing the high-flying clouds around Venus from east to west at speeds of 100 meters per second, or 360 kilometers per hour (see Fig. 4.6, and Focus 4A, The clouds and winds of Venus). These winds are similar to the jet streams of the Earth's atmosphere found at similar altitudes. At such speeds, the planet-wide cloud deck races around Venus' equator once every four Earth days. Curiously enough, Venus' surface also rotates westward, but with a much longer period of 243 Earth days. So the winds of Venus blow the entire outer atmosphere around the planet much more rapidly than the planet spins. By contrast, most of the Earth's atmosphere rotates synchronously with its surface, and although terrestrial jet streams reach Cytherean speeds, they are limited to narrow zones high in the Earth's atmosphere.

The remarkable circulation of the Cytherean atmosphere is probably driven by the energy of sunlight absorbed in its clouds. The Sun's rays fall directly on Venus' equator throughout the year, and the poles would remain quite cold were it not for the heat brought to them from the equator.

In each hemisphere, the heat circulates in a single large Hadley cell (Fig. 4.7). Warm air rises at the equator to the cloud tops, where winds propel it poleward. After warming the poles, the circulating atmosphere sinks and flows back towards the equator at lower levels near the base of the clouds. The stronger zonal (east to west) circulation combines with this weaker Hadley (north–south) circulation, giving rise to a wind vortex that carries the clouds in a slow spiral towards the poles.

4.3 The temperature of Venus

(a) The greenhouse effect – trapped heat

A planet's temperature adjusts so that its radiation to space carries away all the energy it absorbs from the Sun. Thus, its temperature depends upon its distance from the Sun and the fraction of sunlight it absorbs. About 70 percent of the incident sunlight directly warms the Earth's surface; the remainder is reflected into space. On Venus the bulk of the incident sunlight is either reflected or absorbed within the clouds and very little reaches the ground.

Simplified calculations indicate that all of Earth's oceans ought to be frozen solid, and that the surface of Venus should now also be quite cold.

Focus 4A The clouds and winds of Venus

Venus is completely covered with opaque clouds that rise 70 kilometers in the thick atmosphere. Space probes that landed on Venus' surface and balloons that floated in her atmosphere have been able to detect three distinct layers in the sulfurous clouds. The top layer contains small droplets of sulfuric acid; the middle layer contains larger but fewer particles. The bottom layer is the densest and contains the largest particles; it is comparable to a bad city smog in visibility. Beneath the lowest layer, the atmosphere is hot enough to vaporize all particles, so it is relatively clear from the surface to a height of about 31 kilometers. The entire region of clouds, including the lower haze, is about 40 kilometers deep, compared to 6 kilometers on Earth.

At all altitudes the dominant retrograde winds blow with a speed that increases with height. From a gentle breeze at the ground, it increases to a speed of 100 meters per second (360 kilometers per hour) at great heights.

An unprecedented international exploration of Venus' atmosphere was accomplished in June, 1985, when two balloons from the Soviet Vega 1 and Vega 2 spacecraft carried Soviet, American, and French equipment for 33 hours while being blown one-third of the way around the planet, moving at an average of 240 kilometers per hour. One of the balloons was occasionally buffeted by descending currents averaging 10 kilometers per hour for several hours at a time.

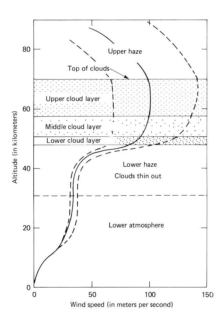

But such calculations do not take into account the greenhouse effect. This is the name given to the process by which the planet's atmosphere traps heat near the surface.

The trapped heat warms the planet's surface to higher temperatures than would normally be achieved by direct sunlight in the absence of an atmosphere. On Earth, the surface temperature is raised about 30 degrees Kelvin by this effect, resulting in the mild climate we enjoy today. But the greenhouse effect has raised the temperature of Venus' surface by hundreds of degrees (Fig. 4.8). Venus' carbon dioxide atmosphere is translucent to visible radiation that contains most of the Sun's energy, but it absorbs most of the heat that is radiated by the surface at infrared wavelengths. Thus, the incident sunlight is allowed to enter, but the emitted heat radiation is trapped near the surface.

In fact, Soviet astronomers once thought that the thick, dense Cytherean atmosphere might not let any sunlight reach the surface. They therefore equipped their earlier spacecraft with floodlights to illuminate the scene when they reached the ground. The floodlights were not necessary, however, for there was enough sunlight to take historic pictures of Cytherean rocks.

(b) Where has all the water gone?

Venus may have once contained vast quantities of water that could have helped initiate the present inferno. When an early, feeble greenhouse effect vaporized some of the ocean water, the increased amount of water in the atmosphere permitted it to trap more heat near the surface. As a result, more water was vaporized, trapping even more heat. Eventually this process went out of control, creating a runaway greenhouse effect that filled the steamy atmosphere with all the ocean's water and heated the Cytherean surface to the high temperatures that are observed today.

Evidence that Venus once had an ocean is found in the fact that there is now an excess of deuterium in Venus' atmosphere. Deuterium is an atom chemically identical to hydrogen but heavier and therefore more likely to be retained in the atmosphere. On Earth, it is found in heavy water, and it comprises about 0.016 percent of the oceans, but there is 100 times as much deuterium in the atmosphere of Venus as in the Earth's oceans. The most natural explanation of this observed excess of deuterium is that Venus once had vast quantities of both water and heavy water. When the liquids subsequently evaporated and became dissociated by sunlight, the hydrogen escaped from Venus, but the deuterium was heavier and less likely to escape. Some of it remained behind as a residue. The amount of remaining deuterium suggests that Venus once had at least as much water as a few tenths of one percent of the Earth's present ocean – sufficient to cover Venus with a global ocean 10 meters deep.

Figure 4.7. Hadley cell. Incident solar energy heats the equatorial regions more than the poles. This drives a Hadley cell circulation which transports heat from the equator to the poles. Heat rises at the equator and flows along the cloud tops to the poles. Heat then sinks at the poles and returns to the equator along the cloud base. The photo is a 72-day average of infrared measurements of the north polar region on Venus taken from the Pioneer Venus Orbiter. It shows a hot "dipolar" structure that straddles the pole and rotates around it in 2.7 Earth days. The two polar hot spots may mark clearings in the Cytherean clouds where air flows downward. (The infrared image was prepared by Dr D.J. Diner of the Jet Propulsion Laboratory, courtesy of F.W. Taylor and NASA.)

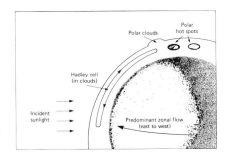

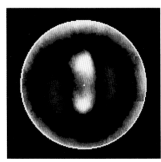

Figure 4.8. Trapped heat. About 70 percent of the sunlight incident on Venus is reflected at the top of its clouds. Most of the remaining energy is absorbed within the clouds (short dashed lines). Only about 2 percent of the incident sunlight actually reaches the surface of Venus, where it is re-radiated as heat (or thermal) radiation at infrared wavelengths (long dashed line). Radiation in the infrared is absorbed by gases in the lower atmosphere and heats up the planet's surface. As a result of this greenhouse effect, the surface and lower atmosphere of Venus are heated to the unusually high temperatures shown by the thick solid line.

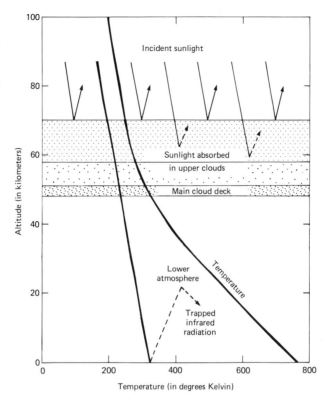

Figure 4.9. What color is Venus? A Venera 13 panorama of the surface of Venus (top) indicates that its thick, dense atmosphere turns sunlight to an orange, peach-colored hue, primarily by absorbing the blue and violet components of the Sun's light. When the data are processed to remove the atmospheric interference, the visible surface becomes dark and nearly colorless (bottom). Spectral data taken at near-infrared wavelengths suggest that the dark surface rocks are oxidized. The high surface temperature literally bakes the color out of the oxidized rocks. (Courtesy of Carle M. Pieters through the Brown/Vernadsky Institute to Institute Agreement with the USSR Academy of Sciences.)

But liquid water would quickly vaporize at Venus' high temperature, and the present store of water in its atmosphere is only 0.001 percent of the water found in the Earth's oceans. Where has Venus' primeval water gone? One plausible mechanism for removing the water is the dissociation of water vapor by solar ultraviolet radiation. It would break apart the water vapor molecules in the upper atmosphere into hydrogen, H, and oxygen, O, atoms, according to the reaction

$$\text{Photon} + H_2O \rightarrow 2H + O$$

The lighter hydrogen atoms could escape into outer space, while the oxygen could combine with crustal rocks. Spectral analysis of Venus' dark crustal rocks suggests that they are oxidized, and they may have absorbed the oxygen associated with the ancient oceans (Fig. 4.9).

If this process of dissociation and escape operated over long periods of time, most of the water vapor in Venus' atmosphere would be lost. But why doesn't the same process remove substantial amounts of water from the Earth today? The Earth's interior releases about a million tons of water into its oceans each day; but most of the Earth's water remains in the oceans. The water that does evaporate soon produces clouds and falls back to the ground as rain. It never reaches the atmospheric levels where intense ultraviolet sunlight can disrupt it.

4.4 Beneath the clouds

(a) Venus rotates backwards

Although no human eye has ever seen the surface of Venus, radio waves can penetrate its obscuring veil of clouds and touch the landscape hidden beneath. By bouncing pulses of radio radiation off the surface, radar astronomers have discovered that Venus spins in a direction that is opposite to that of its orbital motion. Sunrise on Venus occurs in the west instead of in the east.

The radar observations also show that Venus spins with a period longer than any other planet, 243 Earth days. This rotation period is even longer than Venus' 225 day period of revolution around the Sun. Why is the body of the planet rotating so slowly?

(b) Tidal effects

The slow rotation of Venus may be primordial – a result of conditions during its formation. Alternatively, a gentle tidal interaction with the Sun may have slowed the planet, because the Sun's gravitational force produces two tidal bulges on Venus. The rotating planet drags its tidal bulges along with it, causing them to twist out of alignment with the Sun. As a result, the Sun's gravitational attraction tends to oppose Venus' rotation. In much the same way, the Sun may have altered Mercury's spin.

A remaining peculiarity is that the planet's spin is even slower than its orbital motion. Perhaps some other force has slowed the spin to its longer period of 243 days.

(c) Venus has no strong magnetic field

If Venus has any magnetic field, its strength is less than one ten-thousandth (0.0001) that of the Earth's, so an ordinary compass would be useless. (See Focus 4B, Venus' interaction with the solar wind.) The weakness of this magnetic field is rather surprising because Venus and the Earth have a similar size and mass, and they might be expected to have similar interiors and, hence, similar magnetic fields. For instance, we strongly suspect that iron has gravitated to the Earth's center, and heat produced by accretion, core formation or internal radioactivity has melted a portion of its central core. It is thought that the circulation of this molten iron produces the Earth's strong magnetic field. By analogy, Venus ought to possess a similar molten core. However, Venus does not show the magnetic field that would be produced by currents within such a core. One possible explanation for this magnetic difference between the planets is Venus' slow rate of rotation, which may be insufficient to produce the necessary circulation in its core.

(d) Continents, valleys and craters

Our understanding of the surface of Venus was dramatically improved when Earth-based radar telescopes as well as radar systems aboard the Pioneer Venus and the Venera 15 and 16 spacecraft mapped most of the planet's surface. The round trip travel time from each spacecraft radar pulse could be converted into one altitude measurement. As the spacecraft orbited Venus, many thousands of such measurements were obtained, and these were combined to map the surface with a vertical accuracy of 200 meters.

The radar maps show that most of Venus is an extraordinarily smooth world whose surface is quite different from ours. Without its oceans, the Earth would appear to have two main levels, the ocean floors and the continents. In contrast, the surface of Venus is largely at one level, and about 65 percent of the surface lies within one kilometer of the average planetary radius, 6052 kilometers (Fig. 4.10). The high surface temperature associated with a runaway greenhouse effect may have softened the rocks, permitting the peaks to slump down and smooth the surface somewhat.

Although most of Venus' terrain consists of smooth, rolling plains, about 10 percent of the planet's surface is covered by highlands that tower up to 11 kilometers above the plains. There are two large elevated plateaus on Venus: Ishtar Terra in the far north and Aphrodite Terra just south of the equator (Fig. 4.11). Ishtar Terra is named after the Babylonian goddess of love, while Aphrodite was the Greek fertility goddess. Within the western part of Ishtar Terra there is Lakshmi Planum (High Plain), named after the Hindu goddess. Ishtar Terra is larger than the continental United States, and Lakshmi Planum twice the size of the similarly shaped Himalayan

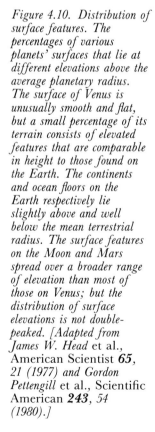

Figure 4.10. Distribution of surface features. The percentages of various planets' surfaces that lie at different elevations above the average planetary radius. The surface of Venus is unusually smooth and flat, but a small percentage of its terrain consists of elevated features that are comparable in height to those found on the Earth. The continents and ocean floors on the Earth respectively lie slightly above and well below the mean terrestrial radius. The surface features on the Moon and Mars spread over a broader range of elevation than most of those on Venus; but the distribution of surface elevations is not double-peaked. [Adapted from James W. Head et al., American Scientist 65, 21 (1977) and Gordon Pettengill et al., Scientific American 243, 54 (1980).]

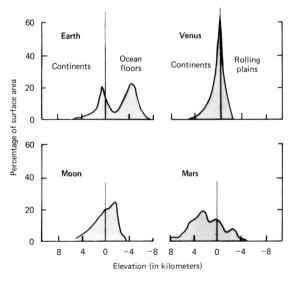

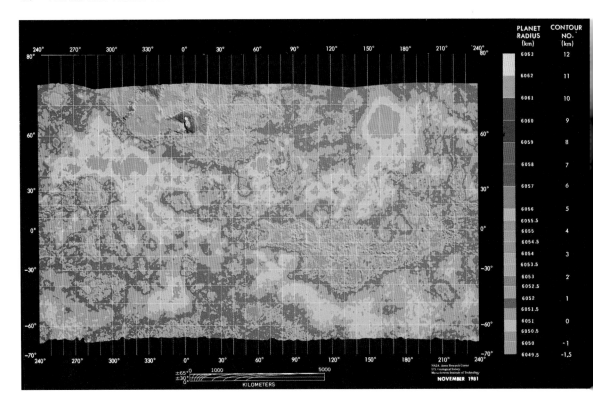

Figure 4.11. Radar map of Venus' surface. Altitudes of the surface of Venus measured by the Pioneer Venus Orbiter. The raised appearance is created by modulating the intensity of the colors, thereby emphasizing the steeper slopes on an unusually smooth planet. Almost 70 percent of the surface does not deviate more than 500 meters from the average radius of 6051.9 kilometers that is coded a blue-green in the altitude color scale. This flat, smooth part of Venus' terrain is called the rolling plains. Some 20 percent of the surface is lowland that lies about 1.6 kilometers below the mean radius. The remaining 10 percent of the surface of Venus is highland. The highlands consist of two large, elevated plateaus: Ishtar Terra in the far north and Aphrodite Terra just south of the equator. Lakshmi Planum is found within Ishtar Terra. This high plain is 2500 kilometers across and rises about 4 kilometers above the rolling lowland plains. The highest elevation of 6062 kilometers is denoted by the pink area at the top of the red Maxwell Montes in the north. The mountain's 11 kilometer height above the average radius exceeds by two kilometers the height of Mount Everest above sea level. (Courtesy of Peter Ford and Gordon Pettengill, Massachusetts Institute of Technology, and Eric Eliason, US Geological Survey and NASA.)

Figure 4.12. Southern hemisphere of Venus. This radar image displays circular features that may be impact craters and long, linear features that could be due to folds and faults in the planet's crust. The smaller crater at the center is about 60 kilometers across. There is still controversy over whether such craters are of impact or of volcanic origin, but the relative sparseness of craters on Venus suggests that old impact craters may have been covered by volcanic outpourings or that Venus has a young, recently formed lithosphere. The image was made at 12.6 centimeter wavelength with the 300 meter radio telescope of the Arecibo Observatory. (Courtesy of Donald B. Campbell, National Astronomy and Ionosphere Center.)

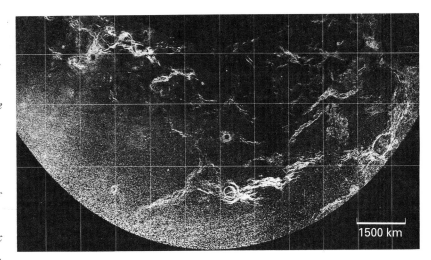

Plateau. Lakshmi is also ringed by towering mountain ranges that resemble the Himalayas. It is tempting to attribute the plateaus and mountains of Ishtar to collision and folding due to large-scale horizontal distortions of the lithosphere of Venus.

Aphrodite Terra is comparable to Africa in size. Like Africa, it is split open by gigantic cracks, or rifts, that are as much as 3 kilometers deep and 1000 kilometers long. The most likely explanation for these long cracks is crustal motions, perhaps similar to those on Earth. At the southeastern end of Aphrodite, there is a huge circular feature that may be the remnant of a giant impact crater. In fact, there are numerous circular or ring-shaped features scattered across the extensive rolling plains. (See Fig. 4.12 and Fig. 4.13.) They often contain central mounds, and resemble the impact craters on the Moon and Mercury. (See Fig. 4.4. and Fig. 4.15.)

Figure 4.13. Northern mountain ranges. Horizontal motions of Venus' crust have created folded mountain ranges like Maxwell Montes (bottom right) and the Akna Montes (left center) and Freyja Montes (top center) that border the Lakshmi Plateau (center). This plateau is 2500 kilometers across and rises about 4 kilometers above the surrounding terrain. This radar image was made at 12.6 centimeter wavelength with the 300 meter radio telescope of the Arecibo Observatory. (Courtesy of Donald B. Campbell, National Astronomy and Ionosphere Center.)

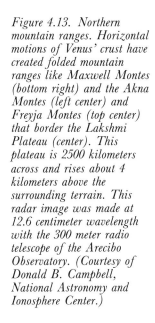

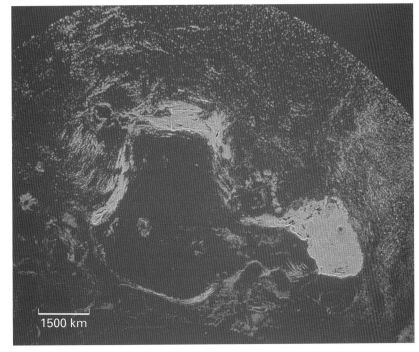

Focus 4B Venus' interaction with the solar wind

Venus has no appreciable magnetic field to fend off the solar wind (see Chapter 5), but its dense atmosphere prevents the solar wind from reaching its surface. Energetic photons in sunlight ionize some of the atoms and molecules in Venus' outer atmosphere, forming an ionosphere. This electrified layer is similar to the Earth's ionosphere and it creates a bow shock and shields Venus from the solar wind. Interactions between the solar wind and Venus' ionosphere may create the electrical noise that resembles lightning discharges on Earth. Lacking a magnetic field, Venus has no belts of trapped radiation such as the Van Allen belts near the Earth.

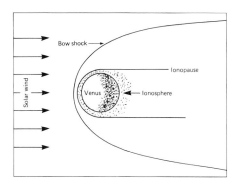

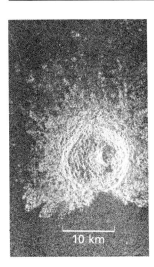

Figure 4.14. Impact crater. Although Venus has a thick atmosphere, many large meteorites have survived the fiery passage and struck the surface, gouging out craters like this one. The impacting object was apparently moving toward the north (top) at a small angle to the surface; this would account for the missing ejecta in the south (bottom) and the small secondary craters seen to the north. The rim diameter of the primary crater is 12.5 kilometers. It has a complex central peak, and its interior exhibits terracing caused by slumping of the inner wall. This radar image was taken from the Magellan spacecraft with a resolution of about 120 meters. (Courtesy of JPL and NASA.)

Figure 4.15. Crater farm. Although fewer impact craters are detected on Venus than on the Moon or Mars, three large impact craters, with diameters ranging from 37 kilometers to 50 kilometers, are found in this radar image of the Lavinia region of Venus. The rough, radar-bright ejecta surrounds smoother, dark crater interiors and bright central peaks. Numerous domes of probable volcanic origin can be seen in the south-east (bottom-right) corner of this Magellan image mosaic; the domes are between 1 and 12 kilometers in diameter. (Courtesy of JPL and NASA.)

(e) Plate tectonics, volcanoes and flowing lava

On Earth, slowly moving plates are driven by the push of the spreading sea floor or the pull of the subducting lithosphere. Internal heat is released through the upwelling and emergence of hot material at the mid-ocean ridges (see Chapter 5). Additional heat is lost by the volcanoes that occur at the colliding boundaries of heavy ocean plates and lighter continental plates. This is called tectonic activity, and similar processes might be expected on our neighbor, Venus.

However, there are no surface features on Venus that are comparable to the Earth's subduction trenches or island arcs. Venus also seems to be missing the volcanic arcs that mark places where a heavy ocean floor would slide under the lighter continental plates. Nevertheless, the vast plateaus, mountain ranges, and large valleys on Venus are similar to those produced by the motions and collisions of continental plates on the Earth. The Cytherean mountain ranges are similar to the Appalachian Mountains and the Himalayas, a series of parallel mountains formed by horizontal movements of the Earth's crust.

Fresh evidence for crustal motion on Venus has been provided by high-resolution radar images of the surface taken by the Soviet Union's Venera 15 and 16 spacecraft. These orbiting spacecraft have surveyed much of the northern hemisphere of Venus, providing pictures that show features as small as 1.25 kilometers in size. They reveal ridges and narrow valleys resembling the folds in a bunched up carpet (Fig. 4.16). These features are probably produced by horizontal motions of the crust. (Also see Fig. 4.17.)

Venus may lose much of its internal heat through regions of volcanic activity, and Beta Regio is considered to be such a region (Fig. 4.18). The volcanoes in this region postdate much of the faulting in its associated rift system, and the apparent lack of impact craters also implies that the activity is geologically youthful. Explosive volcanic activity is also suggested by images taken from the Magellan spacecraft (Fig. 4.19).

Soviet Venera spacecraft provided additional evidence for past volcanic activity on Venus when they landed on the planet's surface and tentatively identified the rocks as resembling a type of lava that is found on the Earth's ocean floors and the Moon's mare basins. This basalt was identified at several landing sites, suggesting that volcanoes once poured lava over much of Venus' surface and perhaps this process continues today.

Photographs of Venus' surface suggest rocks with sharp un-eroded edges as well as loose soil and dust (Fig. 4.20). The fresh-appearing rocks and soil suggest that active volcanic processes may be replacing the old eroded surface with fresh, young material. Moreover, the flat, slab-like appearance of many of the rocks might be attributed to flowing lava. As the molten lava spread across the surface and cooled, it produced a thin, fractured, layer of rock that we see in the Venera photographs.

There is abundant evidence for lava flows on Venus, but the current observations cannot determine whether the planet is geologically alive or dead. Variations in the observed amounts of atmospheric sulfur dioxide have, for example, been attributed to volcanoes that are now belching forth gases, but simple atmospheric circulation might also cause such variations.

Thus, although speculations about current volcanic activity have not yet been proven, there is a wide variety of evidence suggesting that Venus is a dynamic world with many similarities to the Earth. The veiled planet exhibits evidence for crustal motions, volcanic activity, and flowing lava in the recent past, if not now.

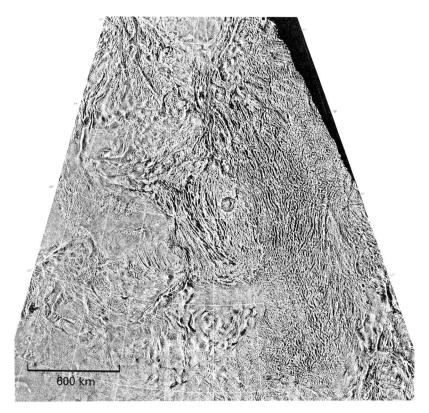

Figure 4.16. Maxwell Montes and surrounding region. This remarkable detailed radar image was taken by the Venera 15 and 16 spacecraft at 8 centimeters wavelength, and it resolves surface features as small as 1 kilometer across. These extremely rugged mountains, called Maxwell Montes, tower up as much as 11 kilometers above the surrounding terrain. The prominent circular crater, Cleopatra Patera, is about 100 kilometers across. The long, linear parallel ridges on Maxwell Montes are attributed to folds and faults in Venus' crust, and the concentric series of cracks or furrows at the south-west (bottom-left) of Cleopatra Patera suggest tensional tectonics. [Adapted from Science **231**, 1271 (1986), Courtesy of Gordon H. Pettengill, Massachusetts Institute of Technology, and Dr. Rzhiga, Institute of Radioengineering and Electronics, Moscow.)

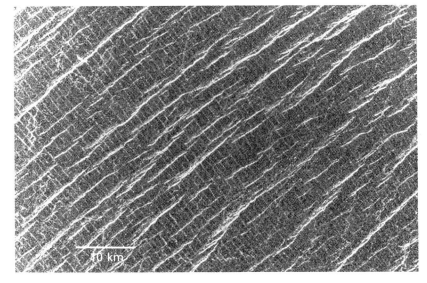

Figure 4.17. Crisscrossing terrain. This Magellan radar image of parts of the Lakshmi region shows a type of terrain that has not been seen previously, either on Venus, the Earth, or any other planet. The fainter lineations are spaced at a regular interval of about one kilometer, and their widths are at the 120-meter resolution of this mosaic. The brighter, more dominant lineations are less regular. The strange, crisscross pattern might be due to intersecting fracture lines. (Courtesy of JPL and NASA.)

Figure 4.18. Volcanism and rift formation in Beta Regio. This radar image of Beta Regio shows two bright features (top and bottom) that are thought to be volcanoes. They are connected by a central depression whose bright features may be the faults of a rift system. The volcano named Theia Mons is the bright circular feature in the south (bottom). It is about 350 kilometers across. The brightest area adjacent to the chasm in the north-west (top left) is the volcano called Rhea Mons. This radar image also portrays variations in surface roughness with a horizontal resolution of approximately 2 kilometers. Bright areas are relatively rough and unweathered, while dark areas are relatively smooth. The image was made at 12.6 centimeter wavelength with the 300 meter radio telescope of the Arecibo Observatory. (Courtesy of Donald B. Campbell, National Astronomy and Ionosphere Center.)

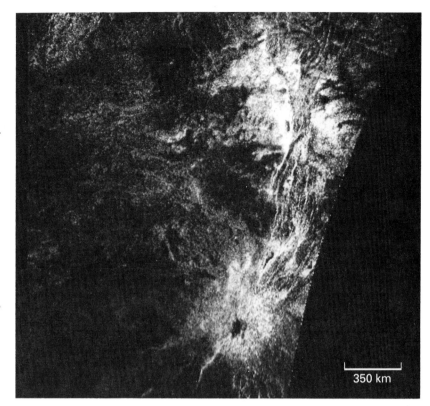

Figure 4.19. Volcanic fallout. Radar-bright surface features broaden and extend away from the one-kilometer crater in the middle of this image, suggesting deposits from an exploding volcano. This Magellan radar image shows an area to the north-east of Ushas Mons on Venus. (Courtesy of JPL and NASA.)

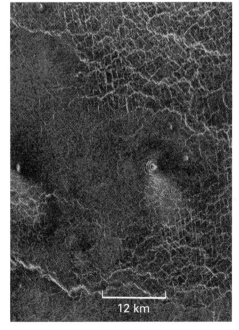

Figure 4.20. Surface rocks. Photographs of the surface of Venus taken from Venera spacecraft. The high temperatures and pressures on Venus and its corrosive atmosphere were expected to melt, deform and chemically weather the surface into a flat, featureless plain. However, these photographs reveal loose soil, dust and rocks. The thin slabs of rock could be due to molten lava that cooled and cracked. A partial chemical analysis indicates that the surface rocks may be composed of basaltic lava. (Courtesy of Iosif Shklovskii.)

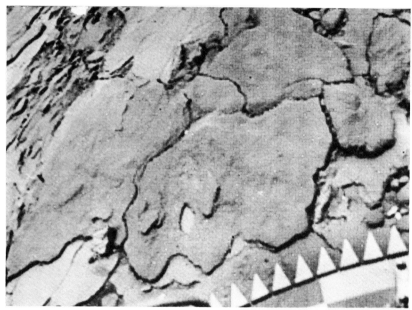

Focus 4C Venus – summary

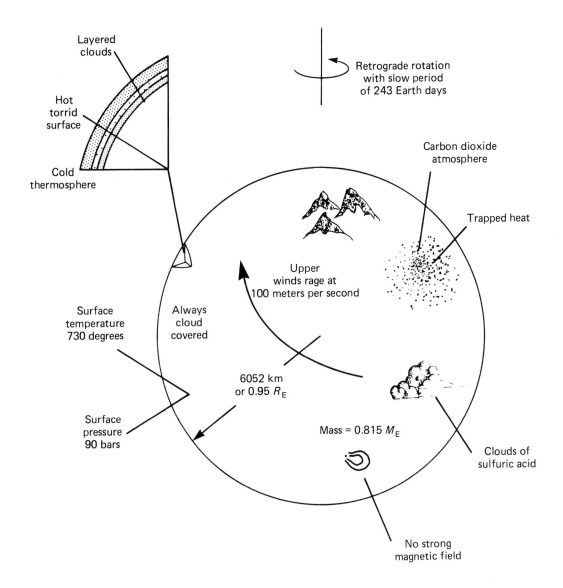

Layered clouds

Hot torrid surface

Cold thermosphere

Retrograde rotation with slow period of 243 Earth days

Carbon dioxide atmosphere

Trapped heat

Upper winds rage at 100 meters per second

Surface temperature 730 degrees

Always cloud covered

6052 km or 0.95 R_E

Mass = 0.815 M_E

Surface pressure 90 bars

Clouds of sulfuric acid

No strong magnetic field

Mass: 4.87×10^{27} grams = 0.815 M_E (Earth = 1)
Mean radius: 6052 kilometers = 0.949 R_E (Earth = 1)
Mean density: 5.25 g/cm^3
Sidereal rotational period: 243 days 36 minutes (retrograde)
Orbital period: 224.701 days
Mean distance from Sun: 0.723 A.U.
Venus has no satellites
Venus has no detectable intrinsic magnetic field

Earth: the water planet. Almost three-quarters of the Earth's surface is covered with water, as seen in this view of the Indian Ocean. The southern ice sheet gleams white at the bottom. (Courtesy of NASA.)

5 The restless Earth

The water planet, Earth, is covered by a thin membrane of air that protects, ventilates and incubates us.

Chemicals are destroying a thin layer of ozone that protects human beings from lethal ultraviolet radiation from the Sun.

The burning of fossil fuels and the destruction of tropical forests are altering the atmosphere's greenhouse effect, perhaps leading to global warming.

All of the planets are continually being bombarded with a lethal wind of energetic particles from the Sun; but we are protected from this wind by an atmosphere and a magnetic cocoon.

The aurora is a spectacular multi-colored light show that is energized by the Sun and shines like a cosmic neon sign.

Ancient magnetic rocks indicate that Earth's magnetic poles keep switching places, and that Earth's magnetic field may now be heading for a flip.

Boston and Italy were once part of Africa, a glacier of ice once covered the Sahara Desert, and the Pacific Ocean once washed against the shores of Colorado.

Continents weld together and create mountain ranges, and they split apart to form new oceans.

5.1 Planet Earth

(a) A delicate balance

Life as we know it, and the oceans of water that probably gave birth to life, exist only on Earth. When we look at our nearest neighbors, we see that Venus is too hot and Mars is too cold. Any water on Venus would be in the form of steam, and water on Mars is now locked beneath the surface in the form of ice and frost. Ours is the only planet whose temperature matches the temperature of liquid water, 273 to 373 degrees Kelvin. (The freezing and boiling temperatures of water are 273 K and 373 K, or 32 and 212 degrees Fahrenheit, respectively.)

The presence of liquid water makes the Earth unique among the known planets. About 97 percent of the water is collected in the oceans, and only a small fraction (1 part in 100 000 each year) is recycled from clouds into lakes as rain and from the lakes back to the oceans by way of rivers and streams. Eighty percent of the world's fresh water is locked in the polar caps, where the salts have been frozen out of the sea ice.

Water is a marvelous substance. We, ourselves, are largely water. As a liquid, it will dissolve almost anything to some extent. And when it freezes it expands and becomes less dense, in contrast to most substances. As a result, ice floats on the surface of lakes and oceans, so they freeze from the top down. This provides an insulating layer that protects animals and plants from freezing solid.

Today, the oceans cover three-fourths of the Earth's surface, and they contain so much water that if the Earth were perfectly smooth the oceans would cover the entire globe to a depth of 2.8 kilometers. During the recent ice ages, the Earth was much colder. The polar glaciers reached half way to the equator, and the oceans were much shallower.

When the solar system was young, the Sun was less than half as bright as it is today. It shed no more light and heat on Earth in those early days than it sheds on the planet Mars today. Would the oceans have existed? Would the Earth have been warm enough to nurture the early stages of life? How did the Earth's atmosphere evolve to its present composition? Is the temperature we enjoy today a permanent feature of our planet?

Astronomers and geologists have long been puzzled by such questions, and they have recently come to realize that the state of the oceans and the atmosphere are intimately linked to each other and to the interior of the Earth. Our atmosphere affects the temperature of the planet. Volcanic activity, as well as life itself, influence the composition of the atmosphere.

The ecosystem of our planetary surface is therefore the result of a delicate balance among many influences. In the past, this system has been a self-regulating one, but industrial activity may be on the brink of causing a shift away from this state of balance. We will discuss this delicately balanced ecosystem in much of the remainder of this chapter.

(b) Our layered atmosphere

On a warm, dry, windless day the air about us is nearly invisible and we are unaware of its touch on our skin. On such days, the air is composed almost exclusively of molecular nitrogen and oxygen (comprising 78 percent and 21 percent) with a 1 percent trace of argon and about 0.03 percent of carbon dioxide. Small as it is, this quantity of carbon dioxide is essential to life on Earth, as we shall see. The atmosphere also contains a variable amount of invisible water vapor, H_2O, up to 5 percent by volume in moist tropical air.

The drift of a cloud occasionally reminds us of the atmosphere about us, and on cold days we feel the air against our skin. The touch of the wind and the sight of birds and airplanes supported by their motion proves there is something substantial surrounding us. If we were to weigh the air in a one liter container we would find it tips the scales at slightly more than one gram; this is about one thousandth the weight of the same amount of water.

We find a further clue in the rise of smoke above a candle or a group of hawks circling above a warm meadow. Hot air rises around the flame of a candle, and the flowing air replenishes the supply of oxygen; without this current, the candle quickly goes out. The current is driven by a layering of the air under the force of the Earth's gravity. A candle would not burn in a spaceship, because there is no layering of the air by gravity, so hot air would not rise. The burned up air would hang near the candle and snuff it out.

Air, like all gases, is highly compressible. Because its atoms are normally far apart, a gas is mostly empty space, and it can be squeezed into a smaller volume by increasing the pressure of its surroundings. Such crowding causes the atoms to collide more frequently with the walls of their container, so the atoms push more vigorously on the walls. This pushing is called air pressure, and it is the force that prevents the atmosphere from collapsing to the ground. (Liquids resist much more vigorously than gases, so gigantic pressures are required to increase the density of a liquid.)

This relationship between pressure and volume – increased pressure brings decreased volume – is one of the keys to understanding the Earth's atmosphere. The other key is gravity that pulls down on the air, tending to create a pile-up. To understand the effect, imagine a hundred mattresses stacked into a pile. The mattresses at the bottom must support the weight of those above, so they will be squeezed thin. Those at the top have little weight to carry, and they retain their original thickness. The Earth's atmosphere imitates a pile of mattresses; the air near the ground is compressed by the weight of the overlying air. At greater heights, the compression is less, and the air gradually tails off into the vacuum of space. At a height of 10 kilometers (slightly higher than Mount Everest), the pressure and density of the air have dropped to 10 percent of their values near the ground. No insects and few birds can fly in such rarified air.

This decrease of atmospheric pressure with height accounts for the rise of balloons. Think of a balloon as a skin full of gas. The outside of the skin is pressed by the air, and the top of the balloon is pressed a little less than the bottom because the air pressure decreases with height. The buoyancy of the balloon is the difference between the upward and the downward force of the atmosphere. If the buoyancy exactly matches the weight of the skin plus its contents, the balloon will remain suspended motionless in the air. If the balloon is filled with a light gas such as helium, its weight will be less than the buoyancy force, and so it will rise.

Not only does the atmospheric pressure decrease as we go upward, the atmospheric temperature also changes, but it is not a simple fall-off. It falls and rises in two full cycles as we move off into space. The temperature of each level is determined by the rates at which it accumulates and loses energy, and the atmosphere can be divided into four regions in which the energy balance is of a different type. The lowest region is controlled by visible and infrared radiation, the next by ultraviolet, and the highest by X-rays from the Sun.

The lowest region is the troposphere, which is the region of our familiar weather patterns (Fig. 5.1). The ground and the surface of the ocean are heated by visible sunlight, and the low-lying air acquires some of the accumulated heat. The average surface temperature is 288 degrees Kelvin (59 degrees Fahrenheit), and radiation of heat provides a natural thermostat that keeps this temperature in a narrow range. If the ground becomes overheated by intense sunlight, the surface will radiate more energy to space and this tends to cool it off, especially on a clear dry night when the infrared radiation can easily escape from the atmosphere. And when the ground becomes cold, it radiates less and tends to conserve its heat.

Another type of thermostat near the ground is provided by currents of air that rise on a summer day. These currents carry heat from the ground and

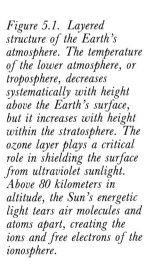

Figure 5.1. Layered structure of the Earth's atmosphere. The temperature of the lower atmosphere, or troposphere, decreases systematically with height above the Earth's surface, but it increases with height within the stratosphere. The ozone layer plays a critical role in shielding the surface from ultraviolet sunlight. Above 80 kilometers in altitude, the Sun's energetic light tears air molecules and atoms apart, creating the ions and free electrons of the ionosphere.

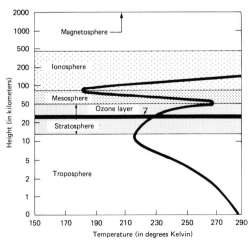

distribute it at higher levels, giving free rides to soaring birds and occasionally producing billowy fair-weather clouds. As the currents move up from the ground, the air expands in the lower pressure and it becomes cooler, at a rate of about 7 degrees Kelvin per kilometer. The lowest temperature occurs at a height of about 12 kilometers above sea level. This is the greatest height achieved by the air currents, and it is the top of the troposphere. Very little air from the ground or the surface of the ocean can rise above the troposphere.

Moving up from the troposphere, we enter the stratosphere. Here the temperature begins to rise again because molecular oxygen (O_2) and ozone (O_3) absorb the ultraviolet portion of the incoming sunlight. If all these radiations could suddenly reach the surface of the Earth, life would probably be wiped out by a terrible case of sunburn. A small portion of the ultraviolet is partly absorbed by O_2 in the stratosphere, and this process dissociates an oxygen molecule into two oxygen atoms, as described by the relation

$$\text{Ultraviolet photon} + O_2 \rightarrow O + O$$

Some of the energy of the photon causing the disruption of the molecule is transferred to atoms, which radiate it as ultraviolet light, producing airglow, which makes the atmosphere glow brightly when viewed in the ultraviolet wavelengths from space (see Fig. 5.2).

Some of the oxygen atoms become attached to another O_2 molecule, forming O_3,

$$O_2 + O \rightarrow O_3$$

Figure 5.2. Earth's airglow. This color-enhanced ultraviolet image was taken by the Apollo 16 astronauts from the airless Moon. It records the red crescent of the ultraviolet airglow emitted by excited oxygen atoms on the Earth's sunlit side. Blue bands of aurorae are also seen on the night side. (Courtesy of NASA.)

This new molecule, ozone, resonates to ultraviolet light and absorbs it from the stream of incoming sunlight. Although only one atmospheric molecule in a million is ozone, these relatively few can absorb almost all of the ultraviolet light before it reaches the ground.

The photons of ultraviolet light from the sun are 2 to 3 times as energetic as the photons of visible sunlight. They can break apart certain biological molecules, such as DNA, and they can increase the incidence of skin cancer. It has been claimed that a severe, 50 percent, depletion of atmospheric ozone would produce cataract blindness in all animals that are regularly exposed to sunlight. Milder depletions could cause systematic crop damage.

Certain chemicals threaten to destroy the ozone. They are man-made gases, invented about a half-century ago and given the euphonious name "chlorofluorocarbons." This name is a giveaway to their composition: chlorine, fluorine, and carbon. They have been widely used as cooling liquids in refrigerators and air conditioners, as foaming agents for insulation, as propellants in hair spray and deodorants, and as cleaning solvents in the process of manufacturing computer chips.

The chlorine is the culprit in the destruction of atmospheric ozone. It breaks apart the ozone molecule, returning it to normal oxygen and removing the absorbing quality of the gas. And it survives the process, so it can strike again and again. A single chlorine atom can disrupt as many as 100 000 ozone molecules before it is captured and locked in some other molecule and removed from the high atmosphere.

We may have already disrupted our ozone shield. An "ozone hole" has been discovered over the South Pole. It develops each polar spring and leaves a broad gap in the ozone layer, wider than the continent of Antarctica. Global winds evidently carry the chemicals away from the tropical and equatorial regions and collect them in a vast polar vortex where they can be released by spring sunshine. This hole appears to have increased over the past decade, and a similar vortex, although weaker and not so cold, is found over the North Pole, but it has not been seen to produce an ozone hole. Such a hole would be closer to population centers and is potentially more dangerous.

The discovery of the ozone hole sparked public awareness to the fragility of the ozone layer and convinced the international community to put a stop to the production of chlorofluorocarbons by the year 2000. An agreement to this effect has been signed, and none too soon, because many of the culprit chemicals are already loaded into the atmosphere. The chlorine atoms are long-lived and they will continue to work their way into the ozone layer for decades to come. We may, in fact, see a continuing increase of the ozone hole into the next century, and we can only hope that the trend is reversed before the damage has become severe.

Above the ozone layer, which is produced by ultraviolet light, is the ionosphere, which is produced by X-rays from the Sun. This is the electrical layer that is crucial for radio propagation. X-rays strip electrons from the atmospheric atoms and the resulting electricity acts as a mirror for longer radio waves, such as those used in amateur and commercial broadcasts. The shortest waves can pass through the ionosphere, because, roughly speaking, they are short enough to pass among the electrons, and these are the wavelengths used in satellite communications from continent to continent.

(c) Continents, oceans and ocean floors

There are two major types of terrain on Earth – the high, dry continents and the low, wet floor of the ocean. Between them, and partially surrounding many continents, is a narrow strip of shallow ocean called the continental shelf.

To those of us who are confined near the surface of the globe, the Earth seems rugged, with towering mountains rising 7 or 8 kilometers above the ocean and deep trenches sinking almost as far beneath the ocean's surface. But a scale model of the Earth would have to be quite smooth. These extremes reach only one-tenth of one percent of the Earth's radius above and below the ocean surface. A basketball this smooth would have bumps no more than 0.1 millimeters high, roughly the size of the dot at the end of this sentence.

The smoothness of the Earth is due to the immense force of its gravity and the weight of its outer layers, which largely overcomes the electrical forces inside the solids making up the Earth and causes them to lie in concentric spherical shells. In smaller bodies, such as asteroids less than a few hundred kilometers in diameter, the interior is strong enough to remain rigid and they retain their original irregular shapes.

(d) Bulk properties of the Earth

Even if we could smooth out the oceans and continents, the Earth would still not be a perfect sphere. The outward force of its rotation produces a slight bulge at the equator – or, as some prefer to think of it, the rotation flattens the Earth's poles slightly. As a result, the equatorial radius is 6378 kilometers, while the polar radius is 21 kilometers shorter. The astronauts could not have noticed this slight bulge with the naked eye, but it can be mapped from the surface by carefully measuring the positions of stars. It can also be detected through its gravitational effects on clocks and Earth-orbiting satellites.

When the Earth's mass, an unimaginable six thousand billion billion tons, is divided by its volume, we obtain a figure for the average density of the Earth. It is 5.5 times that of water, or $5.5 \, \text{g/cm}^3$. To appreciate what this value implies, we need only pick up an ordinary rock. It will typically have a density only one-half as great, about 3 times that of water. A small portion of this difference is caused by the great pressures that squeeze rocks inside the Earth. Most of it, however, must be due to the abundance of high-density material deep in the Earth, and because iron is so abundant in meteorites, most geologists now believe the Earth has a large core of iron.

5.2 Journey to the center of the Earth

The interior of the Earth is layered in the manner of a peach. Its deeper layers are more dense, and they are often separated from one another by sharp transitions. There are three major divisions: (1) the crust, (2) the mantle, and (3) the dense core presumed to be of nickel and iron (Fig. 5.3). They are the skin, pulp, and pit of the Earth, so to speak.

(a) The crust and mantle

The outer skin of the Earth is a thin, rocky crust that varies from 10 to 65 kilometers in depth. The mantle beneath it reaches downward some 2900 kilometers. Most of the rocks of the mantle and crust consist of minerals in which silicon and oxygen are linked with other atoms (silicates). The mantle consists mainly of minerals rich in magnesium and iron. It is solid rock except for a partially fluid zone just below the crust. The crust, on the other hand, consists of lower density minerals such as quartz and feldspars, containing an abundance of silicon and aluminum. The continental granites and oceanic basalts were formed in the fiery melts of volcanism (Table 5.1).

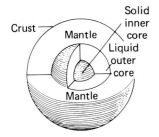

Figure 5.3. Crust, mantle and core. A relatively thin, rocky crust covers a thick silicate mantle. They overlie a liquid outer core, composed mainly of iron, and a solid inner core of pure iron. These nested layers have been inferred from seismic waves that travel through the Earth and become reflected at the layer boundaries.

Table 5.1. *The five most abundant elements in the Earth*

Element	Symbol	Average abundance (percent by mass)
Iron	Fe	34.6
Oxygen	O	29.5
Silicon	Si	15.2
Magnesium	Mg	12.7
Nickel	Ni	2.4

The heat for this melting was generated by radioactivity in the upper crust. The temperature of the rocks increases with depth in the outer layers of the crust, reaching 176 degrees Fahrenheit at 1.5 kilometers and requiring air conditioning in deep mines. If this rate of increase persisted at all depths, the rocks would be melted at a depth of 50 kilometers. But rocks are evidently solid through the mantle to a depth of 2900 kilometers. So the temperature rise must abate, and this suggests that the heat-producing radioactive elements are confined to the outermost layers of the Earth.

Only the outermost layers of the Earth can be directly measured, but the deeper structure can be mapped with the help of earthquake waves. (See Focus 5A, Taking the pulse of the Earth.) The Greek word for earthquake is *seismos*, which is derived from a word meaning "to shake." Today, earthquake waves are often called seismic waves, and seismology is the study of earthquakes.

Most earthquakes occur just beneath the Earth's surface when massive blocks of rock slip and crunch against one another. The reverberations resemble ripples spreading out from a disturbance on the surface of a pond. These waves move in all directions and their arrivals at various places on the Earth can be detected by seismometers. By comparing the arrival times at several seismic observatories, geologists can pinpoint the origin of the waves and trace their motions through the Earth. A large portion of the waves penetrates the deep interior and then reemerges toward the surface on the other side.

Rock layers of different stiffness will propagate the waves at different speeds, much the way that a tightened violin string will sing at a higher pitch. As a result, the paths of the seismic waves are bent and focused by their passage through the interior, as though they had passed through an immense lens. (The lens of the human eye is a rough analogy; its concentric layers focus light onto the retina at the back of the eyeball in much the way that earthquake waves are focused by the interior of the Earth.) By a careful mapping of the patterns of many earthquakes (somewhat in the manner of a catscan analysis of X-rays), it has been possible to outline the layers of the interior of our planet (see Fig. 5.4).

The crust is thinnest under the oceans. The continents are regions of thicker crust, and because the crustal material is less dense, it tends to float on the mantle. High mountains have deep crustal roots that provide buoyancy and keep them afloat, much the way that icebergs float on the ocean.

(b) Lithosphere and asthenosphere

The outer solid region beneath the familiar oceans and mountains is approximately 100 kilometers deep and it is called the lithosphere. Beneath the lithosphere lies the asthenosphere, or zone of weakness, that reaches to a depth of about three hundred kilometers.

The distinction between crust and mantle is one of chemical composi-

Focus 5A Taking the pulse of the Earth

We cannot see the inside of the Earth, and even our deepest mines are tiny dents in its surface. However, scientists have found a way to use earthquakes to illuminate the Earth's interior. Although earthquakes originate no deeper than 700 kilometers below the Earth's surface, they shake the planet to its very center, causing it to vibrate and ring like a bell.

The earthquakes actually produce several types of waves. There are the P and S waves that travel through the solid body of the Earth, and the surface waves that propagate around it. The speeds of these waves depend on the varying density and stiffness of the rock through which they propagate. The P, or push and pull, waves are physically similar to sound waves, although their vibrations are much slower than audible sound. The S waves, on the other hand, set the Earth vibrating at right-angles to the path of the waves. Unlike ordinary sound, the S waves do not travel in a liquid. They propagate only in solid substances that have elastic resistance to twisting. Gelatin is an example of such a material.

Seismometers on the surface of the Earth can record the arrival of these waves in much the way that a stethoscope records the beat of a hidden heart. By combining the arrival times of the different types of waves that have travelled through the Earth to various points on the surface, seismologists can construct a profile of the Earth's interior in much the way that an ultrasonic scanner can map out the shape of an unborn infant in its mother's womb.

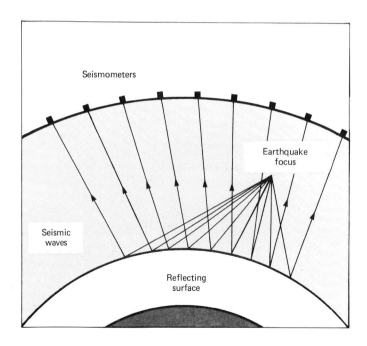

Figure 5.4. Layered structure of the Earth. The Earth's internal structure is delineated by the varying velocity of earthquake waves. A low-velocity zone in the upper mantle marks the hot, plastic asthenosphere that lies at depths of between 100 and 300 kilometers. The cold, rigid lithosphere lies above the asthenosphere. The boundary between the mantle and core is marked by a precipitous drop in the velocity of the P waves at a depth of about 2900 kilometers. The S waves do not propagate beyond this boundary. The liquid outer core is separated from the solid inner core at a radius of 1216 kilometers where the P waves increase in velocity.

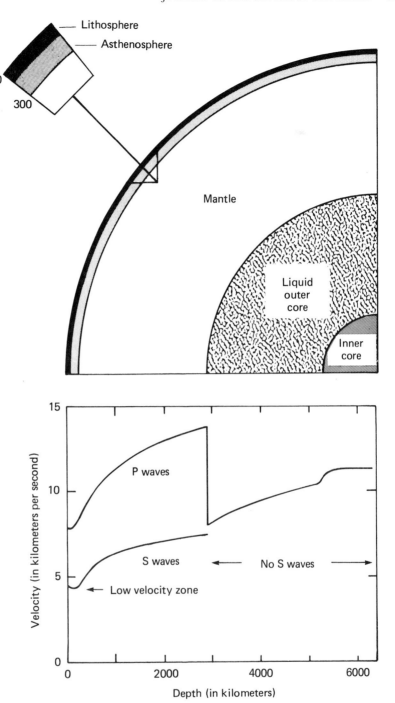

tion, while the distinction between lithosphere and asthenosphere is one of stiffness. The lithosphere takes the root of its name from the Greek *lithos*, for stone. The word asthenosphere comes from the Greek *a-*, without, + *sthenos*, strength. The material of the asthenosphere, which is warm and plastic, is revealed by the slowness with which it propagates seismic waves.

The radioactive elements responsible for the warmth of the asthenosphere were probably concentrated in that region during the formation of the Earth's core. They are too weakly concentrated to melt the rock, but they cause it to soften and behave like putty. Rock in the asthenosphere flows slowly when strained for a long time, but it responds like a solid when it is struck by an earthquake. (The behavior of "Silly Putty" is very similar.)

Continents are embedded in the lithosphere and they float on the asthenosphere. This gives them their mobility and permits them to drift over the face of the globe. These motions are driven by deep-seated, and very slow, currents of plastic rock in the outer core and the asthenosphere (Fig. 5.5).

(c) The core

The Earth's core reaches about half way to the surface, implying a volume that is one-eighth that of the entire Earth. If the density of the Earth were uniform, the core would have an equal share, one-eighth, of the mass of the Earth, but its actual mass is nearly three times greater. So the core's density is very high, and this points to iron as the most likely material. An analysis of seismic waves indicates that the core is far from a smooth sphere and contains troughs and swells on its surface, deeper than the Grand Canyon and higher than Mount Everest. These formations may be related to upwellings of convective motion deep in the Earth.

The temperature of the deep core is difficult to determine, but it appears to be about 6900 degrees Kelvin, which is a bit hotter than the surface of the Sun. At first glance, this would seem to imply that the center of the Earth must be liquid, but this is contradicted by seismic evidence, which indicates

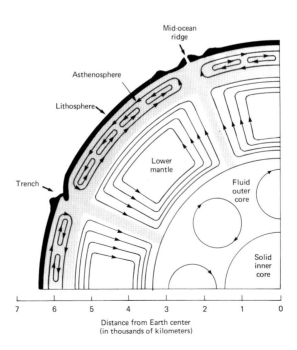

Figure 5.5. Convection cells. Elongated convection cells in the asthenosphere may be aligned in long cylinders that drive the overriding lithospheric plates along like a conveyor belt. A larger-scale circulation transports heat from the volcanic mid-ocean ridge to a deep-ocean subduction trench. Heat-driven convection in the fluid outer core probably generates and maintains the Earth's magnetic field.

a solid region in the deep interior. The clue to the apparent paradox is high pressure. The pressures at the center of the Earth (3.6 million times the pressure at sea level) have been imitated in laboratory experiments, and they lead to a remarkable conclusion. At these pressures, iron can persist as a fairly rigid solid even at a temperature of thousands of degrees.

Heat has apparently been bottled in the deep interior since the early days of the Earth, 4.6 billion years ago. As this heat gradually flows outward from the cooling core to regions of lower pressure, it maintains a liquid layer that slowly turns inside out, like a bowl of thick cereal on a stove. The circulation of this melted rock moves at the speed of a growing fingernail (a few centimeters per year) and carries with it the roots of the overlying continents. The action of these currents on the asthenosphere and the bottom of the lithosphere provide the forces that drive continental drift.

The origin of the layered structure of the Earth's interior is still a geological mystery, but there are two extreme alternatives. According to the hot accretion theory, the Earth was formed from a hot nebula and was molten at a very early stage. This theory supposes that the iron-rich core formed first and then the silicate materials were added to the exterior as the nebula became cooler. If this theory is correct, the layered structure has been a feature of the Earth since its infancy.

A more elaborate process is described by the cold accretion theory. This process begins with the formation of a homogeneous solid Earth. This globe then became heated by radioactivity that was uniformly distributed through the interior, and its temperature gradually rose to the melting point. When the planet melted, its gravitational field caused the heavy elements to sink toward the center, forming a dense core, while the lighter elements rose toward the surface producing a series of chemically distinct layers. After this process of differentiation, the Earth began cooling from the outside. The solid crust was formed, and then the mantle.

5.3 Remodelling the Earth

(a) Continental drift

The Earth has a curiously asymmetric face. Ocean waters dominate in the southern hemisphere while the continents dominate in the northern hemisphere. The outlines of the continents themselves exhibit a number of remarkable symmetries, especially along the shores of the Atlantic Ocean. For example, the eastern edge of South America would fit snugly into the western edge of Africa. In fact, much of the east and west shores of the Atlantic Ocean are as well matched as the shores of a river (see Fig. 5.6). Not only are the superficial shapes similar, but there is also a striking match of geological formations and fossils on opposite sides of the ocean. It now appears that the continents were once part of a single land mass that fragmented and drifted apart. This hypothetical continent is called Pangaea (meaning all lands and pronounced *pan-gee-ah*); it broke apart about 200 million years ago. This is fairly recent in geological terms (see Fig. 5.7).

The theory of moving continents, or continental drift, was proposed near the beginning of the 20th century by Alfred Wegener. It was initially rejected by most geologists, who thought the trans-oceanic matching of shapes was no more significant than the matching of the opposite banks of a river. And geologists could not understand how the continents could plow their way through the crust, especially the ocean floor that seemed so strong. Without a plausible mechanism, the idea of continental drift was disparaged and occasionally ridiculed. Exploration of the ocean floor by sound waves provided the first evidence that Wegener was on the right track after all.

(b) Sea-floor spreading

Only fairly recently, some marine geologists sailed back and forth across the oceans dropping explosives overboard to generate pulses of sound. This "echo sounding" was their method of determining the depth of the ocean and probing into its floor. The first echo determined the travel time to the floor and back. Gradually, these explorers pieced together a series of profiles and built up a contour map of the floor. By the middle of the 20th century a new method had developed, sonar, using a continuous train of sound pulses. This technique was far more efficient and permitted making a highly detailed map of the entire ocean. Nowadays, satellite data obtained with similar techniques allow us to, in effect, empty the oceans and see their floors. The resulting maps have rivalled the most imaginative charts of the lost island of Atlantis that Plato says was swallowed by the sea (see Fig. 5.8).

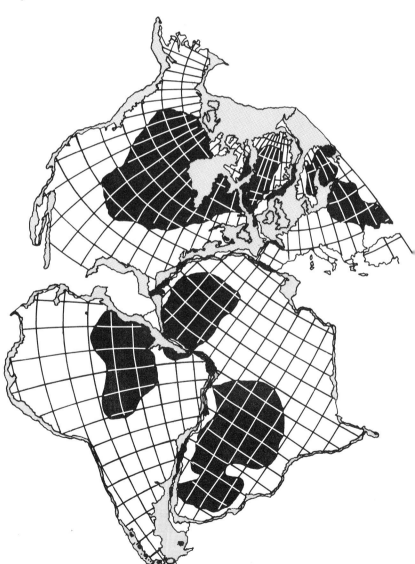

Figure 5.6. Continental fit. The continents fit together like the pieces of a puzzle. Here the fit has been made along the continental slope at the depth of 0.91 kilometers (500 fathoms) below sea level (gray areas). The white areas between continents are gaps, whereas the small black areas between continents are places where they overlap. The large black areas within each continent are terrains that are between 1.7 and 3.8 billion years old. [Adapted from Philosophical Transactions of the Royal Society (London) **A258**, 41 (1965).]

Figure 5.7. Continental drift. Two hundred million years ago all of the continents were grouped into a single supercontinent called Pangaea and the world contained only one ocean (top). The continents then drifted away from Pangaea, riding on the backs of plates to the positions they now occupy (middle). The bottom diagram depicts the world geography 50 million years from now.

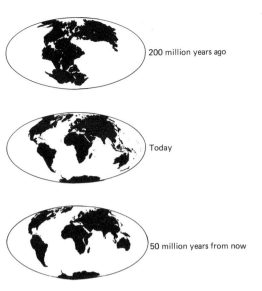

200 million years ago

Today

50 million years from now

Snaking through all the oceans of the world is the mid-ocean ridge, a gigantic network of mountain ranges, 74 000 kilometers long. This is long enough to accommodate the Alps, Andes, Himalayas and Rockies. Where the undersea mountains reach to the surface, they form chains of islands (see Fig. 5.9). They are strings of undersea volcanoes, built of lava that has emerged from deep in the crust or the upper mantle.

Even more remarkable are the deep valleys, called rifts, that run along the mid-ocean ridge splitting it as though it had been sliced with a knife. The mid-ocean ridge is more accurately described as a tear in a sheet of paper rather than a cut, because this rift represents the line from which the spreading ocean floor creeps outward on both sides. Hot lava emerges into the gap, and as it moves away from the mid-ocean ridge it cools and solidifies, becoming more dense and finally tending to sink beneath the continents on either side.

Remarkable creatures live in the eternal darkness of the deep sea rifts. They are warmed and fed by superheated water saturated with minerals emerging from the mid-ocean ridge. Giant clams, tube worms, and crabs thrive without light by digesting minerals and bacteria that flourish in the volcanic effluvia.

The actual spreading of the sea floor is too slow to be seen by present techniques but there are several types of evidence that leave no doubt about its reality.

First, the sediments of the floor of the ocean are relatively young. None are more than 200 million years old, and they are thinner than they would be if they had accumulated over the lifetime of the oceans. Young ocean floors have evidently replaced the older ones.

Second, fossils recovered from the seabed can be dated, and they show that the floor is youngest in the middle and grows progressively older with increasing distance from the mid-ocean ridge. Both the average age and the thickness of the sediment increase away from the ridge.

Third, the ocean floor shows a pattern of magnetic field reversals that is symmetric on both sides of the mid-ocean ridge. Such field reversals have also been found on the flanks of volcanoes on land, and they indicate that the Earth's magnetic field changes its direction from time to time. When fresh molten lava pours out of a land volcano or out of the mid-ocean ridge,

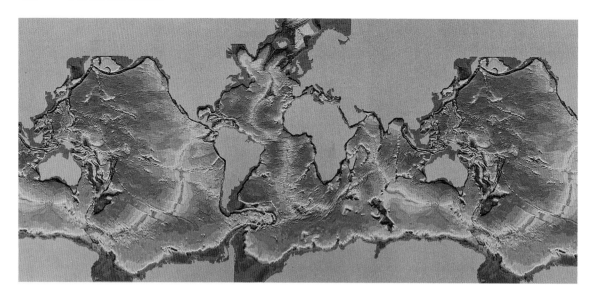

Figure 5.8. Sea-floor map. Map of the world's ocean floors as acquired by the SEASAT satellite. The mid-Atlantic ridge runs down the middle of the ocean floor separating Africa from North and South America. As shown here, a succession of great ridges runs through all the world's ocean floors, although not always in the middle. (Courtesy of William F. Haxby, Lamont-Doherty Geophysical Observatory, Columbia University.)

Figure 5.9. Volcanic islands. Three lava vents erupting from the volcanic island Surtsey on August 19, 1966, almost three years after it rose out of the sea near the coast of Iceland. The volcanic island of Jolnir is in the background. It disappeared back into the sea about one month after this picture was taken, but Surtsey is still visited for research purposes today. All of these volcanic islands, including Iceland, mark points where a mid-ocean ridge has risen out of the ocean. (Courtesy of Hjalmar R. Bardarson, Reykjavik, from his book Ice and Fire.*)*

its iron particles are magnetized by the Earth's field, and this magnetization remains as a fossil when the rock solidifies.

By radioactive dating of volcanic rocks, it is possible to tell when they solidified and to build up a chronology of the magnetic changes. This chronology can then put dates on the reversals found in the sea floor, and from the distances travelled it is possible to compute the rate of sea-floor spreading. (See Focus 5B, Earth's magnetic tape recorder.) The results are remarkably precise. The ocean floor moves away from the ridge at rates of 2 to 20 centimeters per year, or between 4000 to 40 000 kilometers in 200 million years – entirely adequate to explain the present widths of the great oceans.

Focus 5B Earth's magnetic tape recorder

Ancient magnetic rocks indicate that the Earth's magnetic field has not always been the same as it is today. The evidence is found on the flanks of volcanoes on the land and under the sea. When the molten volcanic lava flows to the surface and hardens into rock, its internal magnetism lines up with the Earth's magnetic field and freezes into position. These magnetic fossils record the direction and intensity of the Earth's magnetic field where the lava solidified.

An inspection of magnetic fossils of differing ages from all parts of the world resulted in an amazing discovery. The Earth's magnetic field has flipped, or reversed its direction, many times in the past.

The ages of rock on land can be used to determine the time scale of the magnetic reversals (left). Over the past 3.6 million years, the Earth's magnetic field has reversed itself at least 9 times. The data describe normal epochs (gray) when compasses would have pointed toward the geographic north, and reversed epochs (white) when compasses would have pointed south.

Magnetic detectors dragged in the sea behind ships have revealed alternating bands of normal and reversed magnetism in the ocean floor. The pattern of magnetic reversals on both sides of the mid-ocean ridge is identical – each side is a mirror reflection of the other (right). When lava pours out of the ridge and solidifies, the rock on either side freezes in the magnetic field alignment. Seafloor spreading then carries the rock away from the ridge, creating symmetrical bands of alternating magnetism.

The Earth's magnetic field has weakened by more than 50 percent during the past 4000 years, and it seems headed for a magnetic flip in the next few thousand years. Compass needles will reverse their directions. Animal species that depend on magnetic fields for guidance must continuously adapt to the changing field.

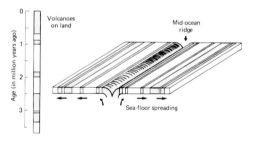

(c) Plate tectonics

As the magma wells up along the mid-ocean ridges and solidifies to form the new lithosphere, it breaks into a mosaic of large plates, thousands of kilometers across and vaguely resembling the cracked pieces of an egg shell (Fig. 5.10). The plates move horizontally over the plastic mantle, carrying the continents with them. In fact, the continents are embedded in the moving plates, and continental drift is a consequence of the motion of plates carried along by the sea-floor spreading. The motion is rapid enough to be measured over the course of a few years, because changes of intercontinental distances can now be measured with an accuracy of better than 1 centimeter. The Pacific Plate is, for example, carrying Los Angeles northward by 3 to 6 centimeters per year. At this rate, Los Angeles will be a suburb of San Francisco in 10 million years. (See Fig. 5.11.) For another example, Massachusetts and Sweden are separating by 1.7 centimeters per year, so the Atlantic Ocean was about 9 meters narrower when Columbus crossed it.

The spreading ocean floors are eventually pushed down inside the Earth at subduction zones where two plates meet. Along a subduction zone, the lithosphere plunges steeply into the Earth, like a down-going escalator, producing a deep ocean trench. The floors of several ocean trenches sink farther below sea level than Mount Everest climbs above it. Because the continental rock is less dense than the Earth's interior, it remains on top while the oceanic floor slides underneath. The continents are persistent floating islands and are much older than the ocean floors.

The plates are fairly rigid and become deformed only at their edges. Each plate may be regarded as a spherical cap that slowly rotates about an axis passing through the center of the Earth. As the individual plates rotate, they grind against each other, occasionally building up stresses. When the stresses exceed the strength of the rock or the force of friction, the plates slip and an earthquake occurs. Thus, most of the plate boundaries are marked by a ring of earthquakes and other types of geological activity, such as volcanoes and mountain building.

If two plates collide head on, they may become welded into a single larger plate, and continents may grow by the accumulation of material along their edges. As an example of the latter, the Pacific Ocean once reached to Colorado, but the western United States was subsequently grafted onto the shore.

But what moves the plates? The hot material beneath the plates is a thick liquid and it rolls over very slowly in a wheeling motion called convection. As the material rotates, the plates are dragged along by convection currents. The heated rock moves upward, spreads sideways, dragging portions of the lithosphere with it, and then cools and sinks again, to be reheated and pushed upward again in an endless cycle.

Convection occurs when molten rock becomes swollen by heat and rises through the cooler overlying material, like the currents in a pot of heated water. The moving material carries heat upward and then it cools and begins to sink again. Convective stirring thoroughly mixes the rock of the upper mantle in several hundred million years. This accounts for the similarity of basalts recovered from mid-ocean ridges throughout the world.

Thus the energy that drives the continents, spreads the sea floor, sets off

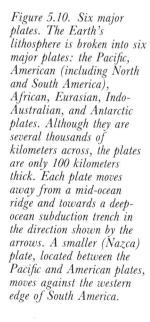

Figure 5.10. Six major plates. The Earth's lithosphere is broken into six major plates: the Pacific, American (including North and South America), African, Eurasian, Indo-Australian, and Antarctic plates. Although they are several thousands of kilometers across, the plates are only 100 kilometers thick. Each plate moves away from a mid-ocean ridge and towards a deep-ocean subduction trench in the direction shown by the arrows. A smaller (Nazca) plate, located between the Pacific and American plates, moves against the western edge of South America.

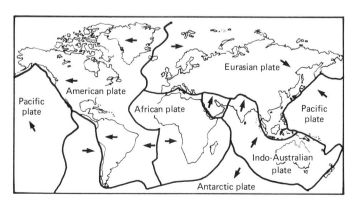

earthquakes and ignites volcanoes is derived from the hot interior of the Earth. When the heat provided by radioactive decay becomes totally depleted, the Earth will become a geologically dead planet. Erosion will gradually flatten the mountains.

(d) Mountain building

As massive as it is, a range of mountains cannot resist eventual destruction by wind, water, and ice. Old mountain ranges, such as the Appalachians and the Urals, once stood as high as today's Himalayas, but they have eroded into gentle undulations and rounded knobs. Given the great age of the Earth, this erosion acting by itself would have worn away the continents, and the globe would have long ago been covered by oceans. But the mountains are constantly being rebuilt. One of the methods of mountain-building is by the collision of plates. (See Fig. 5.12.)

If two plates carrying continents on their backs collide, the continents will buckle upward, heaving land and sea sediment slowly toward the sky, creating a range of mountains. The magnificent Himalayan range was formed this way, when India (formerly a piece of Africa) rammed into Asia. Today the plate that carries India continues to slide beneath the Asian plate, widening the Indian Ocean and pushing the Himalayas upward. Another example is provided by the Alps of central Europe, formed where Italy moved up from Africa and collided with Switzerland's former ocean shore. Today the African plate continues pressing Italy northward and raising the Alps.

Another mode of mountain-building occurs in the ocean floor, above an isolated volcanic hot spot. A hot spot is an outlet that permits lava to flow up through a lithospheric plate. They are formed when a plume of lava beneath a plate melts its way through, piercing it like a welder's torch, and a volcano erupts. The source of lava is anchored deep in the Earth, and the plate may continue to slide past, creating a string of volcanic islands and undersea mountains. (See Fig. 5.13.) The Hawaiian islands were formed this way as the Pacific plate moved eastward; the youngest island is at the western end of the string. According to Hawaiian legend, Pele the fiery

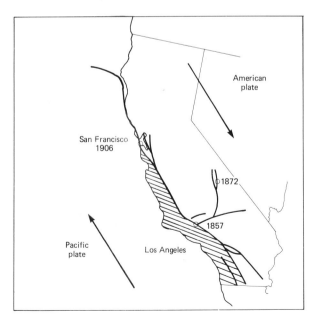

Figure 5.11. Los Angeles is moving. The Pacific plate is carrying Los Angeles towards San Francisco at a rate of about 5 centimeters per year. San Francisco is caught between the American and Pacific plates that move in opposite directions like immense grindstones. The boundary between these plates is the San Andreas fault, which runs right through California. The dated circles denote places where very major earthquakes have occurred with magnitudes of 8 and over on the Richter scale. An earthquake of magnitude 7.1 occurred near San Francisco on October 17, 1989.

Figure 5.12. Colliding plates. Two plates meet head on, one bearing a continent near its leading edge. As the lithosphere of this plate plunges under the overriding continental plate, volcanes are produced (top). But when the two continents collide, new mountains are generated (center). In some situations, the advancing plate may become disrupted, and plate motion may stop. The two continents then become welded together, forming a larger one, and a new subduction zone is formed elsewhere (bottom).

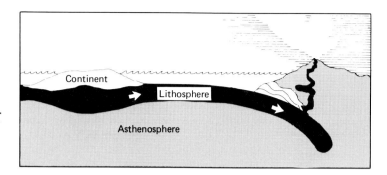

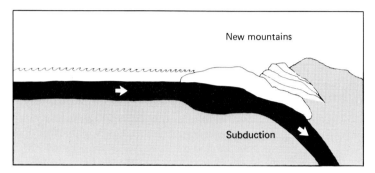

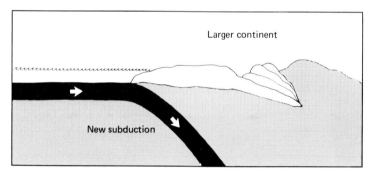

Figure 5.13. Hot spot. A hot spot that is anchored deep within the Earth now feeds molten lava through a long pipe to Mauna Loa on the island of Hawaii. The moving Pacific plate carries volcanic islands away from the hot spot. As the plate moves on, wind and water erode each peak in turn, sometimes reducing them to ocean-covered seamounts.

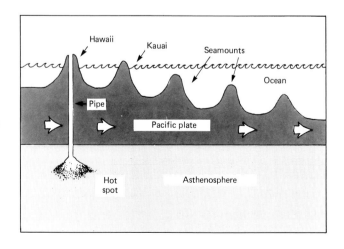

volcano goddess wandered along the chain of islands, digging pits of fire in each before settling on the large island of Hawaii.

If a plate carrying a continent comes to rest over a hot spot, the upwelling lava can lift the continent and cause it to split in two, forming a rift. (See Fig. 5.14.) If the upwelling is short-lived, the result is merely a rift scar, such as the Rhine Valley. If it persists, the rift widens, eventually reaching a coastline and permitting sea water to flow in (Fig. 5.15). The Red Sea and the Gulf of Aden were formed this way, and they are developing into an ocean that will further separate Africa from Asia (see Fig. 5.16).

Thus, a new dynamic picture of the Earth has emerged. Colliding continents can squeeze an ocean out of existence, and new oceans can appear over a hot spot in the midst of a continent. Today, the Mediterranean Sea is narrowing as Africa moves toward Europe, while the Atlantic Ocean and the Red Sea are becoming wider. The ocean floor remains in eternal youth as new floor spills out of the mid-ocean ridges and old floor is destroyed in the deep trenches. This subduction churns up lava and volcanic gases from the interior. These gases have a profound effect on the atmosphere and on the Earth as an abode for life.

5.4 The evolving atmosphere

(a) Clear skies and stormy weather

The Earth is the only place in the solar system where we can stand naked and survive. The air brings oxygen to our lungs and refreshes our bloodstream; sunlight provides just enough heat to prevent our fluids from freezing or boiling. Astronomers once thought that this happy situation was

Figure 5.14. Nyiragongo. Africa rides on a plate that has lingered in one place for about 30 million years. The continent is beginning to fall apart as it slowly disintegrates under the pent-up pressure of numerous underlying hot spots. It is becoming unstitched along the Great Rift Valley, a long forking gash that crosses 4500 kilometers of the continent. Volcanic outpourings like Nyiragongo fill the valley as the rift slowly widens and makes way for a future sea. (Courtesy of Bruce Coleman.)

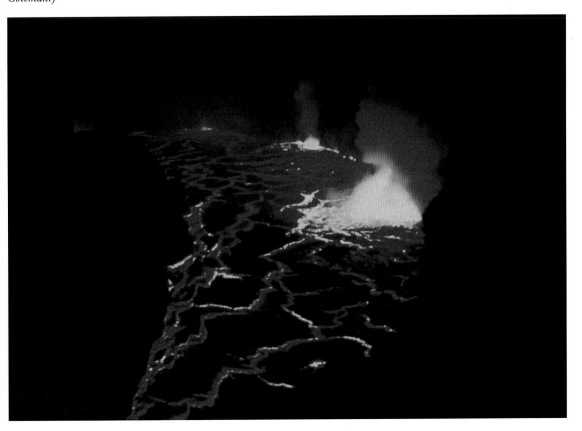

simply a result of our planet's being at just the right distance from the Sun, but we now know that our abode is "air-conditioned" by several types of interdependent cycles that lead to the dynamic balance that makes life possible. The balance appears to be fine-tuned by life itself.

The most obvious of these cycles is the water cycle that brings us our daily weather and produces the climatic differences that starkly distinguish one part of the globe from another. The water cycle begins in sunlight. Of the energy that falls on the Earth in the form of sunlight, 30 percent is reflected directly back into space so it has no effect on our climate. Nineteen percent is absorbed by the air on its way toward the ground, producing the ozone layer and heating the upper atmosphere. The remaining 51 percent is absorbed by the oceans and land. Ultimately, this energy is returned to space as infrared radiation, so the books are balanced, and the Earth emits almost exactly as much heat as it receives from the Sun.

About one-third of the solar energy reaching the Earth's surface is expended on the evaporation of sea water. This evaporation releases warm fresh-water moisture into the air and cools the surface of the ocean. The moisture rises high into the atmosphere, often appearing as clouds, and is transported great distances by winds before condensing again to water and falling to Earth as rain, hail, or snow. Where it falls on the salty ocean, this water produces a slight and temporary drop in the salt content; where it falls on land, it refreshes lakes and streams and eventually finds its way back to the sea.

All the water in the oceans passes through this water cycle once in 2 million years. Yet, the ocean waters are at least 3.5 billion years old, so they must have completed more than a thousand such cycles. These are average

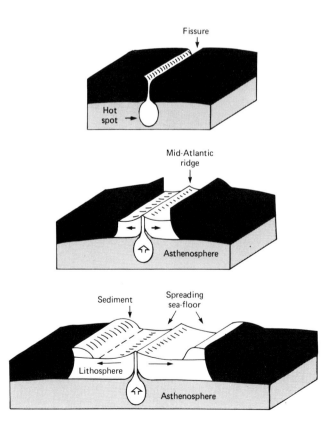

Figure 5.15. The rifting of a continent. A continental rift begins when molten lava rises up from deep in the Earth's interior and splits a continent open. As the fissure grows and widens, a future ocean floor spreads away from the ridge.

*Figure 5.16. A new ocean.
An ocean is being born
where the Arabian peninsula
and the African continent
are moving apart, a process
that began about 20 million
years ago. In a few hundred
million years the Red Sea
could be as wide as the
Atlantic Ocean is now.
(Satellite photograph
courtesy of NASA.)*

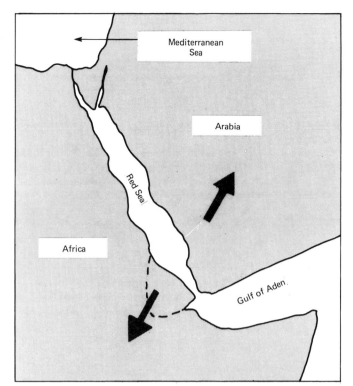

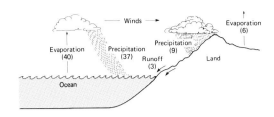

Figure 5.17. The water cycle. Water vapor is released into the air by evaporation. The rising air cools and condenses into clouds. These clouds are moved over great distances by winds, but they eventually return water to the land and oceans by the precipitation of rain, hail or snow. Most of the fresh water that is returned to the land runs back into the oceans. The numbers denote the amount of water that is evaporated or precipitated in units of 10 000 cubic kilometers, or 2.5 million billion gallons.

figures; some water is caught in the cycle and will recycle more rapidly; some water is trapped at the bottom of the sea and will recycle more slowly.

Our weather is largely a consequence of this water cycle (Fig. 5.17). Great quantities of heat are absorbed when water is evaporated and this heat is transported by winds, to be released when the vapor recondenses. These winds are generated by the unequal distribution of sunlight on the surface of the globe. The equator, for example, receives more sunlight than the poles and this difference establishes a pattern of winds that tend to carry heat to the colder places, thus reducing the initial temperature differences.

Air currents in the tropical zone tend to be marshalled into rotating patterns known as Hadley cells. These cells rotate primarily in the north–south direction, although they are deflected by the Earth's rotation. Along the equator they converge to form the trade winds that blow westward almost every day of the year. The temperate zones are dominated by high-altitude jet streams blowing eastward in a sinuous path which resembles the meandering of a river. Between the trade winds and the jet streams, the pattern is primarily a series of high pressure cells rotating clockwise in the north and counter-clockwise in the south. In the polar regions the pattern is primarily a series of low pressure cells rotating counter-clockwise in the north and clockwise in the south. (See Focus 5C, The windy planet.)

Our word "cyclone" is taken from the Greek word for wheel. In the northern hemisphere it indicates a low-pressure cell of air rotating counter-clockwise; an anti-cyclone is a high-pressure cell rotating clockwise. Cyclones are associated with stormy weather, and they occasionally develop into hurricanes (Atlantic Ocean and Caribbean Sea) and typhoons (Pacific Ocean and China Sea). In the southern hemisphere the directions of rotation are reversed.

(b) The breath of life

Almost all of the oxygen and carbon dioxide molecules in our atmosphere are the breath of plants and animals, and they are continually being recycled in photosynthesis and respiration processes. Animals require oxygen for respiration processes, and when they exhale they release carbon dioxide and water vapor. Green plants, on the other hand, absorb carbon dioxide and water, use them in the photosynthesis of nourishment and then release oxygen into the atmosphere. This symbiotic relationship is one of the most remarkable features of life on Earth.

The living and the non-living matter on Earth must affect each other and they may influence each other's evolution to some extent. It has been suggested that they form a global system that behaves somewhat like a great organism. This idea has been called the Gaia (pronounced *GUY-ah*) concept, after the Greek goddess of the Earth. According to this concept, the physical world of rocks, oceans and atmosphere is affected by plant and animal life in such a way as to maintain the conditions that are conducive to life itself.

In fact, the atmosphere was originally poor in oxygen, but blue-green algae began to put oxygen into the air 2 billion years ago. By about 400

Focus 5C The windy planet

The Windy Planet

```
                        pole
            polar    easterlies    pola
      r easterlies polar easterlies polar easterli
      es polar easterlies polar easterlies polar easter


        roaring  forties  roaring  forties  roaring  forties  roarin
      forties  westerlies  roaring  forties  roaring  forties  roar
      ing  forties  roaring  forties  roaring  forties  roaring  forties


      horse   latitudes   horse   latitudes   horse   latitudes   horse   latitudes


      the trades the trades the  trades the  trades the  trades the  trades the
      trades the  northeast trades  the trades  the trades  the trades  the trades
      the trades  the trades  the trades  the trades  the trades  the trades the tr
      ades the  trades the  trades the  northeast trades  the trades  the trades t

      doldrums doldrums doldrums doldrums doldrums doldrums doldrums doldr

      the trades the  trades the  trades the  trades the  trades the  trades the t
      rades the  southeast trades  the trades  the trades  the trades  the trades
      the trades  the trades  the trades  the trades  the trades  the trades the
      trades the  trades the  trades the  southeast trades  the trades the trade


      horse   latitudes   horse   latitudes   horse   latitudes   horse   latitud


        roaring  forties  roaring  forties  roaring  forties  roaring  fo
      rties westerlies roaring forties roaring forties roaring forties roar
        ing  forties  roaring  forties  roaring  forties  roaring  f


          polar  easterlies  polar  easterlies  polar  easterli
            es  polar  easterlies  polar  easterlies  pol
            ar easterlies polar easterlies pol
                        pole
```
                                                            —*Annie Dillard*

million years ago, there was sufficient oxygen in the air to allow life to move out of the water and onto the land.

If plants did not continuously replenish the oxygen in our air, animals and humanity would exhaust the available supply in a mere 300 years. All the water on the Earth is split by photosynthesis and reconstituted by respiration every 2 million years or so (Fig. 5.18). For millions of years, our ancestors have breathed the same oxygen and drank the same water, binding them temporarily in their bodies and then releasing them again to the atmosphere.

(c) Volcanism and the Earth's carbon dioxide cycle

The body of the Earth, as well as plants and animals living on the skin of the Earth, play a role in determining the atmosphere's composition. The early air emerged from ancient volcanoes that spewed molten rock and released gases that were once trapped in the upper mantle. The volcanoes brought forth water vapor, H_2O, carbon dioxide, CO_2, with a bit of nitrogen, N_2, and sulfur dioxide, SO_2. (See Table 5.2.)

The carbon dioxide cycle continues today, and it is crucial to the supply of carbon dioxide in our atmosphere (Fig. 5.19). In addition to taking part

Table 5.2. *Predominant gases emitted by typical Hawaiian volcanoes*

Constituent	Percent by volume
Water vapor (H_2O)	77.0
Carbon dioxide (CO_2)	11.7
Sulfur dioxide (SO_2)	6.5
Nitrogen (N_2)	3.0
Hydrogen (H_2)	0.5
Carbon monoxide (CO)	0.5
Sulfur (S_2)	0.3

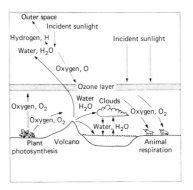

Figure 5.18. Earth's oxygen cycle. Plants continuously replenish the atmosphere with oxygen during photosynthesis while animals consume it during respiration. The oxygen in the air is also used up in forming the ozone layer and weathering crustal rocks, but oxygen is liberated when energetic sunlight tears atmospheric water vapor apart.

in photosynthesis and being converted to oxygen, the atmospheric carbon dioxide dissolves into sea water and it contributes to the bodies of plankton and shell fish. When the animals die, their shells rain down onto the sea floor and a layer of calcium carbonate, $CaCO_3$, builds up, often in the form of limestone deposits such as the white cliffs of Dover. These deposits are so extensive that there is now about 100 000 times as much carbon dioxide in the form of rocks as in the atmosphere. In fact, the amount of carbon dioxide in the Earth's rocks is comparable to that in Venus' atmosphere.

Were it not for sea-floor spreading, this carbon dioxide would be lost from the air. What actually happens is the following. The rocks in the floor of the sea are carried deep into the asthenosphere by the processes that create continental drift. There the heat breaks down the calcium carbonate and releases the carbon dioxide as a gas. This gas then works its way back to the surface, often through the vents of volcanoes, and it is returned to the atmosphere.

The carbon dioxide resides in the atmosphere until it either becomes broken by photosynthesis or becomes bound once more into the rocks. While in the air, it plays a crucial role in determining the temperature of the Earth's surface.

(d) Warming and cooling the Earth's atmosphere

The planet Venus, whose surface is a scalding 730 degrees Kelvin, provides an example of the greenhouse effect of carbon dioxide. This is a severe case, and a similar effect occurs on Earth. In our case it is much milder because our carbon dioxide is, at the moment, largely bound up in rocks. If the amount of carbon dioxide in our atmosphere were to increase or decrease, the greenhouse effect will change with it.

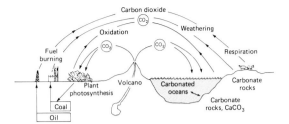

Figure 5.19. Earth's carbon dioxide cycle. Carbon dioxide, CO_2, moves through the atmosphere, plants, organic sediments, oceans, and carbonate rocks in a vast global cycle. The carbon dioxide in the air is dissolved in the ocean waters and taken up by plants in photosynthesis. Dead plants provide the carbon dioxide found in deposits of natural gas, coal and oil, and the seas provide the carbon dioxide that goes into calcium carbonate rocks. The oxidation of organic matter, the weathering of exposed carbonate rocks, the burning of fuel, animal respiration and volcanic exhalations return carbon dioxide to the air, completing the cycle.

The greenhouse effect depends on the fact that direct sunlight can transport heat deep into the atmosphere. On the other hand, escaping heat radiation is blocked by water vapor and carbon dioxide. When sunlight enters, the temperature rises and the rate of escape increases until a balance is established. With more water and carbon dioxide in the atmosphere, the rise must be greater. This is why cloudy nights tend to be warmer than clear nights. The water vapor in the clouds prevents the escape of infrared heat radiation.

In fact, many geologists believe that the swings of temperature in our planetary history – the warmer climates of some eras and the chilly ice ages in others – may be the result of swings in the amount of carbon dioxide and water vapor in the atmosphere. If the carbon dioxide increases, the atmospheric temperature is pushed upward, and this tends to increase the evaporation of the oceans, putting more water vapor into the air. The increased vapor tends to block more infrared radiation, and this may further enhance the greenhouse effect.

Variations in the Earth's greenhouse effect may explain several mysteries. The most profound of them has to do with the very early days of the Earth, when the Sun was only half as bright as it is today and gave only half as much heat to the Earth. (This is inferred from the theory of stellar development, and it is also inferred from the appearances of young stars in the Sun's neighborhood.) At that time, the Earth would have been quite chilly – in the absence of a substantial greenhouse effect – and much of the ocean would have been locked in ice. The Earth would have begun with a deep ice age, but there seems to be no geological record of such an inauspicious start. Now it appears that the early Earth may have been warmed by the presence of carbon dioxide (see Focus 5D, Pacemakers of the ice ages).

And the mystery of the warm climate during the age of the dinosaurs, when tropical plants flourished near the south pole, may also find an explanation in the greenhouse effect combined with variations in the rate of sea-floor spreading. Suppose the spreading was enhanced during an interval of increased activity among the liquid rocks of the upper mantle. This might temporarily increase the rate at which carbon dioxide is released into the air – a consequence would be a warmer climate.

The Earth's greenhouse effect may portend an ominous future, for two reasons. First of all, the Sun will slowly become brighter as it ages. This will heat the atmosphere slightly and may cause more carbon dioxide to be released from the surface rocks. If so, the increased greenhouse effect will heat the atmosphere even further, and the result may be a steam bath. This will take hundreds of millions of years, but there is an equally acute threat in a much shorter time interval.

Since the time of the industrial revolution, humans have pumped carbon dioxide and other greenhouse-producing gases into the atmosphere at an increasing rate. In this brief interval, a mere blink in the eye of time, there

Focus 5D Pacemakers of the ice ages

The Earth has undergone a series of warm and cold periods. During the cold periods, called ice ages, huge ice sheets build up on the continents and the polar seas. The growing layer of continental ice flows towards the equator, scouring and covering large areas of land. Then the climate warms and the ice retreats.

Temperatures recorded in the ancient fossilized shells of deep-sea sediment indicate that the major ice ages (white) during the past half million years recurred at intervals of 25 000, 40 000 and 100 000 years, with a dominant 100 000 year recurrence. These periodicities are caused by subtle changes in the Earth's orbit around the Sun and in the orientation of the Earth's rotational axis. They periodically change the amount and distribution of sunlight on the Earth, producing ice ages separated by warm intervals.

Gravitational forces exerted by the other planets, the Moon, and the Sun produce the three cyclic changes in the Earth's reception of solar energy. Planetary perturbations produce the 100 000 year climatic component by periodically changing the eccentricity of the Earth's orbit, so the orbit shape oscillates between a slight ellipse and a circle. The 40 000 year component is attributed to a periodic change in the tilt of the Earth's rotational axis. The third cyclic change is due to the gravitational pull of the

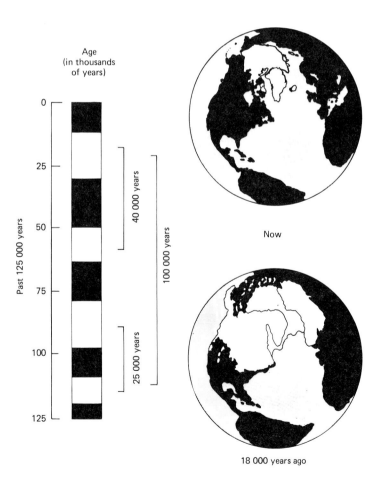

Age
(in thousands
of years)

Now

18 000 years ago

Sun and Moon on the Earth's bulging equator. It produces the 25 000 year precessional motion of the rotational axis on the celestial sphere.

We are now in a warm period, but the Earth could eventually enter the grip of another ice age. It all depends on whether or not the warming trend caused by an increase in atmospheric carbon dioxide can overcome the cooling effects caused by the inevitable future decrease in incident solar energy produced by the three astronomical rhythms.

has been a 25 percent increase in the carbon dioxide content of the atmosphere – partly the result of burning fossil fuels, such as coal, oil, and gas. The burning of forests, whose trees hold much carbon dioxide, has also contributed to the rise. A further doubling in the next century is likely at the present rate. The overall effect of such a doubling is uncertain, but climatologists are, to put it mildly, concerned about the possibility of a global warming.

Some computer models predict a temperature rise of a few degrees within the next century. Such a rise will cause ruinous droughts inland and, ironically, seacoast flooding as the oceans rise from polar runoff. Some scientists go so far as to predict an apocalyptic extinction of life in the searing heat of a runaway greenhouse. Others say these are exaggerated fears and that no computer can adequately simulate the complexities of the real world.

Has the global warming already been detected? There has recently been evidence for a warming trend, but the data are inconclusive for two reasons. First the records are not sufficiently complete or homogeneous to be totally trustworthy; second, the trend is still too short-lived to be ruled out as an accidental fluctuation. Perhaps it is a temporary swing that will be moderated by plants.

This may be wishful thinking; there can be no doubt that the heat is on. Despite the uncertainty, most experts agree that the greenhouse gases are increasing rapidly and global warming is in our future if the trend continues. Only the details – how much and when – remain debatable. Reversing the trend will not be easy. It will require measures that have long-term environmental benefits, such as a drastic cutback in the burning of fossil fuels, eliminating the large-scale clearing of tropical forests, and perhaps even the planting of extensive forests. (Southern California has already moved in the direction of reducing smog by banning the use of lighter fluid for barbecues, outlawing gasoline-powered lawn mowers, eliminating free parking in cities and requiring that all cars be converted to electricity or clean fuels by the year 2007.)

We live in an unsteady balance that is controlled not only by sunlight but also by the moisture and carbon dioxide of our atmosphere. Neighboring Venus and Mars have not sustained the balance, and their climates have become hostile to life as we know it. Venus, with a superabundance of carbon dioxide, has become feverishly hot. Mars, as we shall see, has little carbon dioxide and its surface remains uncomfortably cold. But the frozen surface of Mars may, in a billion years or so, be transformed into a warm, comfortable place with liquid water when the Sun becomes brighter and the Earth's oceans are boiling away. Will we live to see interplanetary real estate agents?

5.5 Aurorae and the Earth's magnetic cocoon

(a) The solar wind

Light and heat are not the Sun's only contributions to our environment. The entire solar system is bathed in a hot gale that blows from the Sun's atmosphere. This is the solar wind, a stream of electrically charged particles (protons, electrons and helium ions) that are driven outward from the Sun. The particles speed past the Earth at 400 kilometers per second – a million miles per hour – so they make the trip from the Sun in only a few days. But at Earth's distance, the solar wind contains only about 5 atoms per cubic centimeter. By terrestrial standards, this is nearly a perfect vacuum, and as it moves outward and spreads into a greater volume its density decreases even further and blends with the gas between the stars.

The existence of this remarkable wind was postulated to account for the motion of comet tails, which often behave like solar weather vanes pointing away from the Sun. It has since been proven and measured by space probes, and it has been called upon to explain one of nature's greatest shows, the aurorae.

(b) The auroral light show

Curtains of light far above the highest clouds flit across the night sky in the polar regions. This light is called the *aurora* after the Roman goddess of dawn. The aurora is a shimmering multi-colored light show that usually lasts for a half hour or so. It is energized by gusts in the solar wind and can be seen at any time of year, although it is seen more often during winter, when the nights are long and dark. (See Fig. 5.20.)

The auroras seen near the north and south poles are called the aurora

Figure 5.20. Aurora Borealis. Bars of bright prismatic color light up Frederic Church's 1865 painting of the fluorescent aurora. (Courtesy of the National Museum of American Art, Smithsonian Institution, gift of Eleanor Blodgett.)

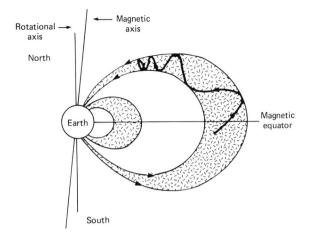

Figure 5.21. Van Allen belts. The charged particles in the Van Allen belts are trapped prisoners of the Earth's magnetic field. The particles spiral back and forth along the magnetic lines of force. As they approach the polar regions, the increasing magnetic intensity turns their motion more and more sideways, until finally they are turned around. The energetic particles return from this mirror point, travelling in just a few seconds to a similar turning point at the opposite pole.

borealis (northern lights) and aurora australis (southern lights). Ancient Vikings thought they were the spirits of fallen warriors being carried to Valhalla, the home of the gods. Later they were ascribed to sunlight reflected off the polar ice, but that idea was rejected when they were found to be 100 to 400 kilometers above the Earth's surface, well above most of the atmosphere.

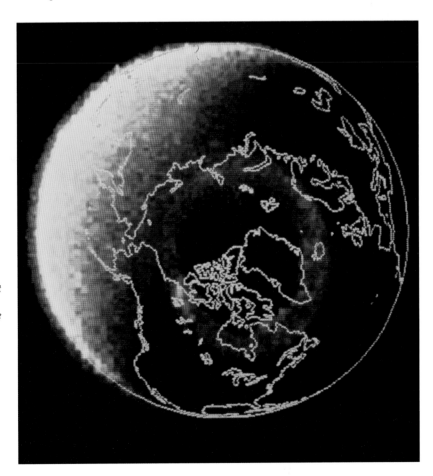

Figure 5.22. The auroral oval. The huge luminous ring of this auroral oval is centered on the north geomagnetic pole. The oval, some 4500 kilometers across, is due to an electrical current discharged at the boundary of the Earth's magnetosphere and the solar wind. The bright crescent at the upper left is the illuminated daylight side of the Earth. This image was made by a University of Iowa team using data taken by the satellite Dynamics Explorer 1 at an altitude of 20 000 kilometers. (Courtesy of Louis A. Frank.)

The auroral colors are produced by glowing oxygen and nitrogen, the principal components of the atmosphere. At great heights (100–200 kilometers) the oxygen is in atomic form and it is excited by an electrical discharge consisting of a stream of high energy electrons cascading down onto the upper atmosphere. The electrons in this discharge collide with molecules and atoms in the rarefied atmosphere, giving up some of their energy. This energy is quickly re-radiated in a process called fluorescence that is similar to that which lights a neon sign. The more energetic electrons excite the oxygen to glow green, while the less energetic electrons cause it to glow red. Ionized nitrogen molecules glow with a bluish color.

This much had been surmised from ground-based observations, and geophysicists thought for many years that the auroral display was the direct result of clouds of electrons emitted from sunspots and funneled down along the Earth's magnetic field. Observations from spacecraft have shown, however, that although the auroral particles come from the Sun, they are held in the Earth's magnetic tail and in the Van Allen radiation belts before being injected into the atmosphere. The Van Allen belts comprise two doughnut-shaped regions of electricity surrounding the Earth high above the equator, and reaching out into space a distance of several times the Earth's radius (see Fig. 5.21).

Electrons and ions from the solar wind are magnetically trapped in the belts, which act as a holding tank occasionally releasing bursts of auroral particles into the upper atmosphere. Each display produces rings surrounding the north and south magnetic poles. These are the auroral ovals (see Fig. 5.22). An observer on the ground sees only a small piece of an oval, which resembles a curtain hanging down from the sky, but from space the aurora can be seen to produce auroral arcs in both hemispheres at the same time. To understand these displays, we need to know how the solar wind interacts with the Earth's magnetic field.

(c) Earth's magnetism

In 1600, William Gilbert, physician to Queen Elizabeth I of England, published a small treatise, *De magnete magnus magnes ispe est globus terrestris*, which may be translated "the terrestrial globe is itself a great magnet." Gilbert's purpose was to explain why compass needles point north, and he suggested that the planet is surrounded by a field resembling the field of a bar magnet. In 1838 the German mathematician Karl Friedrich Gauss showed that the dipolar magnetic field must originate inside the Earth. The name dipolar implies that the field has two poles, north and south, and the lines of magnetic force emerge from the south pole and re-enter at the north pole, after looping outward in a symmetric pattern (see Fig. 5.23).

Electric currents in the Earth's interior generate a magnetic field that stretches into space around our planet. (Magnetic fields in a doorbell, for example, are similarly generated by a current in a coil of wire.) The Earth's currents might be generated by rotation of a semi-liquid core of iron. The approximate north–south alignment of the magnetic axis suggests a connection with the Earth's rotation, and analogy with laboratory magnets indicates that if the current flows eastward, in the direction of rotation, the north magnetic pole would be in the geographic north. Variations in the strength and direction of these currents would then account for the historical variations in the Earth's magnetic field that are so useful in studying sea-floor spreading and continental drift.

The Earth's field at the surface is several hundred times weaker than a toy horseshoe magnet, but it extends far out into space, and it is strong enough to stop the solar wind in a magnetic shock wave at a distance of 10 Earth radii upstream. This bow shock is at the head of the magnetic cocoon

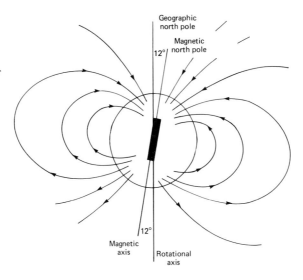

Figure 5.23. Dipole field of Earth. The Earth's magnetic field looks like that which would be produced by a bar magnet at the center of the Earth, with the lines of force looping out of the south magnetic pole and into the north magnetic pole. The magnetic axis is tilted at an angle of 12 degrees with respect to the Earth's rotational axis. This dipolar (two poles) configuration applies near the surface of the Earth, but further out the magnetic field is distorted by the solar wind (see Fig. 5.24).

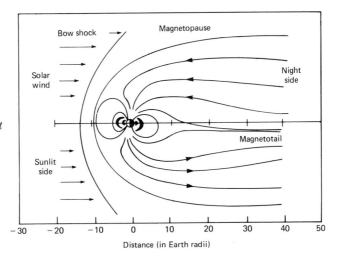

Figure 5.24. The Earth's magnetosphere. The magnetic field of the Earth forms a huge, protective shield called the magnetosphere. The solar wind compresses the magnetosphere on its sunlit side, forming a bow shock at about 10 Earth radii. The opposite night side of the magnetosphere is picked up by the solar wind and drawn out into a long magnetotail. The Earth and its dipolar magnetic field lie within this magnetic cocoon that protects us from the Sun's fierce wind; but small quantities of solar wind particles penetrate our magnetic defenses from the rear. The Van Allen belts and the aurorae are fed by particles coming through the magnetotail.

that surrounds the Earth. This cocoon has a magnetic tail that extends at least 200 Earth radii downstream, away from the Sun, and tapers to a point at the so-called "distant magnetic neutral-line." The tail is produced by the solar wind blowing on the Earth's magnetic field and pulling it into an elongated shape (see Fig. 5.24, and Fig. 5.25).

It is the distortion of this tail that leads to the auroral displays. If the tail becomes stretched by a gust in the solar wind, it may pinch down and create a "substorm magnetic neutral line" nearer to the Earth. When this neutral line forms, the magnetic field near the Earth collapses around the Van Allen belts, squeezing them down toward the atmosphere and squirting particles into the auroral zones. The result is a pair of bright auroral ovals, one in the north and the other in the south.

Figure 5.25. Bow wave of flying sphere. A shadowgraph catches a small sphere in free flight through the air. It forms a bow wave that is similar to the bow shock formed by the interaction of the solar wind with the Earth's magnetic field (see Fig. 5.24). The flying sphere also has a turbulent wake that may resemble the Earth's magnetotail. (Courtesy of Alexander C. Charters, Marine Science Institute, University of California, Santa Barbara. Made at U.S. Army Ballistics Research Laboratory, Aberdeen Proving Ground, Maryland.)

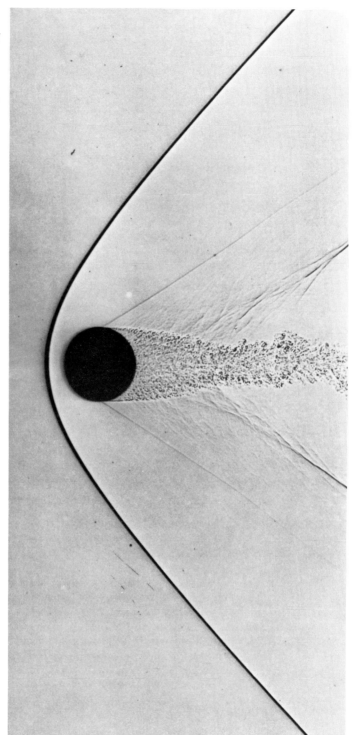

Focus 5E Earth – summary

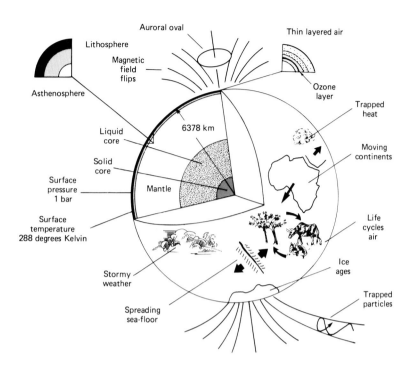

Mass: 5.975×10^{27} grams
Radius: 6378 kilometers
Mean density: 5.52 g/cm^3
Rotational period: 23 hours, 56 minutes, 4 seconds
Orbital period: 1 year = 365.26 days
Mean distance from Sun: 1.00 A.U. = 149.6 million kilometers
Number of known satellites = 1
Surface magnetic field strength = 0.35 gauss

Dawn on Mars. White clouds of water ice flow down the western flank of the volcano Olympus Mons (the dark round spot). Broad, white snows of frozen carbon dioxide (dry ice) have accumulated at the south polar cap in the winter. The great equatorial canyon system, Valles Marineris, makes a long faintly-visible gash in the red equatorial region of Mars. (Courtesy of NASA.)

6 Mars: the red desert

Mars is internally active with towering volcanoes that once erupted.

It cannot now rain on Mars.

Vast quantities of water may be frozen into the Martian ground, and in its polar ice caps.

Catastrophic floods and deep rivers once carved channels on Mars.

Mars may have had a warmer, denser, wetter atmosphere at various times in its history.

Was there life on Mars?

One Martian moon is heading towards eventual collision with Mars.

6.1 An Earth-like planet

(a) Seasonal effects

Mars, fourth planet from the Sun, ranks third in brightness as seen from Earth – after Venus and Jupiter. But Mars is brighter than most stars, and it has intrigued humans since prehistoric times because of its reddish color. The ancients associated the blood-red color with warfare, and the Babylonians, Greeks and Romans named the planet after their gods of war – Nirgal, Ares and Mars, respectively.

However, it is the Earth-like appearance of Mars in a telescope that has intrigued humanity during the past few centuries. Mars has polar caps that wax and wane with the Martian seasons, and mysterious dark markings that seasonally distort its ruddy face. There are also relatively permanent bright and dark areas that were once thought to respectively mark continents and oceans. White clouds repeatedly form at certain locations, and clouds are not possible without an atmosphere. Yellow clouds of dust are often seen, and they occasionally grow and coalesce to envelope the entire planet.

In 1659, Christiaan Huygens sketched the dark triangular feature now known as Syrtis Major, named for the quicksands (major and minor) near the northern coast of Africa. However, the Martian Syrtis Major is actually a lofty plateau. From his observations of this feature, Huygens concluded that the rotation period of Mars is about 24 hours. The exact Martian day is 24 hours 37 minutes 22.6 seconds.

The length of day on Mars is therefore nearly the same as the Earth's day, but the Martian year is almost two Earth years. The rotational axes of both the Earth and Mars are tilted by about the same amount. Both planets have four seasons – autumn, winter, spring and summer – although the Martian seasons last about twice as long.

Thus, Mars has an atmosphere, clouds, polar caps and seasons. These Earth-like qualities led astronomers to speculate that Mars might harbor life, and seasonal variations in bright and dark features were once considered as evidence for the growth and decay of vegetation. (See Focus 6A, Seasonal winds.) Toward the end of the 19th century, a few astronomers even thought they had discovered long straight markings on the surface of the planet indicating construction work by intelligent beings. Most of these

Focus 6A   Seasonal winds

During spring in the northern hemisphere, the north polar cap shrinks and material in more temperate latitudes darkens in color. These seasonal dark features were formerly attributed to the spring growth of vegetation produced by melting ice. They are now known to be caused by strong Martian winds that strip away light-colored dust in some areas and deposit it in others, producing a variable face whose bright and dark features vary with the seasons.

The winds of Mars create a restless world of constant change with wind patterns that vary with the seasons. Bright and dark streaks, often tens of kilometers long and a few kilometers wide, point in the direction of the strong prevailing winds and reveal the global wind patterns. The bright streaks consist of light-colored dust particles that have been deposited on the downwind (leeward) side of craters and hills. The dark streaks can be explained by the erosive action of wind scouring the Martian surface. Strong winds can remove the thin layer of bright dust and expose the dark underlying surface.

When all the streaks in a given area are superimposed by the human eye, they form larger bright or dark features in much the same way as dots in newsprint combine to make a picture. The darkening of Syrtis Major, for example, is apparently due to the growth and coalescence of dark streaks. The global pattern of dark and bright features on Mars varies with the seasons as the winds change direction and become more or less intense. Thus, the seasonal winds of Mars can account for the seasonal growth and decay of dark areas.

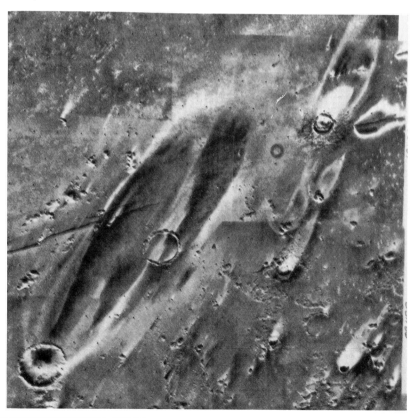

Focus 6B The canals of Mars

In 1877, Giovanni Schiaparelli astonished the astronomical community by reporting that dark, narrow straight lines traverse the Martian surface for long distances. He called them *canali*, or channels, noting that they criss-cross the surface of Mars and sometimes double. (Here we show Schiaparelli's Mercator projection drawn in 1881.) Camille Flammarion subsequently wrote that the channels are actually canals that redistribute scarce water across a dying Martian world. Flammarion was convinced that the Martian inhabitants might be more advanced than terrestrial humans. A few years after Flammarion's 1882 publication of these ideas, Percival Lowell, a wealthy Bostonian aristocrat, enthusiastically reproduced them in his own book on Mars. Lowell convinced much of the American public that the Martian channels are irrigation canals that draw water away from the polar caps, while many astronomers claimed that the channels simply did not exist. As it turned out, none of Lowell's canals coincided with canyons photographed by spacecraft. The "canals" are an illusion created when the eye arranges minute, disconnected details into lines.

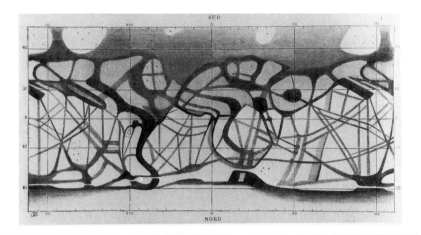

markings ("canali") were never photographed, and recent exploration by unmanned probes has failed to find any evidence for such structures on the planet. (See Focus 6B, The canals of Mars.)

(b) The space-age odyssey to Mars

Even the most powerful Earth-based telescopes provide only a blurred vision of Mars. Motions of the air in the terrestrial atmosphere limit telescope resolution to no better than 100 kilometers on the Martian surface. The details of Mars therefore remained hidden from view until spacecraft flew past the planet, and were then sent to orbit, and finally landed on and analysed samples of its surface.

These voyages to Mars were technological marvels of the 20th century. Large computers and complex calculations were used to track the positions of the spacecraft with an accuracy of 50 kilometers at a distance of 300 million kilometers. Huge Earth-based radio antennae and sensitive detectors were employed to detect weak spacecraft signals whose radiated power was equivalent to that of a match lighted on the surface of Mars. These were no simple accomplishments!

A new Mars was rapidly revealed. It was, and may still be, alive with volcanic eruptions and interior convulsions that reshape its face. An enormous bulge and vast cracks distort its shape, and the planet is divided into two dissimilar hemispheres. Rivers flowed across its surface many times; but large quantities of ice are now frozen into polar caps. These discoveries, made during our space-age odyssey to Mars, will be discussed in the rest of this chapter.

6.2 The Martian atmosphere

(a) Composition of the atmosphere

Astronomers have known for centuries that Mars has an atmosphere. Its presence had been inferred from the alternate growth and recession of its polar caps, as well as from the morning formation of white clouds. An atmosphere was also required to support the global dust storms that were seen to obscure the entire surface from time to time. However, the most fundamental attributes of the Martian atmosphere, its composition, its temperature, and its surface pressure remained vaguely known until the space-age exploration of Mars.

The chemical composition of the Martian atmosphere was finally obtained when the Viking 1 spacecraft landed on the surface of Mars, in 1976. The atmosphere at the surface is primarily composed of carbon dioxide (95.3 percent), nitrogen (2.7 percent), and argon (1.6 percent). Oxygen molecules are present in the Martian atmosphere to the extent of only 0.13 percent. The small amount of free oxygen that is present on Mars is probably the by-product of the destruction of carbon dioxide by energetic sunlight. (See Table 6.1 and Fig. 6.1.)

Very little water vapor is present in the Martian atmosphere, making it drier than the driest of the Earth's deserts. If the water vapor were collected and condensed, it would fit into a cubic kilometer, enough water to make a good-sized lake. But because Mars is cold, the thin air is capable of holding very little water, and in fact the Martian atmosphere is about as wet as it can be. It is saturated with water vapor, always on the verge of snowing. (See Fig. 6.2 and Fig. 6.3.)

Mars also lacks a thick ozone layer, such as has been found high in the Earth's atmosphere. The Earth's ozone layer only allows harmless visible sunlight and small, relatively harmless amounts of ultraviolet light to pass

Figure 6.1. Atmospheres of Earth and Mars. The Earth is covered with billowing white clouds, while Mars, with a relatively thin atmosphere, has only a few high-flying clouds.

Table 6.1. *Comparison of the main atmospheric constituents of Mars and Earth (percent by volume)*

Constituent	Mars (Viking 1 Lander)	Earth (Sea level)
Carbon dioxide (CO_2)	95.3%	0.03%
Nitrogen (N_2)	2.7	78.08
Argon (A)	1.6	0.93
Oxygen (O_2)	0.13	20.95
Water (H_2O)	0.03	Up to 5%
Ozone (O_3)	0.03 per million	1 to 10 per million

Figure 6.2. Cyclonic storm. Clouds of water ice form near the peaks of volcanoes and in storm fronts like the cyclonic storm shown here. These cyclones occur when the cold polar air flows under warmer air at a lower latitude. (Courtesy of NSSDC.)

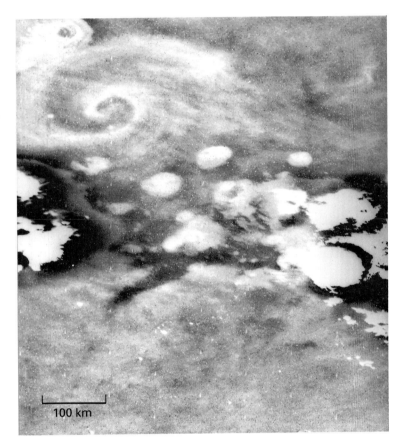

100 km

through to the surface. In contrast, much of the Martian surface is bombarded with lethal amounts of ultraviolet sunlight.

Studies of the present constituents of the atmosphere of Mars indicate that, like Venus and the Earth, the planet Mars now has a secondary atmosphere of gases released from the planet's hot interior. However, the Martian atmosphere contains less carbon dioxide than Venus' atmosphere, and much less nitrogen than the Earth's air. One explanation might be that volcanic activity on Mars was not as complete as on Venus or Earth; but it also seems likely that at one time, in the distant past, the Martian atmosphere was more dense than it is today.

Mars' gravity is sufficient to retain carbon dioxide in the planet's atmosphere. Hence, Mars could have an atmosphere as thick as Venus', but it does not. The thinness of the Martian atmosphere might be due to several factors: the incomplete release of gases locked up in Mars' interior, the weak gravity of the planet, and the chemical make-up of Mars (the planet's low density suggests it is made up of light elements that could escape relatively easily from the atmosphere). Some scientists speculate that Mars once had a dense atmosphere that slowly evaporated into space, under the combined effects of energetic sunlight and the planet's weak gravitational field; ultraviolet light from the Sun breaks up the atmospheric molecules into lighter, more energetic atoms that can escape the relatively weak gravity.

(b) The thin cold Martian air

Spacecraft probes have revealed that the average Martian surface pressure is a mere 1/160th of the Earth's surface pressure. So the ground-level pressure on Mars is less than the pressure outside the highest-flying jet airplane on Earth.

When the surface temperature drops to the condensation temperature of carbon dioxide (about 150 degrees Kelvin), the atmospheric carbon dioxide condenses and then begins to freeze on the surface. This occurs

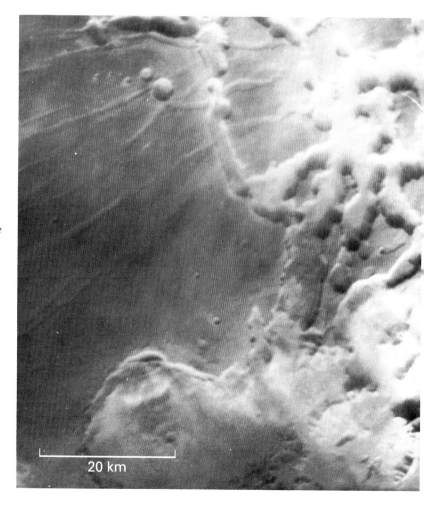

Figure 6.3. Labyrinth of the Night. Clouds of water ice form at low altitudes in the early morning hours. The low-lying mists of water condense and freeze out of the saturated Martian atmosphere. Here the pale clouds of ice crystals are seen against the plains of Noctis Labyrinthus (Labyrinth of the Night). The clouds and mist are mostly within the canyons, but some spills over into the surrounding plains. A few hours after sunrise the clouds all sublimate and vanish back into the atmosphere. (Courtesy of NASA.)

20 km

during the winter at the poles. As a result, broad white polar caps of frozen carbon dioxide form alternately at the poles as the winter comes first to one and then to the other pole. These seasonal polar caps occupy as much as 30 percent of the hemispheric surface area of Mars in the winter, and then either disappear entirely or shrink to less than 1 percent by the end of spring. The winter polar caps of carbon dioxide frost are about one meter thick, and there may be thicker deposits at the south pole that remain throughout the year.

As much as one-fourth of the atmospheric carbon dioxide is recycled between the polar caps and the atmosphere during a Martian year. This interchange produces variations in the surface pressure of about 20 percent as the gas is alternately depleted and replenished.

(c) The winds of Mars

Occasionally, local storms generate powerful winds, stirring up dense, billowing clouds of dust, eroding and severely scouring the Martian surface, carrying dust from one place to another. These are called aeolian effects (after Aeolus, god of the wind). As on the Earth, the winds of Mars have daily and seasonal cycles. (See the earlier Focus 6A, Seasonal winds). These winds have been sculpting the face of Mars for billions of years, but because the Martian air at the ground is about 2 percent as dense as the Earth's, the wind strength required to raise dust on Mars must be 7 or 8 times that required on the Earth. While terrestrial winds of 24 kilometers per hour can raise dust, particle movement on Mars requires winds faster than 180 kilometers per hour. Moreover, high-altitude jet streams whip across Mars at speeds of about 400 kilometers per hour, nearly 3 times the speed of the terrestrial jet streams.

Sand dunes provide excellent examples of aeolian effects on Mars. The Martian winds are continually picking up sand from the upwind side and dropping the larger particles on the downwind side, so the dunes roll along, like dunes on the deserts and beaches of Earth. (See Focus 6C, Wind-blown sand and dust.)

But global dust storms offer the most dramatic evidence for aeolian effects on Mars. As Martian winds roar on an otherwise silent world, they often stir up small, local dust storms, in much the same way that winds occasionally whip the terrestrial soils into towering columns called dust devils. As the local dust storms become more frequent, they sometimes coalesce and eject large quantities of dust high into the atmosphere, triggering a globe-encircling dust storm. When sufficient dust has been tossed aloft, the storm can sustain itself by converting the Sun's energy into wind energy. The airborne dust absorbs the Sun's radiation and heats up the atmosphere, generating strong winds. The entire planet then becomes wrapped in an opaque yellow veil.

Eventually, the storm quenches itself by preventing the Sun's heat from reaching the ground. As the surface cools, the winds die down and the storm subsides. Nevertheless, because there is no rain to wash out the dust, it lingers in the atmosphere for weeks as it slowly settles out of the air.

6.3 The surface of Mars

(a) Craters and impact basins

There is an asymmetric distribution of the major surface features on Mars, as there is on the Moon, Mercury and even the Earth. Mars is divided into two strikingly different hemispheres: in the south there are the older, elevated, heavily cratered highlands that resemble the lunar highlands. In the north there are younger, lower-lying, volcanic plains.

Focus 6C Wind-blown sand and dust
Particles that are moved directly by the Martian wind jump across the surface of Mars, and help move particles that cannot be moved by the wind alone. When the Martian winds become intense, sand-like particles that are about 0.016 centimeters (160 micrometers) in size begin to jump across the surface. This movement is called saltation. Because smaller particles have greater cohesion with the ground, they cannot be moved by wind alone, but when struck by a jumping, or saltating, particle they are tossed into the wind and they may remain suspended in dust clouds. The jumping particles also strike larger particles and push them along the surface.

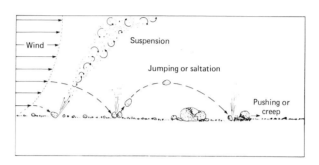

Most of the north is depressed by a few kilometers from the mean, while most of the south is elevated by a few kilometers. (There are exceptions. The north also contains towering volcanoes that rise as much as 25 kilometers above the surrounding terrain.)

The craters and multi-ringed impact basins found in the heavily cratered terrain are probably the scars of an ancient bombardment similar to that recorded on the face of our Moon, Mercury, and satellites of the giant planets (Fig. 6.4). The history of this bombardment, and the subsequent lower rate of cratering, is revealed by counts of the number of craters as a function of their size. Although the number of craters on Mars increases with decreasing size, there are fewer small craters on Mars than there are on the Moon; the larger Martian craters are worn and old looking. This suggests that extensive erosion modified the large craters and wiped out many of the existing small craters during an early period in Mars' history.

(b) Volcanoes on Mars

When the Mariner 9 spacecraft neared Mars in 1971, the planet was engulfed in a global dust storm. The eyes of the spacecraft – its cameras – could only peer at a disappointing, featureless yellow ball, but as the dust storms began to settle four dark, nipple-like spots poked out of the yellow gloom (Fig. 6.5). When the cameras focused in on these spots, central crater-like depressions were found. Even the thick blanket of dust could not cover these towering volcanic mountains, each capped with the coalesced craters that comprise a volcanic caldera.

The tallest of these volcanoes, Olympus Mons, is over 700 kilometers across at its base. This is broader than the distance from San Diego to San Francisco. The summit reaches 25 kilometers above the surrounding plains – about 3 times higher than Mount Everest. The largest volcano on Earth is

Mauna Loa, Hawaii, a mere 120 kilometers wide and 9 kilometers above the ocean floor.

Why should Martian volcanoes be so much higher than their terrestrial counterparts? Perhaps it has to do with the thickness of the crust of Mars. Because Mars is smaller than the Earth, it probably cooled faster, and its lithosphere became relatively thicker. As a result, the Martian lithosphere became too strong to break up into moving plates. This gives volcanoes a longer chance to grow in one spot. In contrast, the thinner lithosphere of the Earth has broken into plates which move with respect to one another and slide over the deep-seated sources of magma. This motion limits the growth of individual terrestrial volcanoes and produces chains of smaller volcanoes, such as the Hawaiian chain in the Pacific Ocean.

No Mars-quakes (Martian ground tremors) have been detected during thousands of hours of observations and this is consistent with the fact that there are no chains of mountains on Mars.

Shield volcanoes like Olympus Mons are formed by the repeated eruption of lava that flows from the summit down the flanks. Like their Hawaiian counterparts, these shield volcanoes have gently sloping flanks and roughly circular caldera at their summits. The paucity of impact craters near the top of Olympus Mons suggests a relative youth for the latest lava flows, whereas the higher crater density on the volcano's flanks and outer edges indicates a ripe old age for the edifice itself. While the most recent lava flows may be only a few million years old, lava may have been flowing down the flanks of Olympus Mons for billions of years before that, but we have no absolute age determinations for the Martian rocks.

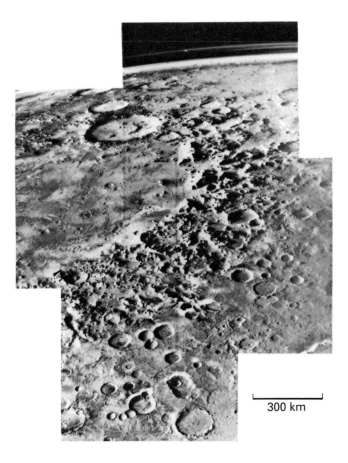

Figure 6.4. Argyre impact basin. This oblique view of the Argyre impact basin shows about half of its flat, lava-filled floor (at the center left). The basin is about 800 kilometers in diameter and is surrounded by a rim of mountains. The adjacent terrain shows the heavily cratered surface that is typical of the southern hemisphere. The parallel white streaks above the horizon are clouds floating 30 kilometers above the planet's surface. (Courtesy of NASA and NSSDC.)

300 km

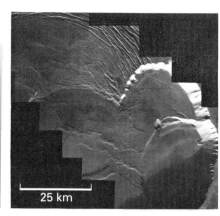

Figure 6.5. Zooming in on Olympus Mons. The summits of four Martian shield volcanoes stand out as four dark spots (left). They appear dark partly because they stand above bright clouds and haze. Notice that three of the volcanoes are lined up. The tallest volcano on Mars, Olympus Mons, is the dark spot seen above an imaginary line joining these three volcanoes. A closer view of Olympus Mons (center) shows a shield volcano that stands 25 kilometers above the broad plateau on which it rests, and volcanic flows that are up to 700 kilometers across. All of the large Martian volcanoes strongly resemble those in Hawaii in that they have large summit craters and lava tubes. The largest collapse crater at the summit of Olympus Mons (right) measures about 25 kilometers across and is nearly 3 kilometers deep. It marks the point where lava has withdrawn from a chamber within the volcano. (Courtesy of NASA.)

Relative ages can be estimated from impact crater counts, for young surfaces will have fewer craters. Ancient volcanoes, called patera, resemble gigantic shield volcanoes that have collapsed and have been eroded (see Fig. 6.6). Volcanoes of the Tholus type, as well as shield volcanoes, have few craters and are relatively young.

Mars is probably still undergoing such volcanic activity at the present time. Even though no spacecraft have yet detected a Martian volcano erupting lava, it is very unlikely that the planet has turned off its internal furnace. There is no telling when another eruption will occur.

(c) A swollen and cracked surface

When Mars was being intensely bombarded and lava flowed out along its surface plains, the planet experienced tremendous internal adjustments. Two huge lumps formed, the most pronounced of which is called the Tharsis bulge. As Tharsis expanded, the Martian surface spread apart and long, wide cracks opened up within the surface, forming parts of Valles Marineris (Valleys of the Mariner). The stretching associated with this surface expansion and consequent stresses also produced the intricately fractured Noctis Labyrinthus (Labyrinth of the Night). The expansion, stretching, and fracturing of the Martian surface may even be continuing today.

The vast system of canyons forming Noctis Labyrinthus and Valles Marineris cuts across the equatorial zone and stretches nearly halfway across the planet (Fig. 6.7). It was as if some cosmic sculptor was trying to split Mars in two. The system of canyons stretches one-quarter of the way around the planet (Fig. 6.8). The difference in temperature at the two ends sends strong, cold winds down its length.

6.4 Water on Mars

(a) Frozen water

Both polar caps are repositories for water on Mars. In the winter, carbon dioxide freezes out of the atmosphere to form the seasonal polar caps, but these are only temporary deposits. Much of the dry ice in the northern cap sublimates back into the atmosphere during the heat of summer, revealing a massive underlying cap, evidently composed of water ice. The presence of

Figure 6.6. Volcanic types. There are very few craters on the flanks of very young shield volcanoes like Arsia Mons (top left). Volcanoes of the Tholus type (top right) have partly convex slopes due to eruptions of a viscous lava, and relatively few impact craters. By contrast, the older volcanoes like Apollinarsis Patera and Tyrrhenia Patera (bottom left and right) have several craters superimposed on their flanks. (Viking Orbiter photographs courtesy of Michael H. Carr and NASA.)

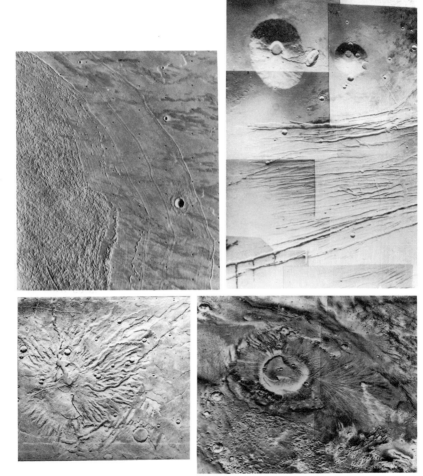

water ice has been inferred by temperature measurements and confirmed by the high amounts of water vapor above the remnant north polar cap in the summer. The southern cap does not sublimate to the same extent, but it probably contains water ice as well.

Although the thin Martian atmosphere contains very little water vapor, great quantities of water ice must lie hidden in the permanently frozen soil. The water might be held in a zone of permafrost, like those in Earth's arctic tundra. The layer of subsurface permafrost on Mars could be as much as one kilometer thick! If melted, the ice in it could produce an ocean at least 10 to 100 meters thick covering the surface of Mars.

Most of the water on Mars is probably to be found in the polar regions, where it resides as permafrost and contributes to the polar caps or is chemically bonded to minerals in the soil. Abundant water may be trapped beneath the surface, even though the surface appears dry today.

There are several types of evidence for a vast reservoir of water. The unusual shapes of some of the craters on Mars can be explained by supposing the ground contained abundant water or ice when they were formed. Craters that are larger than a few kilometers in size are distinctly different from those of the Moon or Mercury. On Mars, the heat of impact either melted and vaporized the water ice that was frozen into the ground,

Figure 6.7.   Color mosaic of Mars. This computer-generated mosaic of the northern hemisphere of Mars shows three volcanoes as dark spots to the west (left), while the bottom center of the scene shows the entire Valles Marineris canyon system from Noctis Labyrinthus to the chaotic terrain. Outflow channels are found in the north (top), and a variety of clouds and hazes are also visible, especially in the violet filter near the limbs. (Viking Orbiter photographs courtesy of Alfred S. McEwen, US Geological Survey.)

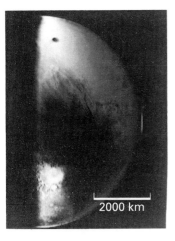

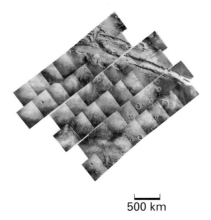

2000 km

500 km

25 km

Figure 6.8. Zooming in on Valles Marineris. The global view (left) shows that the canyons of Valles Marineris stretch nearly halfway across the red planet. The central portion (center) contains two vast chasms, while a close-up view of one canyon (right) reveals a huge landslide. The landslide was probably lubricated by melting ground ice. It moved 100 billion cubic meters of material downhill at speeds of up to 100 kilometers per hour. (Courtesy of NASA and NSSDC.)

or it released liquid water from the ground beneath the permafrost. The steam and liquid water then acted as a lubricant for the flowing debris, and the muddy material sloshed outward like a wave until it dried and stiffened, or became cool and refroze (Fig. 6.9).

(b) The channels and ancient lakes of Mars

Liquid water cannot now exist on Mars. Over most – and perhaps all – of the surface, the temperature is below freezing, and where the surface becomes sufficiently warmed, the water would immediately boil away. So it is not surprising that Mars is now an arid, barren world. Yet there are unmistakable signs that water flowed over the surface in the distant past. The tracks of flowing water in the form of river-like channels are carved and etched into its surface (Fig. 6.10). The amount of water required to gouge out the outflow channels is enormous – probably more than ten thousand times the annual discharge of the Amazon river, and much greater than the amount of water released by a bursting dam.

These outflow channels often extend from the chaotic terrains, where they are widest and deepest, but they are rarely joined by tributaries. The twisting riverbeds contain streamlined hills, scoured floors, interconnected channels, and large teardrop-shaped islands, where flowing water encountered an obstacle. In short, they have many of the properties of terrestrial riverbeds.

The origin of the outflow channels is linked to the origin of the chaotic terrain, for the channels emerge full grown from this terrain. The rapid melting of subsurface ice in the chaotic terrain could have completely filled the outflow channels with raging floods. The permafrost in these regions may have acted as a seal, confining ground water beneath. This water may have been abruptly released by external impact or volcanic activity, and when it rushed out, the ground would have fallen into the cavities left behind. Thus, the formation of the chaotic terrain was probably associated with the sudden release of millions of tons of water in catastrophic floods that carved the outflow channels deep into the Martian surface.

Another category of Martian channels is the sinuous channel – one that is not associated with chaotic terrain. As these sinuous channels wind and meander their way downhill, they coalesce with several well-developed tributaries (Fig. 6.11). These tributaries may be due to slow and prolonged erosion by running water either on or beneath the Martian surface.

There is also another type of evidence that water once stood on the

Figure 6.9. Crater Yuty. The material surrounding Crater Yuty was set loose when an impacting object melted the permafrost, or frozen ground. The ejected sludge sloshed like a wave across the surface and then refroze. These uniquely Martian features are known as rampart craters. (Courtesy of NASA and NSSDC.)

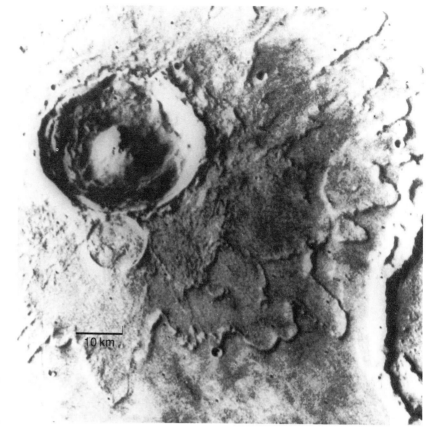

Figure 6.10. Outflow channels. The Ares Vallis region of outflow channeling exhibits the effects of catastrophic flooding. The large channels and teardrop-shaped islands indicate that water once flowed on the Martian surface. (Courtesy of Michael H. Carr, US Geological Survey, and NASA.)

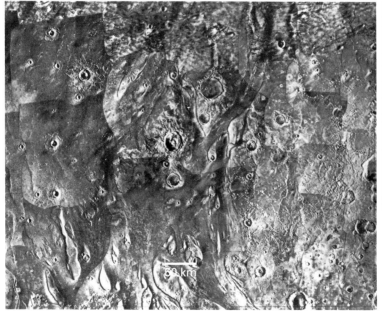

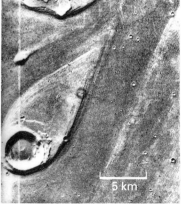

Figure 6.11. Nirgal Vallis. The sinuous channel, Nirgal Vallis, shown together with a high-resolution image of part of its well-developed tributary system. The Nirgal Vallis is more than 500 kilometers long. The deep and steep-walled tributaries suggest an origin by collapse into cavities formed by underground water. This high-resolution image is about 80 kilometers across. (Courtesy of NASA and NSSDC.)

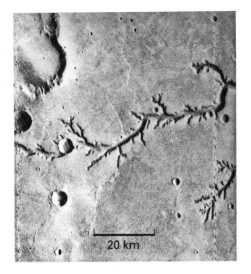

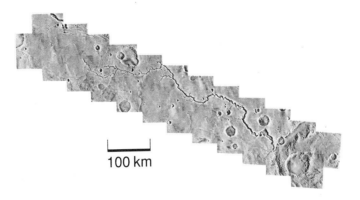

surface of Mars. Sequences of layered deposits within canyons appear to have been deposited by water. They resemble features found near the shores of terrestrial seas (Fig. 6.12).

(c) The wet phase of Mars' history

Evidently, running water existed on Mars in its early history, and may have flowed at irregular intervals over an extended period of time between 0.5 and 3.7 billion years ago. If so, the atmosphere must have been warm enough and dense enough to allow water to remain liquid. Several processes may have favored the presence of liquid water. Early volcanic outgassing may have released large amounts of carbon dioxide and created a warm, dense atmosphere (Fig. 6.13). Such an atmosphere could have been responsible for a warm, wet epoch on Mars (Fig. 6.14).

Higher temperatures may also have been produced by a rather bizarre behavior of the planet's tilt with respect to its orbit. It has been shown by lengthy calculations that Mars' spin axis changes its tilt with respect to its orbit in a variety of cycles, ranging from hundreds of thousands to millions of years in duration. Some scientists think that these changes of tilt had a profound influence on the climate of Mars, perhaps leading to repeated ice ages intermingled with warmer periods. The Earth's climate shows similar changes, but the Earth's spin axis changes much less than that of Mars.

But if Mars once had a warm dense atmosphere, where did its gases go?

The originally thick atmosphere must have somehow evolved into the thin, dry carbon dioxide atmosphere that we now see. Perhaps much of the carbon dioxide dissolved in the primeval water and became fixed in carbonate rocks. As the early dense atmosphere became thinner and colder, Mars entered an ice age. The rivers dried up and the liquid water froze into the ground and the remnant polar caps.

During its youth, Mars may have had liquid water and all the other ingredients for life. This brings us to the search for life on Mars.

6.5 Prospects for life on Mars

(a) Could life originate on Mars?

The question of whether or not life exists on Mars is intimately related to theories for the origin of life on Earth. Many scientists now support the theory of chemical evolution which is based upon a detailed study of the chemical components and evolution of terrestrial life. The chemical evidence indicates that every living thing is composed of the same molecular building blocks – amino acids and nucleotides – while fossil evidence suggests that the diversity of life evolved from single-celled organisms that swam in the Earth's primeval oceans more than 3.5 billion years ago.

Figure 6.12. Candor Chasma in Valles Marineris. This enhanced-color image of Candor Chasma shows blue, flat-lying layers of sediment that appear to have been laid down in liquid water. They may be the remnants of huge, ice-covered lakes that once existed on Mars. (Courtesy of Alfred S. McEwen, US Geological Survey.)

50 km

Figure 6.13. Polar cap. Gracefully curved cliffs of ice spiral around a remnant polar cap. The regular succession of sediments probably means that Mars has its own version of the ice ages that have affected the Earth. (Courtesy of NASA and NSSDC.)

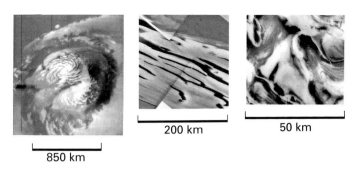

850 km

200 km

50 km

Figure 6.14. Channels with tributaries. Some scientists argue that rainfall was required to produce these tributaries, and that they provide the best evidence for a dense, warm atmosphere in the early history of Mars. Others counter by arguing that the acute-angled tributaries could have been formed when local concentrations of ground ice melted. (Courtesy of NASA and NSSDC.)

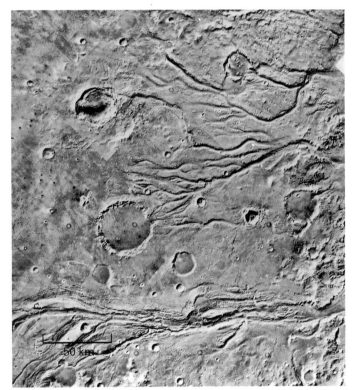

50 km

According to one hypothesis of chemical evolution, life arose spontaneously during the first billion years of chemical interactions in the Earth's primeval oceans. Simple molecules of the cosmically abundant elements were continuously formed, rebroken, chemically combined, and transformed into more complex molecules by the influence of sunlight and lightning. Ultimately, molecules capable of reproducing themselves were formed in the watery soup and life began!

Of course, life could have originated in another way. The hypothesis of chemical evolution is unproven. It supposes that life originated when the Earth had a primitive hydrogen-rich atmosphere without any free oxygen; but we do not really know if the Earth, or Mars, ever had such an atmosphere. If the hypothesis is correct, then there is nothing unique about the origin of life on Earth. Given enough time and the right physical conditions, the interplay of energy and non-living matter could result in life on other planets.

But could life originate on Mars? The Earth and Mars were formed out of similar material at about the same time, and at a similar distance from the Sun. As we have already seen, there is evidence for an ancient period of flowing water on Mars, and some astronomers have found controversial evidence for ancient lakes or seas on the red planet. Martian life may have breathed carbon dioxide, and the Martian air may even have once contained substantial amounts of oxygen that is now locked into its rusty soil. In short, all the basic ingredients for life *may* have once been present on Mars, and the hypothesis of chemical evolution suggests life *could* have arisen there.

If life did arise on Mars, how could we find evidence for it? Martian life ought to be based on the chemistry of the cosmically abundant atoms, including carbon, which is a key substance in building complex molecules. Carbon atoms can form large molecules by combining with other atoms, including other carbon atoms. Complex molecules based upon carbon are called organic molecules. The amino acids are an example.

Organic molecules provide living things with the capacity to evolve, adapt, and replicate themselves. They provide the basis for life, and the discovery of organic molecules on Mars would provide evidence that life might exist, or perhaps has existed, on Mars.

There was even a hope that real live organisms might be detected on Mars, but if anything now lives there it would have to be very strong, tough, and very small. After all, an ice age now prevails on Mars.

(b) Can living creatures survive the hostile Martian environment?

Of all the strange and forbidding worlds in our solar system, Mars is the most Earth-like, with its seasons, clouds, past flowing water, ice caps, and similar daily rhythm. Yet today Mars is implacably hostile to almost all forms of terrestrial life. Mars does not have enough air, water, oxygen or heat. Its atmosphere is nearly all carbon dioxide, with very little oxygen or water vapor, and it is a hundred times thinner than our air.

If any hypothetical Martian creature avoided being asphyxiated by carbon dioxide, or frozen to death at night, it would be faced with damaging ultraviolet light during the daytime. Even on Earth, an overexposure to ultraviolet sunlight can lead to a painful sunburn, and many doctors use ultraviolet light, or UV, to sterilize their equipment; it kills most of the micro-organisms that might have been present. The ozone layer in the Earth's atmosphere absorbs most of these damaging rays, but Mars does not have a similar thick ozone layer, so the deadly ultraviolet light reaches the Martian surface. Most organisms found on Earth today would be quickly killed if they were exposed to the Martian levels of ultraviolet daylight. On top of that, there is the lethal lack of liquid water on Mars.

But the Martian environment is not necessarily unfit for all imaginable types of organisms. In fact, life on Earth probably began in an environment that would be hostile to present life. In the early history of Earth, oxygen was probably absent, and there was no protective ozone layer. The single-celled blue-green algae that lived on Earth more than 3.5 billion years ago probably breathed carbon dioxide as plants do today. In fact, the bacteria that produce tetanus flourish in carbon dioxide and die in the presence of oxygen. Some terrestrial microbes have survived when placed within "Mars jars" that simulate the carbon dioxide atmosphere and freezing temperatures on Mars. Thus, there might be life on Mars that breathes carbon dioxide and has found a shield from the ultraviolet light.

However, many scientists were very skeptical about the chances of finding life in the cold, hostile Martian world (see Table 6.2). To send a spacecraft to Mars in search of life was an exciting long-shot gamble.

Table 6.2. *Obstacles to life on Mars*

Obstacle	Adaptation required
No liquid water on surface	Extract water from ice or rock by heat or chemical processes
Very little oxygen in atmosphere	Breath carbon dioxide
Much lethal ultraviolet light	Hide under rocks and sand, or develop protective shell
Surface temperature rarely rises above freezing temperature of water	Develop internal anti-freeze

Figure 6.15. Surface of Mars. The Martian surface resembles the rock-strewn deserts of the Earth. The wind-blown dunes create a gently rolling landscape. It is a frozen wasteland of rock, sand and sky, glowing in soft hues of yellow, red and brown. Drifting dust clings against wind-eroded rocks and fills the sky. Mars is a cold and desolate world in which the silence is broken by the roar of winds, the hiss of dust, the rumble of mammoth landslides and perhaps by outbursts of active volcanoes.

(c) The gamble in space

One of humanity's most daring and imaginative experiments began on July 20, 1976, when the Viking 1 lander came to rest in the Chryse Planitia region of Mars (Fig. 6.15). Six weeks later, Viking 2 landed in the Utopia Planitia region on the opposite side of the planet. Both landers were specifically designed to search for life on Mars.

How did the landers test for life on Mars? The first, most obvious, moving creature test consisted in looking to see if any creatures were frolicking on the Martian surface. The camera lens could detect anything down to a few millimeters in size if it came within 1.5 meters of the landers. Pictures were taken of all the visible landscape, from the stubby lander-legs to the horizon, for two complete Martian years. A careful inspection of these pictures failed to reveal any shapes or motions suggesting life. Conclusion: Unless they look like rocks, there are probably no forms of life on Mars larger than a few millimeters in size (Fig. 6.16).

But how could the landers detect the invisible microbes? Their presence could be inferred if the Martian soil contained organic molecules. Living microbes would be composed of organic molecules, and dead microbes would leave them in the soil. The organic molecule test would search for this debris.

The landers scooped up some Martian soil, deposited it in their sensitive bellies, and heated it to see what gases were released. Carbon dioxide and water vapor were detected, which was hardly surprising since these gases exist in the planet's atmosphere. But no organic molecules were found, down to a level of less than a few parts per billion. There appears to be a total absence of all likely organic molecules.

There is one small loophole in the argument for the absence of organisms. The test for organic molecules might be called a dead-body test, for soil would be expected to contain a higher proportion of organic molecules derived from dead bodies than from living ones. If there are any dead microbes on Mars, the amount is thousands of times smaller than that present in a comparable amount of terrestrial soil. Perhaps there are living organisms on Mars that are very efficient scavengers, consuming the carcasses of their dead relatives like cannibals; but biologists consider this an extremely unlikely possibility.

The other experiments on board the Viking landers searched for the vital signs of living microbes. If there are any microbes living on Mars, they must breath the air, eat the available food, and release the byproducts of digested food. The three tests for life may therefore be called the breathing test, the eating test, and the waste products test.

Let us first examine the breathing (or pyrolotic release) test. Because the Martian air is almost entirely composed of carbon dioxide, the microbes on Mars probably breathe this gas. Plant-like microbes might take the carbon out of the air and use it to build organic molecules. The microbes would then contain carbon that has been extracted from the air. The trick was to use radioactive carbon, ^{14}C, as a tracer.

Figure 6.16. Rocky surface. The rock-strewn Martian surface near the Viking 2 lander. Fierce winds continually erode the surfaces of Martian rocks, exposing their natural dark color. The sponge-like texture of some of the Martian rocks is due to the bursting of bubbles of volcanic gas. A trough filled with fine-grained sediment extends from the upper left to the lower right. Just beyond the trough, in the right half of the picture, there is a large boulder about 1 meter across. (Courtesy of Craig Leff, Washington University Regional Planetary Image Facility.)

So, small samples of the Martian soil were incubated in carbon dioxide and carbon monoxide containing radioactive carbon for five days. Then the soil was analysed and found to contain small amounts of radioactive carbon, suggesting that plant-like microbes had used carbon from the air, with the energy of sunlight, to manufacture organic molecules.

But then the experiment was redone on soil that had been sterilized by high temperatures before being exposed to the carbon. These temperatures would have killed most known organisms on the Earth, and yet the soil showed the same amount of radioactive carbon as in the earlier test. So, whatever was absorbing the radioactive tracer was probably not alive in the first place.

The eating (or labeled release) test also used radioactive carbon as a tracer, but it looked for the release of carbon by micro-organisms. Scientists created liquid food containing radioactive carbon and exposed it to samples of the soil. If there were animal-like microbes in the soil, they would digest the food and exhale carbon dioxide, just as animals on Earth release carbon dioxide when they burn, or oxidize, food. When this eating test was carried out, large amounts of radioactive carbon dioxide poured out from the soil. This certainly suggested that microbes were giving off the gas, but when additional nutrients were added to the soil, there was no additional increase in radioactive gas. Living creatures would have continued to ingest the food.

Scientists soon realized that radioactive carbon dioxide can arise from chemical reactions involving the nutrients and the Martian soil itself, and they concluded that the eating test does not indicate the presence of life on Mars, it only tells us something about the chemical behavior of the soil.

In the waste product (or gas exchange) test, liquid food was fed to the Martian soil, but this time the food was not radioactive. Instead of looking for radioactive tracers, the sensitive equipment sniffed the atmosphere to see what gases might have been released as waste products by the hypothetical microbes. (Terrestrial organisms release all kinds of gaseous waste products when they digest, as is indicated by the smell of rotting food.) A dramatic response occurred when the soil was first exposed to water vapor, and a burst of oxygen flowed from the soil. At first, the surprised scientists thought plant-like microbes were emitting the oxygen, but they soon realized that the release was too fast and brief. Microbes would grow and produce more oxygen as time went on, thereby releasing oxygen at a more steady rate. Moreover, the oxygen was released when the experiment was performed in the dark, and this behavior would not be expected from photosynthesis.

After further testing, the Viking biologists concluded that the rapid burst of oxygen was simply due to a chemical interaction between the Martian soil and water vapor, and was not a sign of life.

The results of all three tests for living microbes were attributed to non-biological chemical reactions, and the only solid outcome of the search for life on Mars seems to be the discovery of the extraordinary chemical reactivity (oxidizing potential) of the Martian soil.

The total absence of any organic material, down to a level of less than a few parts per billion, places very severe constraints on the possibility of life on Mars. It cannot be completely ruled out by such experiments, but the Viking explorations certainly failed to find unambiguous evidence for the presence of life on Mars today. Of course, they do not exclude the possibility of life in the past. The early histories of Mars and Earth were probably similar, so it is possible that life arose on Mars and was snuffed out as the climate cooled. If so, there may be fossil evidence for life on Mars.

6.6 The mysterious moons of Mars

(a) A maverick moon

The two moons of Mars are named Phobos (fear) and Deimos (terror) after the horses that pulled the chariot of the god of war (Ares) in Greek mythology (Fig. 6.17). Phobos moves around Mars at 2.7 Martian radii, and Deimos at 6.9 radii. Both moons move within the planet's equatorial plane; but they orbit so near to the surface of Mars that neither moon could be seen by an observer at the poles of Mars. They were discovered in the 19th century, but by a remarkable coincidence, Jonathan Swift, who wrote *Gulliver's Travels* a century earlier, endowed Mars with two fictional moons with orbital periods close to those of Phobos and Deimos.

Phobos is the maverick moon. It is so close to Mars that it orbits the planet in less than one-third of a Martian day. (Phobos moves around Mars in 7 hours 39 minutes, but the planet's rotation period is 24 hours 37 minutes.) This is a result of the fact that the orbit of Phobos is steadily shrinking and is spiralling towards unavoidable destruction! If it continues to move towards Mars at the present rate, it will either collide with the Martian surface or be torn apart by Mars' tidal forces in about 100 million years. Because Phobos is about 4.6 billion years old, we are, astronomically speaking, catching a fleeting glimpse of the last few moments of its life.

But what explains Phobos' motion toward Mars? Tides in the body of Mars hold the answer. Phobos raises tides on Mars in much the same way that the Moon produces tides on the Earth. Phobos produces two tidal bulges in the solid body of Mars. As the satellite moves ahead of the rotating planet, the closest tidal bulge pulls gravitationally on Phobos, causing it to lose energy and move inexhorably toward self-destruction. Because it orbits so swiftly, this tidal action pulls the moon inward, instead of pushing it outward as it does for the Earth's Moon. (See Focus 6D, Speculations about Phobos.)

(b) Origin of the Martian moons

Where did Phobos and Deimos come from? Either they were formed out of the debris remaining after the formation of Mars itself, or they were captured after forming elsewhere. The two moons may have originated together – perhaps by the capture of a large asteroid that broke up during its close passage to Mars. Phobos and Deimos closely resemble the innumerable objects that are found in the asteroid belt that lies just outside Mars' orbit. Like the asteroids, the moons of Mars are small and irregularly shaped. Moreover, the surfaces of the Martian moons are as dark as some asteroids, and nowhere near as lightly colored as the surface of Mars. Spacecraft observations indicate that Phobos has a low density, only twice that of water, and this density is comparable to that of certain meteorites that may have been chipped off asteroids. Perhaps these moons are adopted from the asteroid belt.

This belt is our next stop on the journey outward from the Sun.

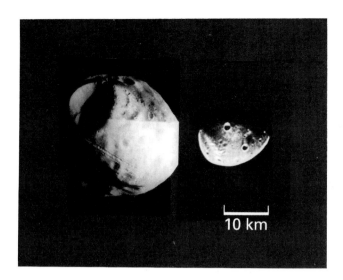

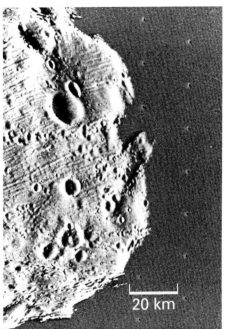

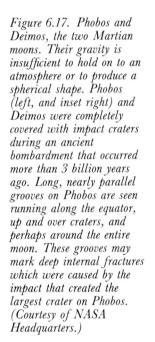

Figure 6.17. Phobos and Deimos, the two Martian moons. Their gravity is insufficient to hold on to an atmosphere or to produce a spherical shape. Phobos (left, and inset right) and Deimos were completely covered with impact craters during an ancient bombardment that occurred more than 3 billion years ago. Long, nearly parallel grooves on Phobos are seen running along the equator, up and over craters, and perhaps around the entire moon. These grooves may mark deep internal fractures which were caused by the impact that created the largest crater on Phobos. (Courtesy of NASA Headquarters.)

Focus 6D Speculations about Phobos

Phobos is speeding up in its orbit as it gradually falls towards Mars. Some astronomers thought that atmospheric friction, or air drag, was responsible, because air drag causes the orbits of artificial Earth satellites slowly to decline. The thin Martian air can, however, only produce the required drag if Phobos has a density of about one-thousandth the density of water. Since no known solid material is that light, the Soviet astrophysicist Iosif Shklovskii concluded that Phobos is not solid. He claimed that it might be a hollow artificial satellite launched by a past Martian civilization. This remarkable conjecture was given the stamp of approval in 1966 when the American astronomer, Carl Sagan, included it in his book with Shklovskii entitled *Life in the Universe*, describing the two moons of Mars as artificial satellites that were sent into orbit by an ancient, dying civilization whose other edifices are today covered by the sands of Mars. Spacecraft observations have shown that Phobos is actually a battered rock, and that tidal interaction with Mars is responsible for the moon's peculiar motion.

Focus 6E Mars – summary

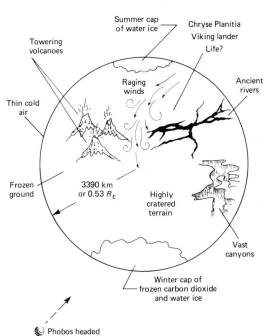

Mass: 6.4×10^{26} grams $= 0.107 \, M_{\mathrm{E}}$ (Earth = 1)
Radius: 3397 kilometers $= 0.532 \, R_{\mathrm{E}}$ (Earth = 1)
Mean density: 3.93 g/cm^3
Rotational period: 24 hours, 37 minutes, 22 seconds
Orbital period: 1.88 years
Mean distance from Sun: 1.52 A.U.
Number of known satellites $= 2$
Magnetic field not yet detected with certainty

Antarctic lode. The midnight Sun illuminates the wind-swept ice at the bottom of the world. Numerous meteorites have been found embedded on the ice in this region near the Allan Hills, Antarctica. These meteorites are probably fragments of asteroids that once had orbits in between those of Mars and Jupiter, but a few of them may have come from the Moon or even Mars. (Courtesy of Ursula Marvin, Harvard-Smithsonian Center for Astrophysics.)

7 Asteroids, meteors and meteorites

The combined mass of millions of asteroids is much less than the Moon's mass.

The asteroid belt is largely empty space.

An asteroid colliding with the Earth may have wiped out the dinosaurs 65 million years ago.

Asteroids may be mined for minerals.

Every year, hundreds of tons of meteorites enter Earth's atmosphere, and many pieces of them are found on the Antarctic ice.

The organic matter found in meteorites predates the origin of life on Earth by a billion years; but the meteoritic hydrocarbons are non-biologic in origin.

A few meteorites may have been blasted off the Moon or Mars, but most of them are chips off asteroids.

7.1 Asteroids in orbit

(a) First discoveries

On the opening night of the nineteenth century (January 1, 1801), Giuseppe Piazzi discovered a new world. While preparing a chart of the stars, Piazzi noticed one point-like star that had moved to a new position since he had last mapped that region of the sky. The moving "star" proved to be a tiny planet, too small and too distant to reveal its disc, even in a powerful telescope. Piazzi gave it the name Ceres, honoring the patron goddess of Sicily. Before other astronomers received word of Piazzi's discovery, the new object had moved too close to the Sun to be observed, and when it returned to dark skies, it could not be located. Hearing of the dilemma, a young mathematician, Karl Friedrich Gauss, devised a method for determining orbits from only three observations, and this led to its recovery about a year after it had first been sighted.

Within a few years of this discovery three more of these tiny planets had been found, and they became known as asteroids because they appeared "star-like." Their apparently rapid motions proved they were inside the solar system and moved in nearly circular orbits between Mars and Jupiter. Estimates of their diameter, based on the amount of sunlight they reflected, suggested they were less than a thousand kilometers across, and often only several dozen kilometers.

Today the orbits of some 2500 asteroids are known, and most of them lie in a great ring at distances of 2.2 to 3.3 A.U., with periods of 3 to 6 years. (The mean distance of the Earth from the Sun is 1 A.U., or one astronomical unit.) This ring is called the asteroid belt. Not all asteroids lie in this belt, but those that do are sometimes said to belong to the main belt.

The brightest asteroids are between 20 and 100 kilometers in diameter. Surveys of the faintest asteroids suggest there may be as many as half a million (500 000) in the main belt larger than one kilometer. Undoubtedly there are even more of the smaller asteroids. Astronomers estimate that more than a billion of these small, undiscovered worlds revolve around the Sun, but the very tiniest (say, less than a few millimeters) would have been swept out of the solar system or drawn into the Sun. Despite their vast number, the asteroids leave plenty of room for space flight, as was demonstrated by the Pioneer 10 and 11 and Voyager 1 and 2 spacecraft, when they passed undamaged through the main belt in the 1970s. The total mass obtained by adding up the contributions of the objects, of all sizes, is far less than the mass of any of the major planets, and is hardly more than 10 percent of our Moon.

(b) The Jovian influence

The asteroids of the main belt move in slightly eccentric orbits. They are scattered around these orbits in a haphazard fashion, much like runners near the end of a long race on a small track. So, at first glance, they appear to fill the asteroid belt quite uniformly. But, if they could be arranged along a line outward from the Sun – as though they had been placed on the starting line – a definite pattern would emerge. Not all distances from the Sun are equally well represented. There are a few prominent gaps, and these are named the Kirkwood gaps after the American astronomer, Daniel Kirkwood, who discovered them in 1866 (Fig. 7.1). In addition, there are several peaks corresponding to groups of asteroids with nearly the same orbital distances. In particular, the Trojan asteroids have orbits that are identical in size to the orbit of Jupiter. (See Focus 7A, The Trojan asteroids and Lagrangian points.)

According to Kepler's laws of planetary motion – which apply equally well to the asteroids as to the major planets – each orbital size corresponds to a specific orbital period. This means that the arrangement of asteroids according to distance can also be considered as an arrangement according to period. As a consequence, the Kirkwood gaps imply that certain orbital periods are missing. A careful comparison with the period of Jupiter, 11.9 years, shows that the missing periods are rational fractions of it. (A rational fraction, such as $\frac{1}{4}$, has integer numerator and denominator.) The most prominent of the Kirkwood gaps are at periods given by $\frac{1}{4}, \frac{2}{7}, \frac{1}{3}, \frac{2}{5}, \frac{3}{7}$, and $\frac{1}{2}$ times Jupiter's period, as can be seen in Figure 7.1.

Kirkwood conjectured that the gaps were the result of recurring gravitational jolts from Jupiter. This seemed reasonable because the large mass and proximity of Jupiter to the main belt would make it much more effective than any of the other planets. Such jolts would dislodge the asteroids from resonant orbits and push them into orbits with slightly different periods. An asteroid with a period two-sevenths of Jupiter's, for example, will receive a similar jolt each time it has made exactly 7 revolutions about the Sun because Jupiter will have made exactly 2 revolutions in this interval. This resonance effect is somewhat analogous to repeated pushes on a swing or a pendulum. If the pushes occur at the same point in each swing, they can increase the energy of the motion. In the absence of such resonance, the

Figure 7.1. Kirkwood gaps. The number of asteroids at different distances from the Sun. Most of the asteroids are found in the asteroid belt that lies between 2.2 and 3.3 A.U. from the Sun. Repeated gravitational interactions with Jupiter have tossed asteroids out of the Kirkwood gaps with orbital periods of $\frac{1}{4}, \frac{2}{7}, \frac{1}{3}, \frac{3}{7}$ and $\frac{1}{2}$ of Jupiter's orbital period. These periods are placed above the relevant gap in the figure.

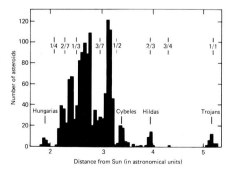

Focus 7A Trojan asteroids and Lagrangian points

The Trojan asteroids precede and follow Jupiter in its orbit by about 60 degrees as measured from the Sun. They lie near the two Lagrangian points that are named after Joseph Louis Lagrange (1736–1813) who predicted their existence. Because the gravitational forces of Jupiter and the Sun converge toward these points, asteroids tend to remain near the Lagrangian points. However, the gravitational perturbations of the other planets produce slight swinging motions, so the Trojan asteroids oscillate within the two shaded lenticular regions. Some of the Trojan asteroids may occasionally move close enough to be captured by Jupiter's gravity, thereby accounting for Jupiter's unusual outer satellites.

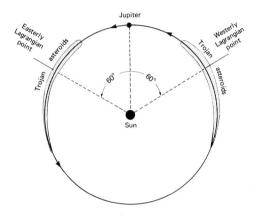

perturbations would be haphazard and would as often increase as decrease the orbit of an asteroid, so no net effect would accumulate over a long time interval.

But, when Kirkwood's idea of resonance in its original form was put to the test, it failed to generate gaps. Simplified calculations involving asteroids moving under the influence of the Sun and Jupiter seemed to show that, contrary to expectation, the orbits would increase and decrease, and the gaps would not appear. Asteroids would move into them as quickly as others moved out, and the distribution of periods seemed to keep its original, smooth shape with no peaks or valleys. So astronomers began to look elsewhere for an explanation. One suggestion was that the gaps might be caused by collisions among the asteroids themselves. Asteroids near resonance, it was argued, would have eccentric orbits and would have a larger than average chance of striking each other. But, as it turned out when the calculations were performed, these collisions would be just as likely to insert asteroids into the gaps as to remove them. So this idea was discarded, and, with the discovery of the interlopers, the asteroids became even more enigmatic.

(c) The interlopers

Although the vast majority of asteroids travel in the main belt lying between the orbits of Mars and Jupiter, there are notable exceptions. Some travel to the outer parts of the solar system; others stray inward toward the Earth. For example, the Amor asteroids cross the orbit of Mars (1.5 A.U.)

and venture almost to the orbit of Earth (1.0 A.U.), where they turn and swing outward again to the inner edge of the main belt at 2.2 A.U.

Two other families of interlopers come into the neighborhood of the Earth. The Apollo asteroids have eccentric orbits extending from the main belt inward, passing the Earth. Another group, the Aten asteroids, never swing out as far as Mars and the main belt, but they have eccentric orbits carrying them in and out past the Earth (Fig. 7.2).

Kirkwood's idea of resonance seemed to have nothing to say about these extreme orbits. Resonances of the type he envisioned would become ineffective as soon as the orbit strayed far from the gap, so many astronomers assumed, for lack of a better idea, that the interlopers were the result of rare collisions, or were debris left from the early days of the solar system.

(d) Chaotic orbits

A satisfactory explanation was finally achieved in the 1980s with the application of an elaborate computer program that permitted dropping some of the simplifying assumptions that had been used before. The result was a partial vindication of Kirkwood's resonance idea, with an important addition, and the theory seems to explain both the gaps and the interlopers.

By following an asteroid near one of the resonances, it was discovered that such an orbit would appear well-behaved for many thousands of years – agreeing with the earlier calculations – but then the orbit would abruptly begin to elongate in a chaotic way that is not completely understood, although it is certainly caused by the gravitational action of Jupiter. This phenomenon created a small number of asteroids with very eccentric orbits, but it still did not produce a prominent gap.

Then the next step, not imagined by Kirkwood, would take place. Over long time intervals, the seemingly regular orbit of an asteroid that strays into one of the Kirkwood gaps grows increasingly elongated until it eventually crosses the orbit of Mars or the Earth, and after a few passages the planet flings it into a totally different orbit, perhaps creating an interloper. (Occasionally the asteroid would crash into the disturbing planet.) The asteroid would be removed from the main belt, and one of the Kirkwood gaps would be slightly enhanced. Only orbits that were near the resonances would show increases of eccentricity that brought them near Mars or the Earth (Fig. 7.3).

So, the explanation lies in the gravitational effects that had been suspected when Kirkwood's ideas were first put to the test. After the advent

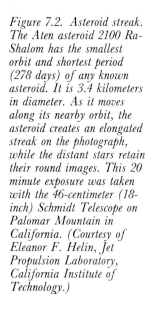

Figure 7.2. Asteroid streak. The Aten asteroid 2100 Ra-Shalom has the smallest orbit and shortest period (278 days) of any known asteroid. It is 3.4 kilometers in diameter. As it moves along its nearby orbit, the asteroid creates an elongated streak on the photograph, while the distant stars retain their round images. This 20 minute exposure was taken with the 46-centimeter (18-inch) Schmidt Telescope on Palomar Mountain in California. (Courtesy of Eleanor F. Helin, Jet Propulsion Laboratory, California Institute of Technology.)

Figure 7.3. Main belt to interloper. Asteroid orbits can become chaotic under the gravitational influence of nearby massive Jupiter. Asteroids at certain locations in the main belt follow a trajectory that becomes increasingly off-center over thousands of orbits. They may eventually become interlopers with orbits that cross the Earth's orbit. Many of these interlopers will eventually collide with the Earth.

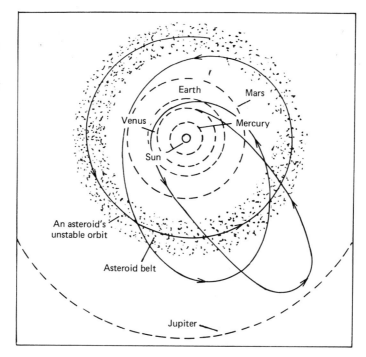

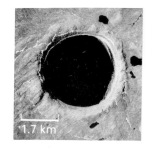

Figure 7.4. Terrestrial impact crater. A colliding asteroid caused this impact crater located in Quebec, Canada. The crater has a diameter of 3.2 kilometers and an estimated age of 5 million years. This type of impact crater has been given the name astrobleme, or star-wound, from the "star-like" asteroids. (Courtesy of Richard A.F. Grieve, Brown University and the Department of Energy, Mines and Resources, Ontario.)

of computers, it was possible to solve the complete equations and simulate the formation of the Kirkwood gaps.

(e) Impacts with the Earth

Altogether there are about 1300 interlopers larger than one kilometer in diameter and with orbits that will cross the Earth's orbit. Most of the time, the Earth will be somewhere else when an asteroid crosses its path, but occasionally the Earth and an asteroid will arrive almost simultaneously at the intersection. When this occurs, the asteroid will strike the Earth's surface and explode with an energy greater than hundreds of nuclear bombs; it will excavate an impact crater much larger than the asteroid itself, in the same way the craters were excavated on the surface of the Moon. The energy associated with the motion of the high-speed projectiles is released explosively – a one kilometer asteroid, for example, releases the equivalent of one thousand 100-megaton nuclear bombs.

So our planet is immersed in a cosmic shooting gallery of potentially lethal projectiles, and on the average, an asteroid with a radius larger than one kilometer will strike the Earth every million years or so. The smaller asteroids are more numerous and will collide more frequently. Because fewer comets cross the Earth's path, impacts with an asteroid are about 50 times more probable than impacts with a comparable comet.

It is not an easy matter to find such craters. With the passage of time, even the largest craters produced by these collisions will be eroded and filled with water and soil, gradually disappearing from sight. And not only are they obliterated with time, they usually occur in remote regions. But scrutiny of aerial photographs has revealed almost 100 terrestrial impact craters ranging in diameter from 1 to 140 kilometers (Fig. 7.4). They are recognized by their circular shapes, uplifted and overturned rims, inverted rock strata and, in some cases, central peaks and multi-ringed structures. (See Focus 7B, Identifying terrestrial impact craters.) Geologists have dated some of the craters by radiometric age determinations of melted rock.

Focus 7B Identifying terrestrial impact craters
How do geologists know that some terrestrial craters are the scars of
external impact? Some of them yield fragments of the impacting body that
can prove an impact, and others contain terrestrial rocks that have been
transformed under conditions of extreme pressure, heat and shock. For
example, the shatter cones found in the vicinity of many terrestrial craters
provide evidence for shocks associated with asteroid impacts. They point
towards the direction of impact, like the cone-shaped plugs of glass that are
often formed when a bullet strikes a window. The shatter cones shown here
are several centimeters in height. They are from the Wells Creek, Tennes-
see basin that is about 9.6 kilometers across.

They find ages from 10 000 to 2 billion years. Two of the largest and oldest
craters (Sudbury, Ontario, and Vredefort, South Africa) are about 140
kilometers in diameter and were formed nearly 2 billion years ago.

(f) Catastrophe from the sky
Evidence for an immense asteroid collision was recently found in a
remarkable layer of clay about one centimeter in thickness. When the layer
was first discovered, a team of geologists headed by Walter Alvarez
determined its age from its position among the geologically dated strata.
They found that the layer had been deposited about 65 million years ago,
and they then set out to estimate how many years had been required to form
the layer. This was to be done by measuring the amount of iridium in the
clay.

Iridium is an element that is rare on Earth but is fairly common in
meteorites, and it can be used as a clock because it steadily rains down
through the atmosphere and settles in the soil. In an average century, a
certain amount of iridium will mix with the soil and become part of any new
layers that are forming. If a layer requires twice as long to form, it will have
twice as much iridium. But the geologists found that the clock had gone
wild for a short interval about 65 million years ago. The amount of iridium
they found in this layer of clay was about 30 times higher than that found in

the fossilized limestone above and below the clay. At first, the finding did not seem highly unusual; most geologists assumed the rate of clay deposition was very slow for some reason in this particular part of the Earth, so the iridium was more highly concentrated. But when the same type of iridium enrichment was found in clay at widely scattered spots on the Earth, a new explanation was needed and there was only one likely conclusion: the entire globe had been drenched in iridium for a short time.

According to one hypothesis, this iridium deluge was brought down by a large asteroid that struck the Earth about 65 million years ago. The asteroid would have exploded and thrown much of its material high up into the atmosphere, where it could be carried by the winds over much of the globe and slowly filtered back down to the ground where it would have produced a thin layer that was rich in iridium. A layer one centimeter thick covering the entire Earth would be deposited by an asteroid about 10 kilometers in diameter, and such objects are estimated to strike the Earth with an average interval of 100 million years.

But that is not the end of the story. According to some paleontologists, the end of the Cretaceous era – and the demise of the dinosaurs and a variety of plants and other animals – was an abrupt event best described as a mass extinction. The cause of this extinction remains a mystery, but it also occurred about 65 million years ago, and many scientists now feel that this agreement of dates is no mere coincidence. They argue the initial impact of an asteroid capable of spreading one centimeter of dust over the Earth will also destroy most of the plants and animals for hundreds of miles around. And the species that survive the initial blast would then face a long period of darkness and cold, as the dust floated about the Earth and prevented much of the sunlight from reaching the ground. A prolonged "winter" would follow, and plant life would wither and freeze, leaving plant-eating animals without food. Such an event might explain the mass extinctions. (Some scientists argue that an analogous "winter" following a nuclear war would wipe out most living species on Earth.)

Recent studies of the clay in this strange layer have produced further evidence for such a catastrophe. Hoping to learn more about the chemical composition of the asteroid itself, a group of scientists at the University of Chicago took clay samples from Denmark, Spain, and New Zealand. When they dissolved the samples, they discovered that most of the elements of the asteroid had been vaporized and the clay held a large residue of carbon. Much of the carbon was in the form of tiny soot particles, irregular and fluffy, which the investigators thought were typical of the particles formed in forest fires. The high concentration of the soot particles suggests that they are the result of enormous wildfires ignited by the intense heat of the impacting asteroid. Soot from such fires might have mixed with the dust from the asteroid and led to a prolonged period of death on Earth.

7.2 Physical properties of the asteroids

(a) Size, shape, and spin

A few asteroids are large enough, and occasionally wander close enough to the Earth, to show their shapes in a large telescope. For example, asteroid 433 Eros has come close enough for astronomers to measure its dimensions. It is a slightly distorted cigar-shaped body. Most asteroids, however, are too far away and their twinkling, star-like images are unresolved. For these, there is no direct way to measure size or shape, but by measuring their brightnesses in visible and infrared (heat) radiation, astronomers can determine their reflective power (albedo). Once the albedo is known, the surface area can be calculated.

We will not go into the details, but the essential point is that the visual brightness of the asteroid may be computed from the product of the albedo and the size of the reflecting surface. If the asteroid's orbit and albedo are known, the surface area and radius can be inferred. For instance, the largest asteroid, Ceres, is covered with dark material and shines less brightly than Vesta, which is somewhat smaller. The number of asteroids increases dramatically as the diameter decreases, as indicated in Figure 7.5.

Many asteroids show varying brightness. The brightnesses of some are seen to change in an interval of a few hours, and this variation is assumed to be caused by non-spherical shapes tumbling through space and rotating in the sunlight. Hence the rotation rates of these asteroids may be inferred from plotting the light curve and seeing how long it takes for the pattern of light variation to repeat itself. For example, if the elongated body of Eros were seen opposite the Sun, its brightness would be greatest when we see it broadside and minimum when we look end-on and see its smaller profile. When it rotates, it will present the broadside twice per revolution, so its light curve would show two maxima (perhaps not exactly the same brightness) for each rotation about its own axis. The elongated shape can be estimated from the ratio of brightnesses at minimum and maximum.

The direction of the rotation axes of asteroids can also be determined by carefully following the light curve as the asteroid moves about the Sun. The results show that, contrary to the orderly rotations of most of the major planets, the rotation axes of the asteroids seem to be rather haphazardly

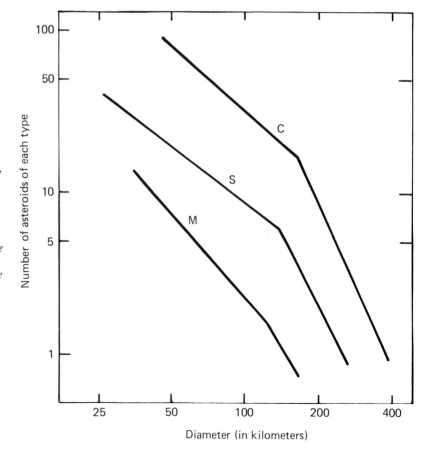

Figure 7.5. Number–diameter relations. The number of carbonaceous, C, silicate, S, and metallic, M, asteroids are plotted as a function of their diameter. The number of all types of asteroids increases with decreasing size. For a given diameter, the C asteroids are more numerous than the S ones, which are in turn more numerous than the M types. The change in the slope of each curve at lower diameters is attributed to repeated collisions between asteroids. (Adapted from Benjamin Zellner and Edward Bowell's article in Comets, Asteroids and Meteorites, *ed. A.H. Delsemme, University of Toledo (1977).)*

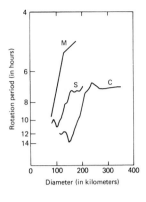

Diameter (in kilometers)

Focus 7C How fast do asteroids spin?

Some asteroids are rotating as fast as they can! The fastest observed rotation periods are a few hours. Theoretical calculations indicate that an asteroid spinning at a faster rate would throw material off its surface.

Within a given class, the largest asteroids rotate the fastest. At a given size, the metallic, M, asteroids tend to rotate faster than the silicate, S, ones, which in turn rotate faster than the carbonaceous, C, types.

If asteroids are homogeneous, then a surface composition of metallic material implies a dense interior, while an asteroid composed of silicate material would be less dense. A carbonaceous asteroid would be even less dense. Thus, the rotation data suggests that larger, denser asteroids rotate faster. (Diagram adapted from Stanley F. Dermott and Carl D. Murray, *Nature* **296**, 418 (1982).)

oriented. Frequent collisions may have been responsible for these random orientations and for the irregular shapes of the asteroids.

The known rotation periods are typically between 5 and 10 hours, shorter than the rotations of most other objects in the solar system. (Jupiter, with a period of 9 hours 55 minutes, is the closest contender among the major planets.) In fact, some of these objects are rotating so quickly that they are nearly torn apart. (See Focus 7C, How fast do asteroids spin?) It is not certain whether they have always been rotating quickly or whether they have been "spun up" by oblique collisions with other asteroids, but the randomness of their orientations suggests that collisions may have played an important role.

(b) Color me different

Detailed examination of the colors and reflectivities of asteroids have given clues to the nature of their surfaces. By comparing their light, wavelength by wavelength, with the light of the Sun, it is possible to map their colors and to distinguish three major groups of surface composition. These groups of asteroids are known as the C (carbonaceous), S (silicate), and M (metallic) groups. The C asteroids are probably composed of dark carbon-rich material; the S asteroids are probably composed of relatively bright and rocky silicate material with a mixture of metals; and the M asteroids, which are relatively rare, reflect sunlight in a way that suggests metallic nickel and iron. (See Focus 7D, Finding the composition of asteroids.)

An intriguing connection exists between composition (as indicated by color) and distance from the Sun (Fig. 7.6). The asteroids with the highest reflecting power tend to lie near the inner edge of the main belt, while the more distant asteroids are, on the average, the ones with the lower reflecting power. The very darkest are found in the remote regions near Jupiter's orbit. Their reddish colors are thought to come from organic compounds.

This progressive decrease in the reflecting power with increasing distance from the Sun is probably related to conditions in the early solar nebula. It may be a consequence of a decrease in temperature with increasing distance when the asteroids were formed. Dark material, rich in carbon and water, could condense only in the colder regions farther from the Sun. On the other hand, the bright rocky material was less volatile and could remain within the hotter regions closer to the Sun. In this way, the temperature of the solar nebula may have led to the pattern of materials and

Focus 7D Finding the composition of asteroids

The sunlight reflected from the surfaces of some asteroids exhibits absorption features, or dips, that can be identified with the absorption signatures of certain minerals found in terrestrial rocks and meteorites. For instance, the sunlight reflected from the S asteroids contains a dip that has been identified with silicate minerals. A prominent silicate absorption feature is shown here for the S asteroid 1685 Toro (dark dots with error bars). Asteroids like Toro may be the source of the stony meteorites recovered on Earth. The shaded spectrum is, in fact, the reflection spectrum of a stony chondrite meteorite.

Rare individual asteroids with well-identified absorption features are also of interest. For instance, the sunlight reflected from the carbonaceous C asteroid 1 Ceres indicates that there is ice on its surface, and the E asteroid 4 Vesta shows the distinct absorption signature of volcanic basalt. (Diagram adapted from R. Chapman *et al.*, *Astronomical Journal* **78**, 502 (1973).)

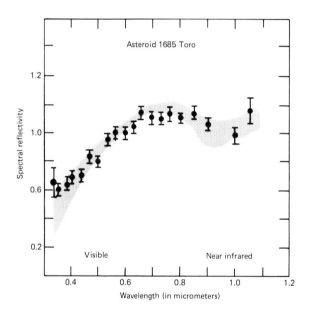

colors now seen in the main belt of asteroids. (See Focus 7E, Mines in the sky.)

7.3 Origin of the asteroids

(a) Former worlds

In the past, there have been two extreme theories for the origin of asteroids. According to the first, the asteroids represent the fragments of a former planet that has been torn apart. According to the second, they are the pieces of a planet that never formed. Today, astronomers favor a theory that lies between the extremes.

It is now known that the combined mass of all asteroids is far too small to make up a major planet. If they were all collected together, they would add

Focus 7E Mines in the sky

Temperature differences within the primeval solar nebula cannot, by themselves, explain the metallic asteroids. These objects are probably the by-products of heating and smelting within a few larger objects. The metals such as nickel and iron might have sunk to their cores, while the silicate rocks floated on top. The outer layers may then have been chipped away by bombardment, leaving their metal-rich cores exposed. There may, in fact, be thousands of these metallic asteroids, richly laden with resources.

We might obtain valuable metals by mining the surface of such an asteroid. Because of its low gravity, material could be easily removed and the metal could be shipped back by space shuttle from an asteroid that came close to the Earth. Some imaginative space-engineers speculate that the asteroid might even be brought closer using a "mass-driver," a device that would chew off pieces and fling them into space, propelling the asteroid like a rocket. They point out that the residual gas and dust would be swept from the solar system by the solar wind, thereby avoiding the problems of industrial pollution.

A metallic asteroid that is one kilometer in diameter contains about 8 billion tons of metal, with a current (1988) market value of about 5 million billion dollars on Earth. The asteroid could supply the world with iron for 15 years, nickel for 1250 years, copper for 10 years and cobalt for 3000 years.

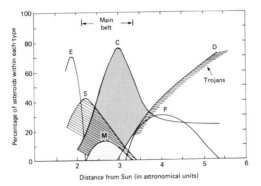

Figure 7.6. Asteroid distribution with distance. The color, or surface composition, of the asteroids is correlated with distance from the Sun. In order of increasing distance, there are the white E asteroids, the reddish S or silicate ones, the black C or carbonaceous ones, and the unusually red D asteroids. This systematic change has been attributed to a progressive decrease in temperature with distance from the Sun at the time the asteroids formed. Simple temperature differences within the primeval solar nebula cannot, however, explain the rare metallic M asteroids found in the middle of the asteroid belt. (Adapted from Jonathan Gradie and Edward Tedesco, Science **216**, *1405 (1982).)*

up to two ten-thousandths (0.0002) of the Earth's mass, and this would make a moon only about one-twentieth the radius of the Earth. So the first extreme must be discarded. On the other hand, the second extreme can also be excluded because there is strong evidence that many of the asteroids were once collected into a relatively small number of slightly larger parent bodies.

About one-third of the asteroids with known orbits can be grouped into 10 families. Within each of these families, the orbits are so similar that the members must have originated from a single object. Hundreds, and perhaps thousands, of small asteroids making up such families are probably the debris of collisions that destroyed the parent bodies. These parents may have been several hundred kilometers in diameter. And not only are the orbits similar within these families, the colors and surface compositions are also alike, and these similarities imply that the families are real physical groupings. Figure 7.7 is a schematic reconstruction of one such family.

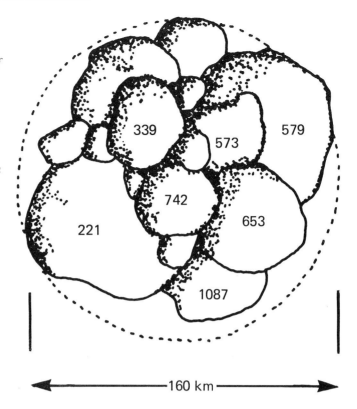

Figure 7.7. Eos family. A reconstruction of the Eos family of asteroids. The number indicates the order of discovery of the asteroids. They may all be the shattered remains of a former larger world that was about 160 kilometers in diameter. [Adapted from Jonathan Gradie, Ph.D. thesis, University of Arizona (1978).]

─────160 km─────

(b) A planet that never formed

But why are there asteroids? Why did the planetesimals between the orbits of Mars and Jupiter fail to coalesce into a single planet and create, instead, the swarm of asteroids? No one knows for certain, but perhaps the formation of a giant planet in the immediate neighborhood prevented the congregation of the asteroids. Jupiter, the most massive of the planets, was forming, and perhaps its gravitational force took charge of the neighborhood, tugging at the planetesimals in the asteroid belt, flinging them into eccentric orbits, causing them to crash into one another and grinding down the larger rocks into pebbles.

7.4 Meteorites – stones from the sky

(a) Space rocks

Some of the ancients recognized that stones have fallen from the sky and they gave the name meteor to the brilliant trail of light flashing across the night sky. We now know that most of the meteors are caused by tiny fragments of comets. They are typically the size of a snowflake and are just as fragile, so they never strike the Earth. (See Chapter 11.) Occasionally an extraordinarily bright meteor will fall, perhaps accompanied by a rumbling sound and what appears to be a great burst of sparks. These are fireballs, and they are produced by tougher chunks of matter from space, resembling pebbles and rocks (Fig. 7.8).

Rocks that survive the fiery descent through the atmosphere and reach the ground were given the name meteorite. And strictly speaking, a meteoroid is the solid object in space that appears as a meteor and can become a meteorite.

Meteorites have long been recognized as celestial objects. The New Testament Acts of the Apostles (19:35) refers to a temple dedicated to Artemis in which there is a "sacred stone that fell from the sky." These black objects have also been found in the Egyptian pyramids with a hieroglyph meaning heavenly iron.

Roughly 40 000 tons of cosmic debris plunge into the Earth's atmosphere each year, but most of it is quickly vaporized and an estimated 200 tons manages to hit the surface – primarily in the form of microscopic grains. About 10 meteorites are recovered each year. They are ones that happened to fall near populated regions of the Earth, and many more must be at the bottom of the ocean, lost in the jungles, or buried in desert sand.

(b) The Antarctic lode

Recently there was a sudden increase in the number of recovered meteorites when a group of Japanese scientists discovered a bountiful source in the ice of the Antarctic. Thousands of meteorites have been recovered there. They rise to the surface of the wind-swept ice and are concentrated in a relatively small area by a combination of the upward motion of the ice sheet and the scouring action of strong winds that erode the ice.

In the course of the past million years or so, these meteorites had been buried deeper and deeper in the snow and ice, where they were preserved against corrosion by the dry cold of the polar region. In time, they were carried by the ice sheet, which creeps slowly toward the coastline in all directions to a mountain barrier, where they were forced to the surface and exposed once more to the air (Fig. 7.9). Radioactive dating indicates that most of the Antarctic meteorites collided with the Earth about half a million years ago, while the meteorites recovered elsewhere on Earth are usually much more recent and have fallen within the past 200 years.

Figure 7.8. Fireball. A great flash of light, called a fireball, is produced when a large meteoroid streaks through the atmosphere. It is often accompanied by sonic booms and rumbling noises. This fireball was photographed by the Prairie Network station at Hominy, Oklahoma. Spaces between the luminous segments of the fireball's trajectory are caused by a chopping shutter used for timing and velocity determinations. The faint background lines are star trails caused by the Earth's rotation during the three-hour exposure. (Courtesy of the Smithsonian Astrophysical Observatory.)

(c) Chronology of the meteorites

If we ask "How old is a meteorite?", the question can have several meanings. Each meaning refers to the time since a significant event in the history of a meteorite, and we shall describe them as follows.

1. Formation

All meteorites date back to the earliest days of the solar system. The majority (chondrites) were formed with the planets. They accumulated directly from the primeval solar nebula and they have compositions similar to that of the Sun (except for their lack of hydrogen and helium). They are non-igneous, that is, they did not go through a hot, liquid stage. A small fraction of meteorites (called the achondrites) formed by igneous processes in parent bodies.

The dates of formation of the meteorites can be determined by radio-active dating in much the same way that the ages of lunar rocks were determined. (The relative concentrations of the decay products of elements such as rubidium and uranium reveal the time since these rocks were formed.) Such measurements indicate that the formation of meteorites occurred about 4.6 billion years ago. The meteorites are hundreds of millions of years older than the oldest rocks on the surface of the Earth, so they reveal the age of the solar system and give clues to its origin.

2. Breakup and exposure

Another type of radioactivity also occurs in meteorites – radioactivity that is continually being caused by cosmic rays in the solar system. These "rays" are not rays in the usual sense; they are atomic particles that bombard the meteorites and penetrate their surface for short distances. This cosmic ray bombardment performs a bit of alchemy and transforms some of the atoms of the meteorite into radioactive nuclei. These radioactive nuclei slowly disintegrate, creating "daughter" nuclei. As the meteorite continues to be exposed to cosmic rays, the daughter nuclei become more and more abundant, and by a careful measurement of the relative amount of daughter nuclei in a meteorite it is possible to estimate the duration of this exposure interval. This gives the exposure age of the meteorite.

Now, the exposure ages of the meteorites that have been recovered on Earth are remarkably short in astronomical terms. Typically they are between 5 and 60 million years – just an instant in the life of the solar system and the meteorites. Evidently the meteorites have spent most of their lives shielded from cosmic rays. Astronomers now believe that most meteorites spent a larger portion of their life inside a parent body that was much smaller than the Earth but larger than a typical meteorite. According to this view, an important event in the chronology of most meteorites was the

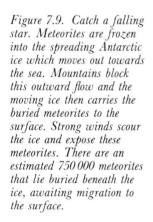

Figure 7.9. Catch a falling star. Meteorites are frozen into the spreading Antarctic ice which moves out towards the sea. Mountains block this outward flow and the moving ice then carries the buried meteorites to the surface. Strong winds scour the ice and expose these meteorites. There are an estimated 750 000 meteorites that lie buried beneath the ice, awaiting migration to the surface.

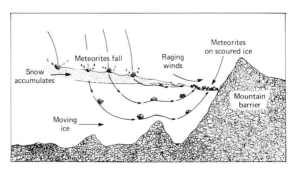

Table 7.1. *Classes of meteorites*

Name	Composition	Density^a (g/cm^3)	Percent of total meteorites
Stones	Silicates and metallic nickel-iron	3.5–3.8	95
Irons	Nickel, iron	7.6–7.9	4
Stony-irons	Silicates and metallic nickel-iron	4.7	1

^{a}For comparison, typical rocks on Earth are largely silicates with densities in the range of 3.1–3.3 g/cm^3, so meteorites are usually denser than terrestrial rocks.

Table 7.2. *The two types of stony meteorites*

Name	Percent of stonys	Appearance
Chondrite	90	Aggregate with spherical bodies up to 5 mm
Achondrite	10	Melted, homogeneous

Figure 7.10. Stony meteorite. Fragment of the stony meteorite (a chondrite) that fell near Johnstown, Colorado. (Courtesy of the American Museum of Natural History.)

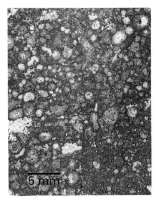

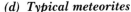

Figure 7.11. Chondrules in Allende. This photomicrograph of a thin section of the Allende meteorite shows numerous round silicate chondrules together with irregular inclusions. The meteorite section was 21 millimeters across and 27 millimeters high. (Courtesy of the Smithsonian Institution.)

breakup of the parent body, exposing smaller fragments to space and to the bombardment by cosmic rays. The exposure ages measure the time that has elapsed since the breakup took place.

3. Collision with the Earth

Finally, when a meteorite falls through the blanket of the Earth's atmosphere, it becomes protected from cosmic ray bombardment. No more radioactive atoms are created, and the ones that already exist inside the meteorite begin to decay, like the slow ticking of a clock.

In this way, the atoms of the meteorite carry a record of their chronology that can be unlocked with radiochemistry.

(d) Typical meteorites

Meteorites, together with rocks returned by astronauts from the Moon and grains of dust collected from the high levels of the Earth's atmosphere, are the only samples we know of extraterrestrial material. Although a meteorite's surface is usually coated with dark, glassy material that melted during its descent through the atmosphere, the heat of friction did not have time to penetrate deeply into the meteoroid. Its interior was unaffected, so meteorites may be cut open and examined with microscopes and may be subjected to chemical analysis.

They are divided into three groups depending on their major constituents (see Table 7.1). The most common ingredient is stone (Fig. 7.10). Most of the stony meteorites are, in turn, classified as chondrites (see Table 7.2). The name "chondrite" is derived from the ancient Greek word, *chondros*, meaning grain or seed (Fig. 7.11).

Most meteorites are denser than terrestrial rock, so if you find a dark, rather smooth rock that you suspect of being a meteorite, it must weigh at least as much as an ordinary Earth-born rock of the same volume if it is to pass muster as a rock from the sky.

But, of course, every rule has an exception. (Except that rule?) There is a rare class of meteorites, with the ponderous name carbonaceous chondrites, that are fragile and have unusually low densities in the range 2.2–2.9 g/cm^3, making them less dense than an average terrestrial rock. They contain appreciable amounts of carbon and water and they are considered to be among the most primitive and least altered samples of solids in the solar system. Their chemical composition suggests that they are daughters of the Sun.

(e) Rare and exotic finds

One tiny greenish-brown meteorite recovered in the Antarctic is strikingly similar to the highland breccias (welded rocks) from the Moon (Fig. 7.12). The abundances of various elements and gases in this meteorite are virtually identical to those found in lunar rocks; at the same time, they are unlike those found in any other known meteorite or terrestrial rock. It now seems certain that this tiny stone was blasted off the Moon, perhaps by an asteroid impact about 100 000 years ago.

The frozen cargo of the ice sheet also includes a few enigmatic achondrites with the code name SNC. The first such meteorites were found near Shergotty (India), Nakhla (Egypt), and Chassigny (France). Meteorites are named for the localities where they are found and the name of this class, SNC, comes from the initials of these locations.

Radioactive dating indicates that the SNC meteorites solidified from molten lava 1.3 billion years ago. This is long after the formation of the solar system, so they must have come from ancient volcanoes. And, looking at

Figure 7.12. Meteorite from the Moon. Polarized light brings out the structure in a thin slice of meteorite that probably came from the Moon. The relative abundances of several elements found in this meteorite are virtually identical with those found in rocks returned from the Moon, and unlike those found in any other meteorite or terrestrial rock. (Courtesy of Darrell Henry, NASA).

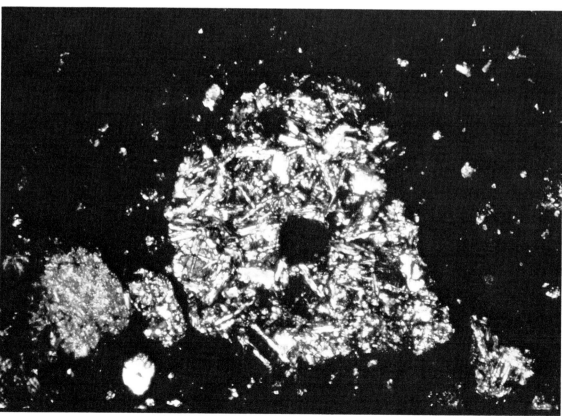

their structure, we see bits of glassy substance of the type produced when a rock is struck a hard blow, such as a meteorite impact. These glassy portions have been dated at 180 million years ago, and taken together these facts suggest that the SNC meteorites solidified from a volcano 1.3 billion years ago and then underwent a powerful shock 180 million years ago. The shock was most likely caused by the impact of a meteorite, perhaps launching the SNCs into space and sending some of them toward the Earth.

Where could the SNCs have come from? Probably not from an asteroid, because most of the asteroids solidified more than 4.5 billion years ago. Probably not from the Moon, because the Moon's lava stopped flowing 3.1 billion years ago. Probably not from Venus, whose thick atmosphere would have prevented the escape of any rocks knocked off its surface.

A process of elimination leaves us with Mars as the nearest planet that might have been volcanically active 1.3 billion years ago. In fact, the chemical makeup of the SNC meteorites is similar to that of the Martian soil sample analyzed by the Viking spacecraft. Thus, the SNC meteorites seem to have come from Mars. Perhaps they were ejected from an ancient lava bed by the impact of an asteroid 180 million years ago.

(f) Organic matter in meteorites

For more than a century, organic molecules have been suspected inside meteorites, although their presence in newly-arrived meteorites led to the suggestion that they were the result of terrestrial contamination. But, their existence recently became a certainty when 20 kinds of amino acids – the building blocks of life – were found in the carbonaceous chondrites from Antarctica. These meteorites had lain in a sterile environment and were collected using sterile procedures. Most convincing of all, they contain a form of right-handed amino acids that are not found in living systems on Earth (which are invariably left-handed).

The discovery implies that the amino acids and other organic molecules probably existed in the solar system a billion years before the appearance of life on Earth. Does this imply the existence of early life before it came to Earth?

Probably not. The molecules found in carbonaceous chondrites are generally thought to be of non-biological origin, perhaps created in early lightning storms. Carbon monoxide and hydrogen can be converted into organic molecules in the presence of an iron catalyst, and when ammonia is present in a laboratory sample, amino acids can be produced by electrical sparks.

So, the organic molecules found in meteorites are not in themselves vestiges of extraterrestrial life. But they are certainly primitive, and their cousins may have been the precursors to living matter.

(g) The asteroid–meteorite connection

What is the source of meteorites? There is little doubt that most of them have come from the asteroid belt. They are probably chips off wayward asteroids, and there are two pieces of direct evidence for this conclusion.

1. Orbits

Photography of meteorites as they descend through the Earth's atmosphere can be used to determine their precise speed and direction of motion when they encountered the Earth. From these data, their orbits may be inferred, and many of the objects came from space beyond Mars, in the main belt of asteroids.

Figure 7.13. Achondrite meteorite. A photomicrograph of the achondrite meteorite that fell near Juvinas, France on June 15, 1821. It contains basaltic material that is the product of the melting and separation of material inside an asteroidal-sized parent body. The section shown here was 3.2 millimeters across. (Courtesy of Martin Prinz, American Museum of Natural History.)

2. Colors

Spectra of sunlight reflected off the asteroids closely match the spectra obtained from light reflected off meteorites in the laboratory. For example, the spectra of S asteroids closely resemble those of ordinary stony meteorites; the spectra of C asteroids closely resemble those of carbonaceous chondrites, and the M asteroids have the colors of the iron meteorites (Fig. 7.13).

In addition, there is some compelling indirect evidence.

3. Crystalline structure

When the majority of iron meteorites are cut and polished and then are etched with acid, a delicate and complex pattern emerges (Fig. 7.14). It is produced by regions of crystalline structure that are more or less susceptible to the acid, depending on the local orientation of the crystals in the iron. The sizes and shapes of these crystals indicate that they grew very slowly, and that the meteorite must have been hot, almost to the melting point, for tens of millions of years. It probably cooled at the rate of a few degrees in a million years!

Such a slow cooling rate is compelling evidence that the meteorites were once inside a sizeable parent body. If a small iron meteorite had been exposed to space when it was still hot, it would have cooled in a matter of days. Small meteorites cool rapidly because their material is close to the surface, through which the heat can escape. Large crystal patterns would not have grown in such small bodies.

The meteorites that retained their heat for millions of years must have

Figure 7.14. Widmanstätten pattern. When polished and etched with acid, the iron meteorites display this distinctive Widmanstätten pattern that results from crystals of two different iron-nickel alloys. It provides evidence that the meteorites were once buried within parent bodies of between 50 and 200 kilometers radius. These sliced specimens are about 5 centimeters across. (Courtesy of the Smithsonian Institution.)

been buried within parent bodies between 50 and 200 kilometers in radius, and this is just the size of the typical asteroids that have been seen from Earth.

So the crystalline patterns of meteorites also suggest an asteroid–meteorite connection. Figure 7.15 shows schematically the relationship among the sizes of asteroids, meteorites and non-cometary meteoroids. The classes are not mutually exclusive, and there is considerable overlap. The simplest explanation for the asteroid–meteorite connection is that the meteorites are the debris of relatively recent collisions among the asteroids.

Whatever their precise history, the asteroids and meteorites are primitive objects that can act as beacons to the past. They represent the frozen tableau of the ancient planetesimals that littered space 4.6 billion years ago. They carry a code for our understanding of the formation of the solar system.

Figure 7.15. Interplanetary debris. Repetitive collisions between interplanetary objects has produced many more smaller meteoroids than larger ones. Some of the larger asteroids are comparable in size to small moons, and ongoing collisions between asteroids have produced numerous smaller meteoroids.

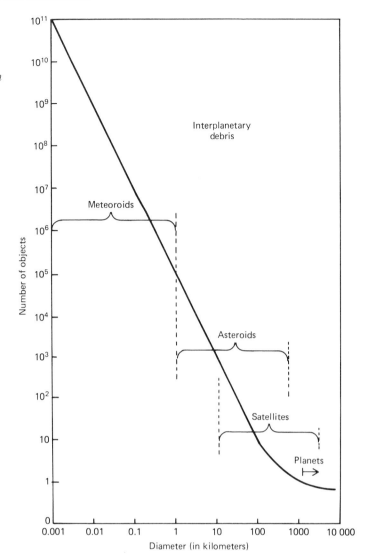

Focus 7F Asteroids – summary

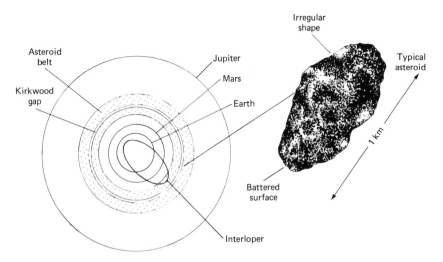

Ceres – the largest asteroid

Mass: 1.2×10^{24} grams = 0.0002 M_{E} (Earth = 1)
Radius: 512 kilometers = 0.08 R_{E} (Earth = 1)
Mean density: 2.3 g/cm^3
Rotational period: 9 hours, 4 minutes, 41 seconds
Orbital period: 4.61 years
Mean distance from Sun: 2.77 A.U.
Number of known satellites = 0

Giant world. Jupiter's clouded world with its alternating structure of light zones and dark belts. The two innermost Galilean satellites are also visible. Bright-orange Io is seen just above the cloud tops, and icy-white Europa lies above it to the right.

8 Jupiter: a giant primitive world

One of Jupiter's cyclonic storms has lasted for more than three centuries.

What gives Jupiter its vivid colors?

Jupiter is a primitive incandescent globe that radiates its own heat.

There probably is no solid surface beneath the clouds of Jupiter.

Jupiter is broadcasting radio waves with about 400 billion watts of power.

A vast current of 5 million amperes flows between the satellite Io and Jupiter.

The volcanoes on Io have turned the satellite inside out.

8.1 King of the planets

(a) View from Earth

With the naked eye, we see Jupiter as the fourth brightest object in the sky, after the Sun, Moon, and Venus. Jupiter's motion among the stars is relatively sedate. The planet orbits the Sun with a period of 11.86 Earth years, so each year it moves one-twelfth of the way around the sky and passes eastward through one constellation of the zodiac.

Its orbital radius is 5.2 times the radius of the Earth's orbit, so the planet's distance from Earth changes relatively little in the course of a year. As a consequence, its apparent size and brightness are fairly constant, unlike the behavior of Mars and the inner planets.

Jupiter surely deserves its title as the king of the naked-eye planets. It is also the largest planet in the solar system, with a radius that is 11 times that of the Earth and a volume that is 1330 times the Earth's. In a large telescope, Jupiter is a splendid sight, rivalled only by ringed Saturn. Jupiter's disc is striped with narrow dark bands parallel to its equator, and the smooth profiles of these bands are interrupted occasionally by darker spots. The largest of these spots – the Great Red Spot – is easily seen in a small telescope, and it has been followed for 300 years.

Despite its great size, Jupiter rotates so fast that its day is less than one-half Earth day. The precise rotation rate is found by tracking radio bursts that emerge from deep within the atmosphere. A spin period of 9 hours 55 minutes 29 seconds is obtained from the repeated passage of storm centers. This rapid rotation can easily be detected in an hour or so with a small telescope if the planet's cloud markings are carefully watched. Jupiter's rapid spin stretches its atmosphere into colorful bands that encircle the planet parallel to the equator. And the spin also reduces the effect of gravity near the equator, giving Jupiter a perceptible bulge around its mid-section.

Four of Jupiter's moons are bright enough to be seen in a pair of binoculars or a small telescope, and they move around the planet so quickly that their positions can be seen to change from hour to hour. (If it were not for the glare of Jupiter, these moons would be visible to the naked eye.) They were discovered by Galileo Galilei, and perhaps the German mathematician Simon Marius, in January, 1610, using the newly invented telescope. Although these objects are now collectively called the Galilean satellites, they retain the individual names given to them by Marius: Io, Europa, Ganymede, and Callisto.

Focus 8A Determining the composition of Jupiter's air

The composition of Jupiter's thin upper atmosphere can be inferred from emission features in the spectrum of its reflected sunlight. Every atom and molecule absorbs and emits energy at specific wavelengths that can be used to identify it, much the way a fingerprint can identify a person.

The infrared radiation from the hot regions of the high atmospheres of Jupiter and Saturn exhibits numerous emission features that have no counterpart in the spectrum of sunlight. As illustrated here, strong features are seen in the spectrum of Jupiter for molecular hydrogen, H_2, ammonia, NH_3, and methane, CH_4. Saturn's atmosphere is also abundant in molecular hydrogen and methane, but the ammonia features are missing and those of acetylene, C_2H_2, and ethane, C_2H_6, are enhanced.

There are no detectable features of helium atoms in the cold outer atmospheres of Jupiter and Saturn. The presence and amounts of helium have nevertheless been inferred from the hydrogen features because the appearances of these features are influenced by collisions between helium atoms and hydrogen molecules. (Courtesy of Rudolf A. Hanel.)

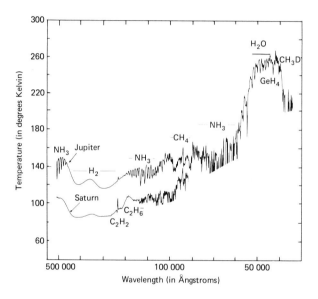

The Galilean satellites provided the first clear example of objects moving about a center other than the Earth, and for this reason they played an important role in the eventual acceptance of Copernicus' model of the solar system (see Chapter 1).

Jupiter's total mass may be determined from the periods and the sizes of the satellite orbits. The result is rather remarkable: Jupiter has 318 times the mass of the Earth, or nearly three-quarters of the total mass of all the planets put together. But compared to the Earth, Jupiter is voluminous for its mass. If we divide the mass by the volume we find an average density that is only one-quarter the density of the Earth. In fact, the density of Jupiter is only slightly greater than the density of water, and this implies that Jupiter, like the Sun, is composed primarily of hydrogen. No other element is light enough to account for the low density of the planet.

Table 8.1. *Composition of Jupiter's upper atmosphere*

Molecule	Abundance[a]
Hydrogen, H_2	79%
Helium, He	19%
Methane, CH_4	0.0007
Ammonia, NH_3	0.0002
Ethane, C_2H_6	0.0004
Acetylene, C_2H_2	0.0001

[a]The abundances of H_2 and He are similar to the solar values, 78% and 20%, respectively.

Jupiter is not alone in having such a low density. The Jovian planets, which also include Saturn, Uranus and Neptune, all have relatively low densities. And Jupiter and Saturn also have extensive satellite systems that resemble miniature solar systems.

(b) Space-age odyssey

Altogether, five spacecraft have been sent to Jupiter's region of the solar system. The Pioneer 10 and 11 spacecraft performed the first reconnaissance, safely traversing regions of space filled with asteroids, energetic particles and previously unmapped magnetic fields.

The Pioneers were followed by the two Voyager spacecraft. The energy received from the spacecraft as they transmitted images from Jupiter was so feeble that it would have to be collected for hundreds of billions of years to make a light bulb shine for one second. And yet, one astronomer happily exclaimed that trying to understand all of the scientific data sent back by these spacecraft was like trying to drink from a fire hose!

After visiting Jupiter, Pioneer 10 was flung by the giant planet's gravitational sling shot on a flight to the stars, while the other three (Pioneer 11, Voyager 1 and Voyager 2) were hurled to a rendezvous with Saturn. In turn, Saturn's gravity was then used to swing Voyager 2 towards its encounters with Uranus in 1986 and Neptune in 1989. In October, 1989, the Galileo spacecraft started on its planned six-year journey to Jupiter.

8.2 Jupiter's upper atmosphere

(a) Poisonous gas and abundant hydrogen

Jupiter is an alien world with a smelly and, to us, lethal atmosphere. The spectrum of the planet's reflected sunlight indicates that its relatively cool atmosphere contains 79 percent molecular hydrogen, and 19 percent helium, with small amounts of noxious gases such as methane and ammonia (see Focus 8A, Determining the composition of Jupiter's air and Table 8.1). Water vapor, and complex hydrocarbons such as acetylene, and ethane, are also found in Jupiter's atmosphere, but in extremely small amounts – one part in a million or less.

The relative proportions of helium, hydrogen, carbon, and nitrogen in Jupiter are similar to the proportions in the Sun, suggesting that Jupiter, unlike the terrestrial planets, is a representative sample of the primeval solar nebula from which it was born 4.6 billion years ago.

Figure 8.1. Jupiter's turbulent atmosphere. Alternating dark belts and white zones stretch around Jupiter (left). The enormous red spot (right) is large enough to cover two Earths. The huge high-pressure vortex swirls in the counterclockwise (anticyclonic) direction with a period of six days. (Courtesy of JPL and NASA.)

(b) Stormy weather

Windswept clouds churn and seethe in Jupiter's colorful atmosphere. Huge storms larger than the Earth in size swirl across Jupiter, while giant cyclones create continent-sized spots. Smaller spots chase each other, whirling and rolling about, and even devouring each other. Zones and belts race around the huge planet, driven by hurricane-speed winds. Lightning bolts illuminate the Jovian night with enough energy to vaporize a city. Everywhere there is stormy weather as clouds billow, churn and surge above a vast sea of hydrogen.

The colorful spots, zones, and belts are weather patterns. In fact, nearly everything we see on this awesome planet is a storm cloud (Fig. 8.1). Jupiter's rapid rotation has pulled the cyclonic storms into alternating bands of light-colored regions, called zones, and dark regions, called belts. The zones and belts surround the planet, running parallel to the equator at different speeds in opposite eastward and westward directions relative to the average eastward flow. They move with speeds of up to 540 kilometers per hour, exceeding those of the fastest jet streams on Earth. The light-colored zones are rising currents of gas, while the dark belts are regions of descending gas.

Red, white and brown spots or eddies stare out of Jupiter's atmosphere like gigantic eyes. These spots resemble huge cyclones or anticyclones in the Earth's atmosphere, but they are larger, more colorful, and often last for months, decades or even centuries. The Great Red Spot is an enormous reddish oval, the vortex of a violent, anticyclonic storm that drifts along its belt, eventually wandering completely around the planet.

The origin of the color of the Great Red Spot is a mystery. Lightning bolts or the Sun's ultraviolet light may be breaking down molecules to liberate phosphorus or a form of sulfur that give the Great Red Spot its color. Alternatively, red organic compounds made lower down in the atmosphere might be brought to the top by rising currents.

The spots of Jupiter are eddies that roll like ball bearings between the oppositely directed east–west winds. The smaller ones are soon torn apart by the counter-flowing winds. Larger spots survive, and the largest of them actually gobble up smaller ones and thereby replenish their own energy. On occasion they spin off a new, smaller eddy (Fig. 8.2).

The lifetime of the spots therefore depends on their size. The smaller spots are created and destroyed in time scales of days, while larger spots last for decades.

But why have the Great Red Spot and the counter-flowing east–west winds persisted for decades on Jupiter when similar storms and winds only last for days or weeks on Earth? Computer simulations indicate that the planet's rapid rotation imparts spin to the atmosphere and stabilizes the rotation of vortices. Further, there is probably no solid surface beneath the clouds to interfere with the flow, so the weather pattern is free to flow in response to the spin, and the flows may penetrate deep into the planet.

Figure 8.2. Turbulent eddies. The Great Red Spot swirls in the counterclockwise direction, sucking in nearby smaller eddies like leaves in a whirlpool of water. Other small eddies roll around the Great Red Spot, probably reinforcing its circulation. (Courtesy of JPL and NASA.)

(c) Layered clouds

All of the visible activity on Jupiter takes place in an atmospheric layer whose thickness is less than one-hundredth of the planet's radius. This layer lies above a vast sea of liquid hydrogen and helium. If we could descend through the clouds, we would find that the temperature and pressure increase with depth, as illustrated in Figure 8.3. At the cloud tops, the temperature is a freezing 114 degrees Kelvin, but in the deeper layers it rises to a balmy 300 K, well above the freezing point of water (273 K). In

these warmer regions, the pressure is comparable to the air pressure at the surface of the Earth.

In contrast to the Earth's atmosphere, which has only one layer of stormy weather, Jupiter may have three distinct layers of clouds. Theorists speculate that, at the low temperatures of the cloud tops, gaseous ammonia freezes to form graceful white clouds that make up the cold, light-colored zones observed from Earth. Below this level the ammonia combines with hydrogen sulfide to form brown crystals that smell something like rotten eggs. Clouds of ammonia hydrosulfide crystals make up the warm, dark belts. The hottest and deepest layer of clouds – which no one has seen – may be composed of bluish crystals of water ice. Balmy temperatures similar to those at the Earth's surface are reached at about 60 kilometers below Jupiter's cloud tops.

Jupiter's poisonous gases would kill most present-day terrestrial life, but its hydrogen-rich atmosphere might be similar in chemical composition to the Earth's primitive atmosphere. Organic molecules have been created in the terrestrial laboratory by passing an electrical discharge through a mixture of methane, ammonia, and hydrogen, a mixture resembling the atmosphere of Jupiter. This success has led to speculations that Jupiter could harbor some sort of primitive life. But Jupiter has no solid surface on which primitive creatures could creep or crawl, and strong atmospheric currents would probably drag them to lethal hotter levels. Nevertheless, imaginative astronomers argue that enormous inflated organisms could be floating in Jupiter's global sea of hydrogen. They might bob up and down like jellyfish in terrestrial oceans, thereby seeking more clement conditions.

8.3 Beneath Jupiter's clouds

(a) An incandescent globe

With the advent of infrared measurements of Jupiter's radiation, astronomers were surprised to discover that the giant planet is an incandescent globe with its own internal source of heat! It radiates twice as much heat as it receives from the Sun.

Jupiter is so far from the Sun that each area of its clouds receives only 4 percent as much solar heat as the Earth's surface. But its deep clouds are much warmer than expected. There is only one acceptable conclusion: The excess heat and the warmth of the polar regions are caused by heat escaping from Jupiter's hot interior. Planets shine by reflected sunlight and do not normally generate energy, whereas stars produce their own light and energy by thermonuclear reactions in their interiors. Jupiter evidently has an excess of internal heat, although it is not massive or hot enough to ignite thermonuclear reactions.

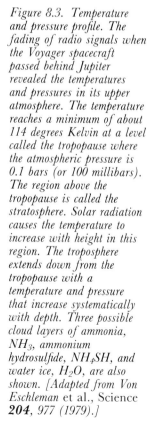

*Figure 8.3. Temperature and pressure profile. The fading of radio signals when the Voyager spacecraft passed behind Jupiter revealed the temperatures and pressures in its upper atmosphere. The temperature reaches a minimum of about 114 degrees Kelvin at a level called the tropopause where the atmospheric pressure is 0.1 bars (or 100 millibars). The region above the tropopause is called the stratosphere. Solar radiation causes the temperature to increase with height in this region. The troposphere extends down from the tropopause with a temperature and pressure that increase systematically with depth. Three possible cloud layers of ammonia, NH_3, ammonium hydrosulfide, NH_4SH, and water ice, H_2O, are also shown. [Adapted from Von Eschleman et al., Science **204**, 977 (1979).]*

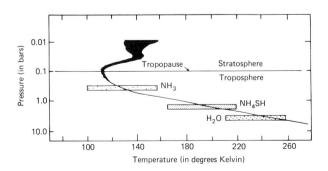

Table 8.2. *Range of pressures*

Location	Relative pressure
Beneath the foot of a water-strider	0.00001
Inside a light bulb	0.01
Earth's atmosphere at sea level	1.0
Inside a fully charged scuba tank	100
Deepest ocean trench	1 000
Pressure at which hydrogen becomes metallic	3 000 000
Center of Jupiter	8 000 000

Much of Jupiter's internal heat may be left over from the planet's formation 4.6 billion years ago. The gravitational collapse and compression that accompanied the growth of new-born Jupiter would have heated the interior. Jupiter's enormous size makes it a much better heat-trap than the terrestrial planets, so this primordial heat would be trapped inside Jupiter and only slowly conducted out of the massive planet.

(b) Enormous pressures and strange matter

Because the weight of overlying material increases with depth, there is an increase in pressure with depth within Jupiter. This increase can be calculated from theory, and the central pressure of Jupiter must be about 80 million times the surface pressure of the Earth's atmosphere at sea level. This is more than 7 times the pressure at the center of the Earth, but if Jupiter were as dense as the Earth, its central pressure would be 125 times as great as that of the Earth. (See Table 8.2.)

Jupiter is mainly composed of hydrogen, which turns liquid at the tremendous pressures and relatively low temperatures inside the planet. Most of Jupiter's interior consists of a vast sea of liquid hydrogen. The planet may be gaseous to a depth of only 1000 kilometers, or 1.4 percent of its radius.

At pressures greater than 3 million bars, which occur at depths greater than 17 000 kilometers (about one-quarter of its radius) below the surface, liquid molecular hydrogen is transformed into metallic hydrogen. The hydrogen molecules are squeezed so tightly that individual hydrogen atoms are no longer bound within molecules, and the electrons are squeezed free of the atoms, to move about and conduct electricity. The hydrogen is said to be in a metallic state because, like a metal, it is an excellent conductor of heat and electricity.

Because Jupiter is composed of Sun-like material, it probably has a core composed of heavy elements such as silicon and iron. The presence of a dense core (either liquid or solid) is suggested by a detailed study of Jupiter's non-spherical shape. A planet of a given size and rotation speed will be more flattened if it has a dense core. Jupiter's shape is consistent with a rocky core that has 15 times the mass of the Earth but squeezed into a volume that is only twice the Earth's. This core is overlaid by an enormous shell of liquid metallic hydrogen, and above this is a global ocean of liquid molecular hydrogen (Fig. 8.4). Electrical currents within Jupiter's liquid metallic shell generate a strong magnetic field.

8.4 Jupiter's magnetic field

(a) Radio broadcasts from Jupiter

Jupiter is broadcasting billions of watts of noise at short radio wavelengths. This radiation is called decimetric radiation because the wavelength is around one-tenth of a meter, or 10 centimeters. The decimetric radiation is generated by high-speed electrons that spiral about Jupiter's magnetic field, so the radiation provides an outline map of Jupiter's magnetic field.

The strength of Jupiter's magnetic field and the energy of its trapped electrons are much greater than those on Earth. The magnetic poles are flipped with respect to the geographic poles, so that a terrestrial compass on Jupiter would point south. The magnetic axis is tilted with respect to the rotation axis by 11 degrees. (The Earth's field is tilted in much the same way, causing our magnetic compasses to point away from the true north pole on most of the Earth's surface.) This axial tilt causes the rotating magnetic field to wobble like a warped phonograph record, and this wobbling is seen in the decimetric radiation pattern.

Observations from ground and space have shown that Jupiter's magnetic field has a strength that is about 12 times that of the magnetic field at the Earth's equator (see Table 8.3). Currents of electricity that are driven by Jupiter's internal heat and rotation probably generate a magnetic field in much the same way that the Earth's field is generated, so Jupiter's rapid rotation and large zone of liquid metallic hydrogen may account for the exceptional strength of its magnetic field.

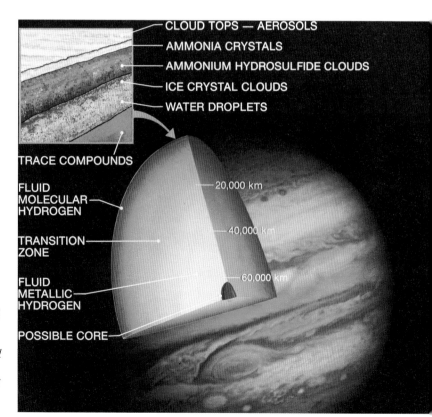

Figure 8.4. Inside Jupiter. Giant Jupiter has a thin gaseous atmosphere covering a vast global ocean of liquid hydrogen. At the enormous pressures that exist deep within its interior, the liquid molecular hydrogen is converted into liquid metallic hydrogen. (Courtesy of JPL and NASA.)

Table 8.3. *Magnetic fields in the solar system*

Object	Strength of field (gauss)[a]	Inclination of axis relative to rotation axis (degrees)
Mercury	0.003	<10
Earth	0.35	12
Jupiter	4.225	11
Saturn	0.2	0.7
Uranus	0.1 to 1.1	60
Neptune	0.3	59
Sunspots	1000 to 3000	0 to 90

[a]The strengths are given at the surfaces of Mercury and the Earth and at the cloud tops of the Jovian planets. Venus and Mars have no detectable fields.

(b) The magnetosphere

Jupiter is surrounded by a vast invisible structure called a magnetosphere. It is essentially a rotating bag consisting of a magnetic field and charged particles, electrons and ions, that are trapped by the magnetic field. Near the planet, the magnetic field has a dipolar shape similar to the Earth's field, but its outer region is greatly elongated into a magnetic tail by the pressure of the solar wind and of the trapped particles (Fig. 8.5).

Jupiter's magnetosphere is the largest enduring structure in the solar system, although it is occasionally exceeded by comet tails. Jupiter's enormous magnetic tail, almost a billion kilometers long, spans the distance between the orbits of Jupiter and Saturn – which is as great as the distance from the Sun to Jupiter, itself. In contrast, the Earth's magnetic tail barely flicks across our Moon's path, less than a half-million kilometers from the Earth.

The outer magnetosphere is buffeted by the solar wind so it expands and contracts between distances of 50 R_J and 100 R_J from the planet. (The radius of Jupiter, written R_J, is 71 492 kilometers.) The inner magnetosphere, on the other hand, is a stiff, permanent structure that is tied to the planet and rotates with it (Fig. 8.6).

Charged particles are trapped near Jupiter within doughnut-shaped regions that girdle the planet in its magnetic equator. These regions

Figure 8.5. Jupiter's magnetosphere. Jupiter's magnetosphere is the largest permanent structure in the solar system, dwarfing the Sun. When the solar wind encounters Jupiter's magnetosphere, it forms a bow shock resembling the wave ahead of a boat. The solar wind flows around Jupiter, stretching its magnetic field out into a long magnetotail. The variable solar wind buffets Jupiter's magnetic field, pushing its bow shock in and out between 50 and 100 Jovian radii from the planet's center. The outward forces and pressures associated with Jupiter's rapid rotation inflate the magnetosphere, and form a thin elongated disc called the magnetodisc. A ring current encircles Jupiter in the magnetodisc.

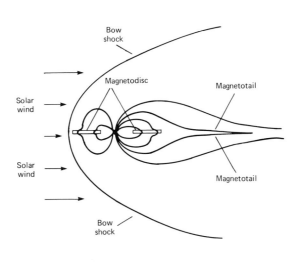

resemble the terrestrial Van Allen belts in shape, but Jupiter's are up to a million times more densely filled with particles than those near the Earth. They contain ions of sulfur and oxygen that come from its innermost large satellite, Io, as well as electrons and protons that come from the solar wind. In contrast, the terrestrial radiation belts mainly contain electrons and protons that are fed into it by the solar wind.

As the powerful magnetic field rotates, it lashes and accelerates the charged particles that are trapped near the planet. The moving particles exert an outward pressure on the magnetic field, inflating it like an air-filled balloon. The forces and pressures associated with the rapid rotation are greatest at the equatorial regions, where the magnetic field is stretched outwards in the form of a thin, elongated magnetodisc. The charged electrons and ions are swept along by the magnetic field within the magnetodisc, producing a ring of current that girdles the planet near the plane of its equator.

When gusts of solar wind compress Jupiter's outer magnetosphere and push it back, some of its energetic particles squirt out into interplanetary space. The particles in this Jovian wind are more energetic than those in the solar wind. The Jovian wind is continually replenished by particle acceleration within the planet's magnetosphere; the wind's particles are sprayed throughout the solar system, some reaching the Earth, and even to the orbit of Mercury.

Figure 8.6. Inside the magnetosphere. Ions and electrons in the solar wind (right) impinge against the magnetosphere, distorting the field lines and creating a turbulent shock front. Io's plasma torus (dark twisted shape) contains energetic sulfur and oxygen ions arising from the moon's active surface. The light disc is a plasma "sheet" along the plane of Jupiter's magnetic equator. (Courtesy of Robert Wolff, JPL.)

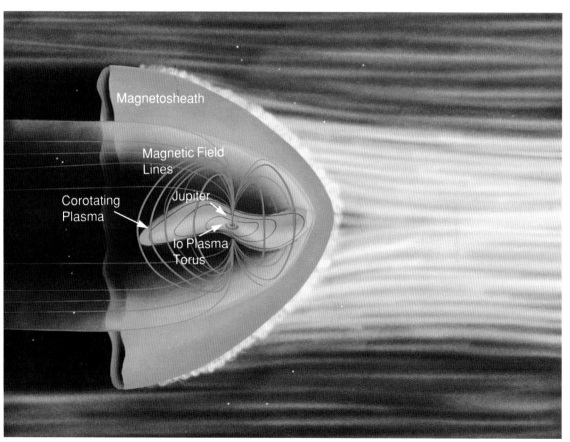

(c) Io's domain

Sodium atoms are spread all around the orbit of Jupiter's innermost large satellite, Io. They remain most concentrated near their source, the satellite Io, forming a glowing sodium cloud that is as big as Jupiter (Fig. 8.7). The cloud stretches forward and backward along Io's orbit, more of it preceding the satellite than following it.

Evidently these atoms have been chipped off the surface of Io by the persistent hail of high-energy particles in the planet's radiation belts. Although the sodium atoms can escape from Io, they are captured by Jupiter's stronger gravity and therefore go into orbit around the planet.

As Jupiter's rotating magnetic field sweeps past and through Io's domain, it generates currents that flow from Io to Jupiter through a magnetic tube (Fig. 8.8). It is a sort of electromagnetic umbilical cord that attaches the moon to its planet. A vast current of about 5 million amperes flows along Jupiter's flux tube, generating power of 2.5 trillion watts. Io and Jupiter may therefore combine to make a natural power station whose output vastly exceeds that of any terrestrial energy-generating plant. The electrical currents in this cosmic power station often produce powerful bursts of radio noise and bright polar aurorae on Jupiter when they strike Jupiter's atmosphere.

Sulfur and oxygen atoms are abundant on Io's surface and in its tenuous atmosphere. When these atoms are ionized, they are lost from Io and are caught up by Jupiter's spinning magnetic field. Carried by the field, which is itself anchored to Jupiter, they rotate about the planet every 10 hours. Io orbits Jupiter in a leisurely period of 42 hours. As the atoms are swept along, they spread into a doughnut-shaped ring. This ring glows in the ultraviolet light produced by electrons colliding with the sulfur and oxygen ions.

8.5 Jupiter's moons and ring

(a) The Galilean satellites

The four Galilean satellites – named Io, Europa, Ganymede, and Callisto in order of increasing distance from Jupiter – move in nearly circular orbits near Jupiter's equatorial plane (Fig. 8.9). They are the four largest of Jupiter's sixteen known moons, and they each keep the same face towards

Figure 8.7. The sodium cloud. Io's neutral sodium cloud (left) as seen from the Earth shown together with Jupiter (center) and a schematic drawing of Io's orbit to scale. Excited sodium atoms that came from Io (location denoted by a cross) emit radiation at optical wavelengths (5890 and 5896 A). This sodium cloud observation was made using a Silicon Imaging Photometer Systems (SIPS) at the Coudé focus of the 24-inch (61-centimeter) telescope at JPL's Table Mountain Observatory in California. (Courtesy of Bruce Goldberg and Glenn Garneau, JPL.)

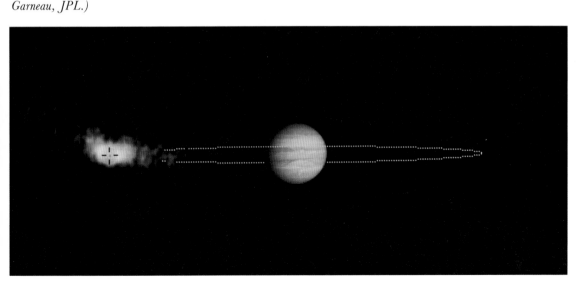

Table 8.4. *Properties of the Galilean satellites*

Satellite	Distance from Jupiter center[a]	Orbital period (d)	Radius[b] (km)	Mass (g)	Density (g/cm³)
Io	5.95 R_J	1.716	1816	8.92×10^{25}	3.55
Europa	9.47 R_J	3.551	1563	4.87×10^{25}	3.04
Ganymede	15.1 R_J	7.155	2638	1.49×10^{26}	1.93
Callisto	26.6 R_J	16.69	2410	1.07×10^{26}	1.83

[a]Mean distances in units of Jupiter's radius, $R_J = 71\,492$ km.
[b]For comparison, the radius of our Moon is 1738 km.

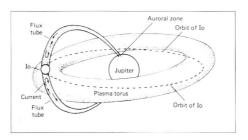

Figure 8.8. Flux tube and plasma torus. An electric current of 5 million amperes flows along Io's flux tube. It connects the ionospheres of Io and Jupiter, like a giant umbilical cord. The plasma torus is centered near Io's orbit, and it is about as thick as Jupiter is wide. It is filled with energetic sulfur and oxygen ions that have a temperature of about 100 000 degrees Kelvin. Because the planet's magnetic axis is tilted with respect to the rotational axis, the satellite Io weaves up and down within the plasma torus. Charged particles probably flow from the plasma torus along magnetic field lines to Jupiter's polar regions where they create glowing aurorae.

Jupiter, as the Moon does to the Earth. They orbit around Jupiter with periods of days (see Table 8.4).

The Galilean satellites are individually named after lovers of Zeus, the Greek equivalent of Jupiter. Zeus changed the mortal Io into a cow to hide her from his jealous wife, and Callisto was punished for her affair with Zeus by being changed into a bear. Europa took the form of a white bull after being carried off to Crete on the back of Zeus, while Ganymede was a Trojan youth carried off by an eagle to be Zeus' cup bearer. Amalthea, the fifth Jovian satellite to be found, is an exception for it is named after the she-goat that suckled the infant Zeus.

Io and Europa, the innermost of the Galilean moons, are about the size of our Moon, and both have densities comparable to that of rock (about 3 g/cm³). Ganymede and Callisto, the larger, outermost of the four, are about the size of Mercury, and they have a significantly lower density of about 2 g/cm³. Apparently they consist of roughly half rock (3 g/cm³) and half water ice (1 g/cm³).

The compositions of the Galilean satellites were most likely affected by their relative proximity to Jupiter. Io and Europa are largely composed of rock because of the radiant heat of hot, new-born Jupiter. Some theorists find that primordial Jupiter was initially about one-hundredth as bright as the Sun, with a central temperature of 50 000 degrees Kelvin, so it heated the region where the inner moons formed. Satellites that formed in this region were too hot to retain substantial quantities of water, and they are therefore mainly composed of rock. (Europa is covered by ice, but its high density indicates that this covering is a relatively thin veneer.)

In contrast, the relative cold of regions farther from Jupiter permitted Ganymede and Callisto to retain their ice and become mixtures of ice and rock. This would explain their low density, as well as the fact that they are more massive than Io and Europa.

As they are comparable to the Moon or Mercury in size, the Galilean satellites were generally expected to have cratered surfaces and to show few

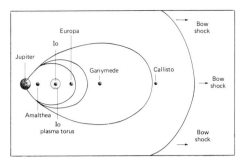

Figure 8.9. The four Galilean satellites. Io, Europa, Ganymede and Callisto are all embedded within Jupiter's magnetosphere. The outermost, Callisto, orbits Jupiter near its bow shock on the planet's sunward side. A small satellite named Amalthea is about one-tenth of the size of the Galilean satellites and orbits Jupiter at a little less than half the distance to Io, the innermost Galilean satellite. All of these satellites are being continuously bombarded with energetic charged particles that are trapped within Jupiter's magnetosphere.

signs of internal heat and geological activity. But the Voyager 1 and 2 spacecraft discovered bizarre and beautiful worlds – each uniquely sculpted – including both the smoothest and roughest surfaces in the solar system, as well as the youngest and the oldest.

(b) Io: a world turned inside out

The innermost Galilean satellite, Io, has a radius and density that are nearly identical to those of our Moon, but contrary to expectation, there are no impact craters on Io. Eruptive plumes, volcanic calderas and fuming lava lakes are found instead! Io's active volcanoes eject plumes of gas and dust that spread out like a fountain, ejecting enough material to cover the satellite's surface to a depth of 100 meters in the short span of a million years (Fig. 8.10). In fact, Io appears to be the most volcanically active body in the solar system. The Voyager 1 cameras recorded 8 major eruptions simultaneously. By comparison, 8 eruptions on Earth might require a century or so.

Geyser-like eruptions send plumes to heights of hundreds of kilometers. The material shoots upward at one kilometer per second (3 times the speed of sound on Earth) propelled by sulfurous gas. Because of the satellite's low gravity and apparent lack of substantial atmosphere, the plumes spread out in graceful fountain-like trajectories; depositing circular rings of material of up to 1400 kilometers in diameter (Fig. 8.11).

Io's volcanoes are literally turning the satellite inside out! Volcanic activity continuously resurfaces the satellite, creating its relatively young surface. All of the material that we now see on Io was probably deposited there less than a million years ago. Evidently Io's mantle and crust have been recycled many times over the span of Io's history.

Volcanic activity is also revealed by large volcanic calderas and their associated flows (Fig. 8.12). Hundreds of the black calderas pock Io's surface. The heat radiated by these active volcanoes can also be detected and monitored from Earth.

But what drives Io's continuous volcanism? The heat released during the moon's accretion and subsequent radioactive heating of its interior should have been lost to space long ago. Io's internal heat is generated by the enormous tidal bulges that massive Jupiter raises on the tiny satellite.

If Io remained in a circular orbit with the same face toward Jupiter, its tidal bulges would not change in height and no heat would be generated; but gravitational forces from the other Galilean satellites pull the orbital path of Io slightly inward and outward.

In effect, Io is caught in a gravitational tug-of-war between Jupiter and the other Galilean satellites, particularly the nearest one, Europa. The resultant variation in the tides flex Io's surface, bending it in and out by as much as 100 meters during each orbit. (The flexing heats Io's interior in much the same way that a wire heats up when rapidly bent back and forth.)

Figure 8.10. Active volcano on Io. Pele, the largest observed volcanic eruption on Io, has been named after the Hawaiian goddess of volcanoes. The erupting plume is visible at the upper right, rising to a height of about 300 kilometers above the surface. It has been ejected from the central blue and white complex of hills. In this enhanced color image we also see concentric brown and yellow rings, that have been deposited around the source of the plume. Here the outermost brown ring runs from the upper left to the lower right, averaging 1400 kilometers in diameter. (Courtesy of Alfred McEwen, Tammy Rock and Laurence Soderblom, USGS.)

This tidal heating melts Io's interior rock and produces volcanoes. The high temperature may have driven off any water that was on Io.

If there is no water on Io, then what acts as a propulsive agent for its volcanoes? According to one theory, Io's volcanoes occur when liquid sulfur dioxide and red-hot molten sulfur come into contact at shallow depths. The liquid sulfur dioxide then begins to boil and vaporize. A mixture of liquid and gas accelerates up a conduit, erupting at the surface and creating volcanic plumes. Alternatively, the large plumes could arise when hot silicates in the crust vaporize sulfur.

(c) Europa: a bright, smooth world

The smallest, and yet brightest, of the Galilean satellites, Europa, has a density comparable to that of rock, but its surface is as bright and white as ice. In fact, it is water ice! Europa's incredibly smooth surface might mark the top of a vast frozen ocean.

Europa has the smoothest surface of any known planet or satellite in the solar system. No surface features extend as high as 100 meters. Its face is marked only by a veined, spidery network of long, shallow dark streaks (Fig. 8.13).

The tidal heating that apparently melted Io's interior operates on Europa as well – to a smaller extent because Europa is further from Jupiter,

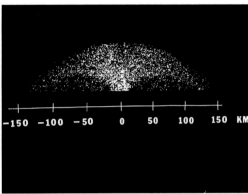

Figure 8.11. Fountain-like eruptions. An active volcano erupts with explosive violence from Jupiter's satellite Io (left). Because Io has practically no atmosphere, the sulfur and frozen sulfur dioxide spread out in graceful, ballistic trajectories. A computer simulation of a volcanic eruption on Io (right) shows the fountain-like trajectories of the volcanic plume. The symmetry and umbrella shape of the volcanic eruption are due to Io's low gravity and the lack of winds or substantial atmosphere on the satellite. (Courtesy of JPL and NASA (left) and Nicholas M. Schneider, Lunar and Planetary Laboratory, Tucson (right).)

but it may be sufficient to keep most of the water on Europa from freezing. A deep, interior ocean of liquid water may be covered by Europa's outer ice shell, which may be only a few kilometers thick.

Europa's icy mantle has numerous fractures, which may be due to global expansion, cracking by impact, or tidal distortion by Jupiter. Dirty water has apparently welled up and frozen between the large ice flows, producing a lacework of dark streaks. There is an interesting resemblance between Europa's surface and the cracked and refrozen ice fields in the Earth's arctic regions, where currents have cracked the floating ice and the fractures have been filled with upwelling water that subsequently froze.

(d) Ganymede: a cratered, wrinkled world

Ganymede, the largest moon in the solar system, has a radius that exceeds that of the planet Mercury, but the satellite's density is so low that it must contain substantial quantities of liquid water or ice. It probably has a thick mantle of water ice (Fig. 8.14).

Ganymede's surface is one of great geological diversity that apparently includes crustal movements and mountain building. The outer shell of ice has been broken into dark blocks (Fig. 8.15). These large polygonal blocks seem to have moved sideways for tens of kilometers along Ganymede's surface. Other regions are covered with ranges of wrinkled mountains, resembling furrows from a giant rake.

The satellite's dark crust probably cracked and spread apart when the surface expanded. Widespread crustal expansion may have resulted from the segregation of Ganymede's rocks into its interior and the ices into the exterior. The ice would expand in the lower-pressure outer regions. Crustal expansion could be responsible for the dark blocks and the parallel mountains.

Sets of intersecting mountain ridges overlap and twist into each other. Some of the ridges cut across craters, while craters appear on other ridges.

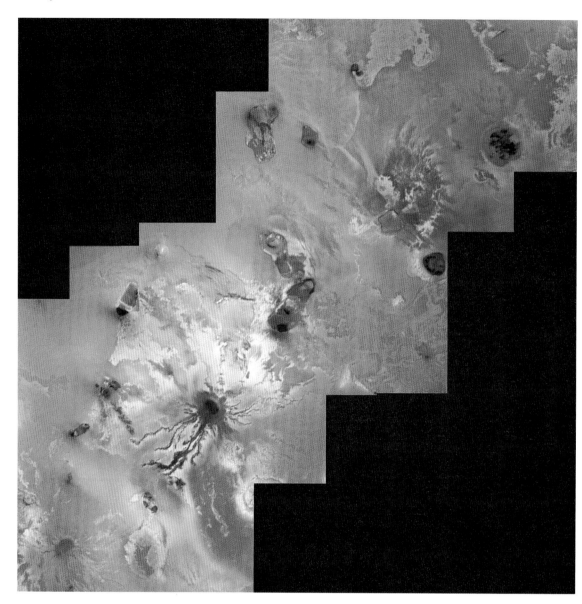

Figure 8.12. Lava flows on Io. Dark snake-like tendrils of red, brown and black lava, thought to be colored by sulfur at different temperatures. They mark the slopes around Ra Patera (lower left), some stretching 200 kilometers from the hot, black caldera into the cooler surrounding terrain. This enhanced color image also shows white deposits that may be erupted sulfur in a different form or in compounds such as sulfur dioxide. Another black caldera (center) contains a bluish crescent that may be attributed to a less energetic eruption. (Courtesy of Alfred McEwen, USGS.)

Figure 8.13. Europa. Dark streaks mark Europa's smooth surface, forming a spidery, veined network. Internal stresses apparently fractured its icy mantle, producing intersecting cracks that extend for thousands of kilometers but reach depths of less than 100 meters. The fractures may have been filled by water or soft ice gushing out from the satellite's warm interior. (Courtesy of JPL and NASA.)

Ganymede evidently experienced several epochs of mountain building. These crustal deformations probably continued for perhaps a billion years, and crater counts suggest that even the youngest mountain on Ganymede is at least 3 billion years old.

(e) Callisto: an ancient, battered world

Remotest of the Galilean moons, Callisto is a primitive world with little sign of internal activity (Fig. 8.16). Because its surface contains more impact craters than any other Galilean satellite, its surface is the oldest one – perhaps 4.6 billion years old. The surface seems unaltered since it was formed, and it is probably a fossil of the origin of our solar system.

Although superficially similar to the crater-scarred surfaces of the Moon and Mercury, Callisto is different in many respects. There is a conspicuous lack of large craters, volcanic plains and mountain ranges on Callisto. Moreover, her ice craters are much flatter than the rocky craters on our Moon. Callisto's icy crust was probably unable to support the weight of their heavy rims.

Figure 8.14. Ganymede.
The face of Ganymede
contains ancient dark areas,
light regions, and bright
spots that are impact craters.
Because Ganymede's icy
surface is somewhat plastic,
its craters are shallow for
their size. (Courtesy of JPL
and NASA.)

Figure 8.15. Polygonal blocks on Ganymede. The surface of Ganymede contains dark polygonal blocks frozen within its icy surface. They resemble brown, frozen-over continents floating on a background of translucent ice. The ancient blocks have apparently separated like the moving pieces of a huge mosaic or giant jigsaw puzzle, perhaps because of crustal expansion. The brilliant white material that surrounds some craters is probably clean water ice that splashed out from inside the satellite. (Courtesy of JPL and NASA.)

Glacial ice flows may have deformed and leveled many craters (Fig. 8.17), because ice, which is rigid to sharp impacts, can flow gradually over long periods of time.

(f) Outer and inner moons

Jupiter's bevy of 16 known moons includes the four Galilean satellites, four satellites that orbit inside Io's orbit, and eight outer moons in eccentric, tilted orbits that are so far from the planet that the Sun competes for their gravitational control. These eight outer moons fall into two widely separated groups (Fig. 8.18). The innermost four move in direct orbits, but the outermost four move backward, in retrograde orbits. The Sun's gravitational perturbations would have dislodged the outermost satellites if they had direct orbits.

All eight of the outer moons are probably former asteroids or planetesimals that once orbited the Sun, and wandered near Jupiter and became captured into its family. The two groups of maverick moons may have resulted from the breakup of two of these objects.

Before the Voyagers, only one small satellite, Amalthea, was known to reside inside Io's orbit. This dark reddish moon is very irregular, about 270 kilometers long and 150 kilometers across, with its long axis pointing toward Jupiter. The Voyagers discovered Thebe, about 80 kilometers across, between the orbits of Io and Amalthea, as well as Adrastea and Metis, 40 and 25 kilometers across, in the outer fringes of Jupiter's ring.

Figure 8.16. Callisto. The Jovian satellite Callisto is a battered world, pockmarked with impact craters. Because Callisto's icy surface is as rigid as steel, it retains the scars of an ancient bombardment. The craters are flat for their size, and there are very few large craters. As a result, the bright limb lacks vertical relief. Many of the existing craters have bright rims that resemble clean water ice splashed upon the dirtier surface ice. (Courtesy of JPL and NASA.)

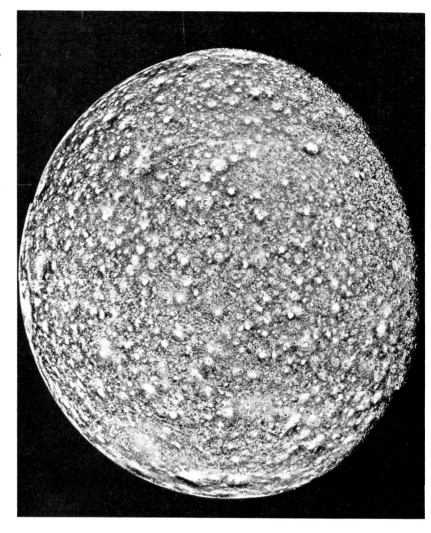

(g) Jupiter's mere wisp of a ring

The rings of Saturn were discovered in the 17th century. In 1977 several faint and unsuspected narrow rings were discovered about the planet Uranus (see Chapter 10). Jupiter was next to join the group of ringed planets, but this time the discovery was not a complete surprise. In 1974, the Pioneer 11 spacecraft had encountered anomalous decreases of radiation in the vicinity of Jupiter. Not much was made of these anomalies although some scientists thought they could be interpreted as weak and unconvincing evidence for "an unknown satellite or ring of particles (surrounding Jupiter) at approximately 1.83 R_J." Finally, in 1979 – after much debate about the likelihood of finding a ring – a search was carried out with a camera on Voyager 1, and a narrow ring was discovered in a single exposure when the Voyager 1 spacecraft passed through the planet's equatorial plane. As anticipated, the brightest part of the ring lies relatively close to Jupiter at a distance of 1.8 Jovian radii (1.8 R_J) from the planet's center. The ring was not previously observed from Earth because it was too faint and close to the bright planet. Since its discovery, the ring has been detected by Earth-based telescopes sensing infrared radiation.

Figure 8.17. Valhalla. The most prominent feature on Callisto is the extensive system of concentric rings that is named Valhalla after the home of the Norse gods. Long ago a large object smashed into Callisto, and like a rock dropped into a pond, it sent waves rippling across the surface. The impacting object apparently punctured the surface and disappeared. Today only the frozen, ghost-like ripples remain. (Courtesy of JPL and NASA.)

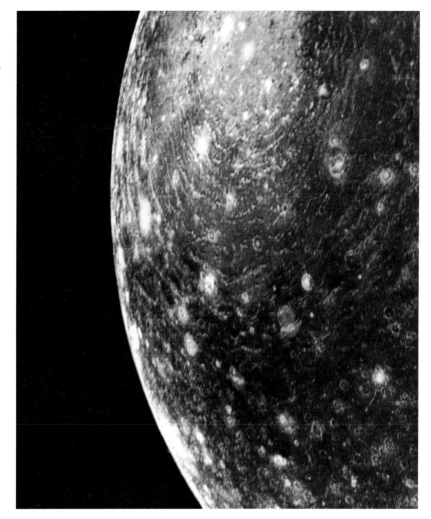

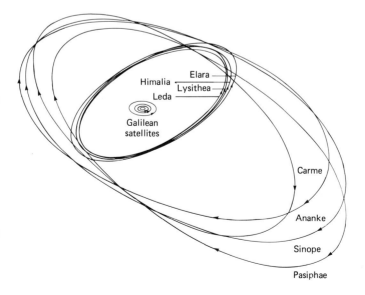

Figure 8.18. Jupiter's outermost satellites. These eight satellites have eccentric orbits that are inclined with respect to Jupiter's equatorial plane. One group of four moves around Jupiter in conventional direct orbits at distances between 11 and 12 million kilometers. Their names all end in the letter "a" – Leda, Himalia, Lysithea, and Elara. The other group of four have backwards retrograde orbits that lie at distances between 20 and 24 million kilometers from Jupiter. The outer group has names ending in "e" – Ananke, Carme, Pasiphae, and Sinope.

Figure 8.19. Jupiter's ring. Sections of Jupiter's tenuous ring (left) look unexpectedly bright from the far side looking back in the sunward direction. Because the small ring particles scatter light in the forward direction, the ring brightens up when sunlight flows through it. Motes of dust or cigarette smoke similarly brighten when they float in front of a light. A close-up photograph (right) shows the main ring's thinness, as well as the diffuse gossamer ring that extends from it. (Courtesy of NASA (left) and Mark Showalter and Joseph Burns, Cornell University (right).)

When the retreating Voyager cameras looked back at the shadowed side of Jupiter, the ring became brighter (Fig. 8.19). This behavior is typical of very small particles that scatter light in the forward direction, like tiny salt grains on the windshield of an automobile or the smoky haze in a movie theater. The size of the ring particles can be inferred from the way they scatter light, and the conclusion is that they are about one ten-thousandth centimeters across, or about the same size as the particles in cigarette smoke. Although Jupiter's tenuous ring is made visible by its dust-sized particles, larger unseen chunks of matter are hypothesized as causing the observed absorption of high-energy radiation.

The individual particles only reside temporarily in the ring, for they rain down into the atmosphere of Jupiter (Fig. 8.20). Such fine particles must be continually replenished.

The ring particles may be renewed by particles that are sandblasted off Jupiter's two innermost moons by meteorites. One of these small satellites, Adrastea, orbits close to the ring's edge at $1.8\ R_J$; the other, Metis, lies nearer to the brighter midpoint.

This ends our space-age visit to Jupiter. Following the Voyager spacecraft, we now step out to Saturn – and beyond.

Figure 8.20. The ring system. Jupiter's ring system is composed of two main parts – the bright ring and a diffuse halo. The bright ring is roughly 6000 kilometers wide and less than 30 kilometers thick. According to one theory, collisions with micrometeoroids shatter some of the ring particles into a fine powder that is blown away by electrostatic forces. As a result, both the bright ring and the faint sheet are enshrouded by a diffuse halo that extends about 5000 kilometers above and below the central ring plane. (Adapted from David Jewitt and G. Edward Danielson, Journal of Geophysical Research **86**, *8500 (1981).)*

View from above ring plane

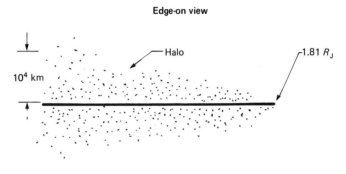

Edge-on view

Focus 8B Jupiter – summary

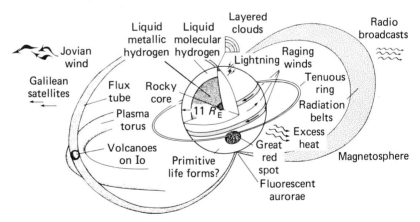

Mass: 1.90×10^{30} grams $= 317.89\ M_E$ (Earth $= 1$)
Radius: 71 492 kilometers $= 11.2\ R_E$ (Earth $= 1$)
Mean density: 1.314 g/cm^3
Rotational period: 9 hours 55 minutes 29.7 seconds
Orbital period: 11.86 years
Mean distance from Sun: 5.203 A.U.
Number of known satellites $= 16$
Magnetic field strength at cloud tops $= 4.3$ gauss

Saturn's realm. Saturn's magnificent rings encircle the butterscotch planet, never touching its surface. The yellow-brown atmosphere of Saturn has a banded structure, but it lacks Jupiter's bright colors. (Courtesy of JPL and NASA.)

9 Saturn: lord of the rings

Although it is much bigger and more massive than the Earth, Saturn could float on water.

Theoretical calculations indicate that helium rain has been falling inside Saturn for the past two billion years.

Saturn's rings are completely detached from the planet and separated from each other; they may contain thousands of icy ringlets.

Why do planets have rings, and why do these rings have sharp boundaries?

Titan's atmosphere is filled with nitrogen. Ethane oceans might lap its shores.

Two of Saturn's satellites shepherd a narrow ring between them, while another two episodically exchange orbits.

9.1 The planet Saturn

(a) Fundamental properties

Majestic Saturn, the sixth planet from the Sun, was the most distant world known to the ancients, and it moved least rapidly around the zodiac. The Greeks identified the planet with Kronus, the father of Zeus; while the Romans named the planet Saturn after their god of sowing. Both the Greeks and the Romans associated Saturn with the ancient god of time, which later became Father Time.

Saturn orbits the Sun at a mean distance of 9.5 A.U., with an orbital period of 29.5 years. Perhaps because of its sedate motion in a distant orbit, the planet's name has been adopted for the word "saturnine," to describe a cool and distant temperament.

At its large distance from the Sun, Saturn receives only about 1 percent as much light and heat as the Earth does. There is a seasonal variation in this feeble sunlight similar to that on Earth, for Saturn's equator is inclined only slightly more than the Earth's – by 29 degrees with respect to the planet's orbit about the Sun. This inclination allows us to see Saturn's fabled rings which encircle the planet while remaining unattached to it.

Saturn's mass is 95 times greater than the Earth's mass, and its radius, without the rings, is slightly more than 9 times that of the Earth. Its volume is great enough to encompass 900 Earths. This giant planet is second only to Jupiter in size and mass, and the two planets are similar in many respects. Together, they dominate the planets.

From Saturn's mass and volume, we calculate its average density to be only 0.71 g/cm^3, the lowest of any planet and less than that of water. If Saturn were placed in a large enough ocean of water, it could float. Like Jupiter, giant Saturn has a low average density because it is primarily composed of the two lightest elements, hydrogen and helium.

Saturn rotates with a day of only 10 Earth-hours – nearly the same as Jupiter's – and the rapid rotation makes Saturn's globe bulge at the equator, although not quite so severely as Jupiter's, whose radius is greater. (The distance from the north or south pole to Saturn's center is 9.6 percent less than the equatorial radius of 60 330 kilometers.)

(b) Saturn's clouds and winds

The clouds of Saturn are composed mainly of ammonia crystals that create a global haze the color of pale butterscotch. The planet's rapid rotation has pulled these clouds into zones and belts that resemble pale versions of those on Jupiter. But the dominant winds on Saturn blow eastward, in the same direction that the planet rotates, unlike those of Jupiter, whose alternating bands of light and dark clouds coincide with reversals in wind direction (Fig. 9.1).

Winds near Saturn's equator reach 1800 kilometers per hour, two-thirds the speed of sound on Earth and almost four times the speed of Jupiter's fastest winds. The equatorial wind extends over more than 70 degrees of latitude, or 80 000 kilometers in width. Jovian-like winds that counterflow in the eastward and westward directions are only found near Saturn's poles.

The temperature of Saturn's outer atmosphere is, like Jupiter's, hotter than that expected from solar heat alone, and it radiates into space about twice as much heat as it receives from the Sun. This implies that Saturn is an incandescent globe with an internal source of heat (Fig. 9.2).

(c) Inside Saturn

Saturn's excess heat is partly a residue from the gravitational collapse at the time of its formation and from its continued gravitational contraction. But this heat is only one-third the amount needed to explain the observations. Additional heat is believed to arise from the gradual separation of helium from hydrogen in the interior of the planet.

At sufficiently high temperatures and pressures, metallic hydrogen and helium are completely dissolved in each other, as they probably are in Jupiter. When the temperature falls, some of the helium separates out as

Figure 9.1. Winds on Saturn. Although Saturn's weather is dominated by its equatorial wind, there are counterflowing eastward and westward winds near the planet's polar regions. Enhanced color brings out the details of these winds, including a dark spot that marks a high-pressure, anticyclonic storm. The white spots are two of Saturn's icy moons. (Courtesy of JPL and NASA.)

small drops. These drops are denser than hydrogen, and they rain down toward the planet's center, stirring the liquid hydrogen as they fall and converting some of their energy to heat. In much the same way, raindrops on Earth become slightly warmer when they strike the ground and their energy of motion – acquired from the gravity – is converted to heat.

Saturn's upper atmosphere contains less helium than Jupiter's (11 percent by mass as contrasted with 19 percent) and this can be understood as a result of helium rain inside Saturn, whose smaller mass would have permitted more extensive cooling and a more complete separation of helium.

Saturn and Jupiter are giant liquid drops surrounded by thick atmospheres. The atmospheres of both planets give way to liquid in the deep interior, but because Saturn is less massive than Jupiter, the pressure will be lower and the transition to a liquid is expected to occur closer to the center of Saturn (Fig. 9.3). This difference probably explains a peculiarity of Saturn's magnetic field.

(d) Saturn's weak magnetosphere

The strength of Saturn's dipolar magnetic field is about one-twentieth that of Jupiter's, despite the fact that both planets rotate in about 10 hours. The weaker field of Saturn may be a result of Saturn's thinner shell – if we are correct in presuming these fields to be created by rotationally driven electric currents in the planets' liquid metallic shell. Saturn's magnetism is almost precisely aligned with its poles of rotation and, like Jupiter's, it is reversed compared to the magnetism of the Earth. (A terrestrial compass would point south.)

Saturn's magnetic bow shock bobs in and out of the orbit of its largest satellite Titan, depending on the fluctuating pressure of the solar wind. The

Figure 9.2. Saturn and its rings at infrared wavelengths. Because Saturn's rings are made of ice, they reflect relatively large amounts of sunlight at an infrared wavelength of 3.8 micrometers (coded blue). Methane in Saturn's atmosphere absorbs solar radiation at this wavelength, but the incandescent globe has its own internal source of heat that makes it shine brightly at the longer infrared wavelength of 4.8 micrometers (coded orange). The heat welling up from within Saturn is probably due to helium raining inside the planet. (Courtesy of David Allen, Anglo-Australian Telescope Board © 1983.)

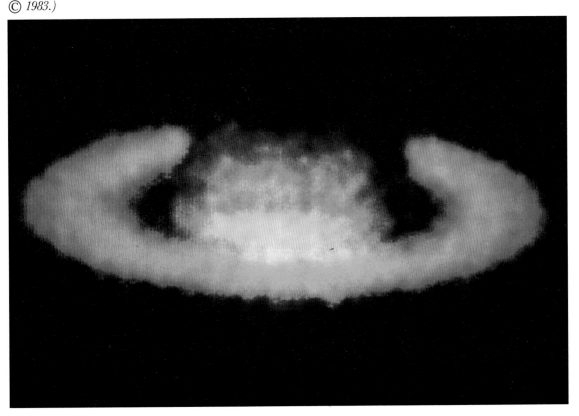

Figure 9.3. Inside Jupiter and Saturn. Jupiter's greater mass results in a higher internal pressure and a larger shell of liquid metallic hydrogen. Saturn's smaller shell of metallic hydrogen may explain its weaker magnetic field. Both Jupiter and Saturn may have Earth-like rocky cores; Saturn's core is estimated to be about 25 percent of its mass, while calculations suggest that Jupiter's core is only about 4 percent of its mass.

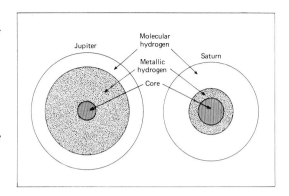

satellite therefore acts as a variable source of particles for the planet's magnetosphere. Atomic hydrogen, released by the action of sunlight on Titan's atmosphere, streams in orbit around Saturn, forming a huge hydrogen torus (Fig. 9.4).

9.2 Saturn's remarkable rings

(a) The main rings

Saturn's beautiful rings are completely detached from the planet, nowhere touching its surface. Three main rings are visible from the Earth: the A, B, and C rings (see Table 9.1). The A and B rings are separated by a dark zone named the Cassini division, first noticed in 1675 by Jean Dominique Cassini. The Voyager spacecraft showed that the Cassini division is by no means empty (Fig. 9.5). Interior to the A and B rings is the crepe, or C, ring. As its name implies, it is the most transparent of the three.

We only partly understand why Saturn's rings have these sharp boundaries. In the absence of other forces, collisions between ring particles should cause them to fall inward toward Saturn and expand outwards away from it. A satellite can evidently help to prevent a ring from spreading if its orbital period is properly tuned to the ring's rotation. For example, the particles in the outer edge of the B ring (the inner edge of Cassini's division) have orbital periods that are exactly half that of the satellite Mimas. The A ring's outer edge is similarly related to the satellite Janus. The influence is a gravitational give-and-take, somewhat like the motion of a swing.

Because the rings are tipped about 29 degrees to the plane of the planet's orbit, they slowly change their appearance when viewed from the Earth. During Saturn's orbit about the Sun, they are successively seen edge-on (when they briefly vanish), from below (when they are wide open), edge-on again and then from above (see Focus 9A, Solving the mystery of Saturn's rings). The complete cycle requires about 30 years, so the rings vanish every 15 years. The next disappearance will take place in 1995.

Figure 9.4. Outside Saturn. Saturn's magnetosphere, satellites and rings. The magnetic bow shock reaches out to the orbit of Titan. A torus of neutral hydrogen atoms lies between the orbits of Rhea and Titan. It is fed by hydrogen liberated from Titan's atmosphere by energetic sunlight. Saturn's rings and satellites absorb energetic particles, and there are therefore no radiation belts in their vicinity.

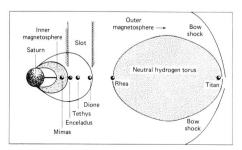

Table 9.1. *The main rings of Saturn*

Name	Distance from planet center (Saturn radii[a])	Orbital period[b] (hours)	Width (kilometers)
C ring	1.2–1.5	5.6–7.9	17 500
B ring	1.5–1.9	7.9–11.4	25 500
Cassini division	1.9–2.0	11.4–11.9	4 500
A ring	2.0–2.3	11.9–14.2	14 700

[a]The radius of Saturn is 60 330 kilometers, nearly ten Earth radii.
[b]Saturn's rotation period is 10.65 hours.

Figure 9.5. Beneath the rings of Saturn. When the Voyagers dove beneath the rings, they could view sunlight transmitted through the rings, a perspective not available from the Earth. When seen from beneath, the rings of Saturn present a reversed image of the sunlit side. Both the C ring and Cassini's division appear bright because they are sparsely populated with particles that efficiently scatter light in the forward direction, whereas the A and B rings appear dark because their densely-packed particles absorb all the incident sunlight. (Courtesy of JPL and NASA.)

The wide rings are incredibly thin. They are nearly 400 000 kilometers across, but they extend no more than a few meters from top to bottom. If the thickness of Saturn's rings is represented by a sheet of paper, then a scale model would require a sheet of paper extending across 40 city blocks. However, the rings also bend like a warped phonograph record, and astronomers measure the larger thickness of this warping when observing the rings edge-on from Earth.

The inner parts revolve faster than the outer parts, in good accordance with Kepler's third law. This means that the rings are composed of innumerable particles, each in orbit around Saturn (see Focus 9B, The motion of Saturn's rings). The particles undoubtedly collide with one another, although the collisions are quite gentle because relative velocities are small.

Sunlight reflected from the rings reveals signs of water ice. The ring particles are therefore thought to be composed of, or at least covered by, frozen water. Typical chunks of ice in the main rings vary in size from a centimeter to tens of meters across. The larger ones, though far less numerous, contain most of the ring mass. Snowflakes, hailstones, snowballs and icebergs make up Saturn's main rings. Occasional components may be as large as a house.

(b) Spacecraft discoveries

When the Pioneer 11 and Voyager 1 and 2 spacecraft entered Saturn's realm, their cameras revealed remarkable and unexpected details. To begin

Focus 9A Solving the mystery of Saturn's rings

When Galileo first observed Saturn through his primitive telescope in 1610, he discovered that the planet was not round; it had blurry objects on each side. When these objects disappeared two years later, Galileo wondered if Saturn "had devoured his own children."

Saturn's puzzling handle-like appendages subsequently reappeared and then disappeared again, causing the planet to appear alternately as an extended ellipsoid and a round disc. The mystery of this slowly changing appearance was not solved until Christiaan Huygens reasoned that it must be caused by a ring surrounding the planet. He supposed that it must revolve as rapidly as a satellite and yet it did not change appearance from one week to the next. Only a ring could revolve without changing appearance.

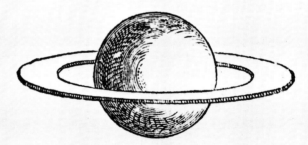

SYSTEMA SATVR... A   47

ea quam dixi annuli inclinatione, omnes mirabiles Saturni facies sicut mox demonstrabitur, eo referri posse inveni. Et hæc ea ipsa hypothesis est quam anno 1656 die 25 Martij permixtis literis una cum observatione Saturniæ Lunæ edidimus.

Erant enim Literæ a a a a a a a c c c c c d e e e e g h i i i i i i i l l l l m m n n n n n n n n n o o o o p p q r r s t t t t u u u u u; quæ suis locis repositæ hoc significant, *Annulo cingitur, tenui, plano, nusquam cohærente, ad eclipticam inclinato.* Latitudinem vero spatij inter annulum globumque Saturni interjecti, æquare ipsius annuli latitudinem vel excedere etiam, figura Saturni ab aliis observata, certiusque deinde quæ mihi ipsi conspecta fuit, edocuit: maximamque item annuli diametrum eam circiter rationem habere ad diametrum Saturni quæ est 9 ad 4. Ut vera proinde forma sit ejusmodi qualem apposito schemate adumbravimus.

Cæterum obiter hic iis respondendum censeo, quibus novum nimis ac fortasse absonum videbitur, quod non tantum alicui cælestium corporum figuram ejusmodi tribuam, cui similis in nullo hactenus eorum deprehensa est, cum contra pro certo creditum fuerit, ac veluti naturali ratione constitutum, solam iis sphæricam convenire, sed & quod annulum

Occurri-turiis quæ de annulo objici possent.

In 1655, Huygens, then only 26 years old, announced his discovery in the form of an anagram, a succession of scrambled letters. In 1659 he deciphered the anagram on this page of his monograph *Systema Saturnium*. They spelled out the solution that can be translated as "Saturn is girdled by a thin flat ring, nowhere touching it, and inclined to the ecliptic."

Because the rings are tipped with respect to the ecliptic, they change their shape when viewed from the Earth, slowly opening up and then turning edge-on as Saturn makes her slow 29.5 year orbit around the Sun. When the rings are opened up, they resemble handle-like appendages, but when they are viewed edge-on the rings virtually disappear.

In 1867 James Clerk Maxwell, also just 26 years old, showed that the rings cannot be solid because a solid ring would be torn apart by Saturn's gravity – similar to the tidal effect of Roche's limit. Maxwell proposed that the rings are instead composed of an indefinite number of particles, each in orbit about Saturn. They act as innumerable tiny satellites that move in accordance with Kepler's laws. This theoretical conclusion was confirmed observationally by James Keeler's spectroscopic observations in 1895 (see Focus 9B, The motion of Saturn's rings). (Figure permission of Houghton Library, Harvard University and courtesy of Owen Gingerich.)

Figure 9.6. Cross-section of rings and satellites. The satellite Mimas helps to gravitationally control the outer edge of the B ring, and the satellite Enceladus probably feeds the tenuous E ring. All of the main rings lie within the Roche limit within which Saturn's gravity will tear a large satellite apart. For clarity, the thickness of the rings has been exaggerated.

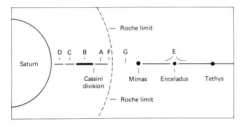

with, there were the tenuous D, F and G rings (Fig. 9.6). These rings are diffuse and nearly transparent. Inside the C ring is the D ring that is so tenuous and transparent that it is probably impossible to see from the Earth using the best telescopes. According to one hypothesis, splintered chips from colliding ice particles drift down into Saturn's atmosphere and form the D ring.

Pioneer 11 discovered the incredibly narrow F and G rings. The F ring lies just outside the A ring and is only a few kilometers wide. It displays an astonishing series of kinks, strands and braids.

The widest and outermost ring is the tenuous E ring, first discovered with ground-based telescopes. Its particle density reaches a maximum at the orbit of Saturn's satellite Enceladus. Watery eruptions from the surface of Enceladus – perhaps produced by meteorite impacts – might feed icy particles into Saturn's E ring.

When the Voyager cameras zoomed in on the main rings, they appeared to break up into countless rings – almost without limit (Fig. 9.7). Thousands of tiny ringlets appeared within the B and C rings, some of them making a complete circle around Saturn. Other ringlets form a tight spiral pattern resembling the grooves on a long-playing phonograph record. The A ring remained broad and free of numerous fine ringlets, while the Cassini division contains more than a hundred of them. The ringlets of Saturn vary with time and vaguely resemble undulating ripples running across the surface of a pond.

Focus 9B The motion of Saturn's rings

The motion of Saturn's rings can be measured using spectral features in their reflected sunlight. When a ring particle moves towards or away from an observer, these spectral features are displaced in wavelength by an amount that depends on the velocity of motion. This phenomenon is called the Doppler effect. Motion towards the observer produces a displacement toward shorter (blue) wavelengths, while motion away produces a shift toward longer (red) wavelengths.

The Doppler effect was used by James Edward Keeler to determine the motion of Saturn and its rings. His results are illustrated here in a figure adapted from Keeler's article in the first volume of the *Astrophysical Journal*, published in 1895. Keeler's observations showed that the inner parts of the rings move faster than the outer parts, implying that the rings are composed of separately moving particles, each in its own orbit around Saturn.

If the rings rotated as a solid, like the planet's body does, the speed at the outer edge would be higher than the inner edge, but the opposite is the case. The observed motion is in accord with Kepler's third law, depicted by the dashed line in the figure.

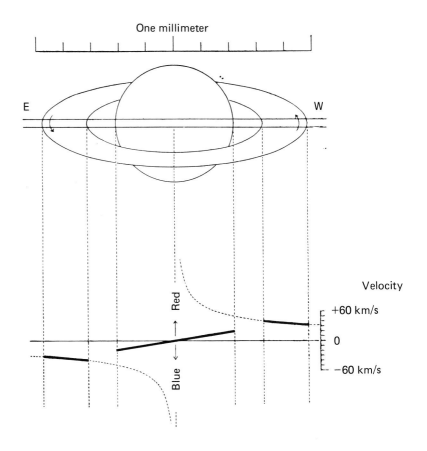

Figure 9.7. Ringlets. When viewed with high resolution, the B and C rings separate into countless rings within rings, while the A ring remains a broad diffuse band. This enhanced-color image shows the blue C ring, and the green and gold of the B ring, suggesting that the particles in these rings are different in composition and have not moved between rings. This suggests that the major structure of the rings may have remained the same since their formation. (Courtesy of JPL and NASA.)

Apparently, the resonating pull of the moons of Saturn creates spiral wave-like patterns. Such concentrations, called spiral density waves, wind about the planet in the form of a watch spring, and they have been found in parts of the main rings. Similar waves on a vastly greater scale are thought to create the arms of spiral galaxies.

Perhaps the most bizarre Voyager discovery was the long dark lanes that project radially across the B ring like the spokes of a wheel. These ephemeral spokes suddenly appear, move around the planet, and then dissipate within a few hours. (See Fig. 9.8 and Focus 9C, Enigmatic spokes.)

(c) Why do planets have rings?

One might expect the particles of a ring to have accumulated long ago into larger satellites. But the interesting feature of rings – and a clue to their origin – is that they do not coexist with large moons. The rings are always closer to the planets than their large satellites and are confined to an inner zone where tidal forces prevent them from coalescing to form a larger moon.

Focus 9C Enigmatic spokes

Dark radial markings across the rings move around Saturn like spokes about the hub of a wheel (Fig. 9.8). They cannot simply be clusters of dark particles, because such clusters would move with speeds that decrease with increasing distance from Saturn. The spokes would quickly stretch out and disappear.

One clue to these enigmatic spokes came from this photograph taken when the Voyager spacecraft passed beyond Saturn. The spokes appear dark when viewed in their reflected sunlight, but they are bright in this perspective when sunlight passes through them. This means that the spokes are composed of very small particles with sizes that are comparable to the wavelength of light (one ten-thousandth of a centimeter).

Another clue comes from the fact that the spokes lie near that distance from Saturn where the orbital velocities are synchronous with the planet's rotation. This means that the particles move with velocities equal to the velocity of Saturn's rotating magnetic field. At larger distances from Saturn the magnetic field sweeps ahead of the ring particles, whereas at smaller distances it lags behind.

Thus, the spokes contain small dust-like particles and they are somehow related to Saturn's rotating magnetic field. In fact, they have been identified with an active sector of the magnetic field that is associated with aurorae and radio bursts on Saturn. This suggests that the spokes are a manifestation of the planet's electromagnetic activity, but no one completely understands their cause.

According to one hypothesis, the spoke particles become charged, perhaps as the result of collisions with energetic electrons. Electromagnetic forces then levitate the particles above the rings, and the spokes are swept around Saturn by its rotating magnetic field. It sounds bizarre, but subtle forces are required to overcome gravity.

The outer limit of this zone is called the Roche limit after Eduard Roche, who described it in 1848. For a satellite with no internal strength and whose density is the same as the planet, the Roche limit is about two and one-half times the planetary radius.

To visualize the physical significance of the Roche limit, try to imagine what happens when two ring particles approach each other slowly while orbiting a planet. As they come closer together, their gravitational attraction for each other increases, and the maximum attraction occurs when the particles are touching each other. Larger particles will feel greater attraction. At the moment of contact, the planet pulls harder on the particle that is closer to it. This is the tidal force, and if it exceeds the mutual gravitational attraction of the particles toward each other, the particles will not stay together. The outcome of the tug-of-war between the tidal force and the mutual attraction is primarily decided by the particles' distance

Figure 9.8. Dark spokes. Dark spokes streak across the central third of Saturn's B ring. They sweep around Saturn with a uniform velocity in apparent defiance of Kepler's laws. (Courtesy of JPL and NASA.)

Figure 9.9. The Roche limit. A large satellite that moves well within a planet's Roche limit will be torn apart by the tidal force of the planet's gravity. The side of the satellite closer to the planet feels a stronger gravitational pull than the side farther away, and this difference works against the self-gravitation that holds the body together. A small solid satellite can resist tidal disruption because it has significant internal cohesion in addition to self-gravitation.

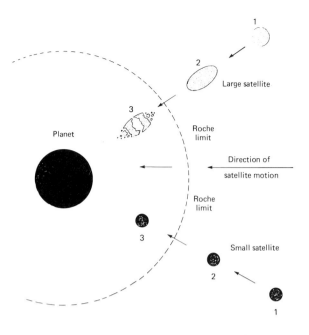

from the planet. At distances less than the Roche limit, particles are pulled apart and this prevents the accumulation of larger moons.

All of Saturn's major rings lie within the Roche limit, as do the rings of Jupiter, Uranus, and Neptune. Large objects coming within this limit will be torn apart by the planet's tidal force, but smaller rocks can remain intact. Because of their great internal cohesion, small solid satellites (less than 100 kilometers) can pass within the Roche limit without being tidally disrupted (Fig. 9.9).

Planetary rings may be the debris of a former satellite, or a group of satellites that passed within the Roche limit and were torn apart or suffered catastrophic collisions. In fact, the combined mass of Saturn's rings amounts to less than that of a medium-sized icy satellite that is 500 kilometers across. Alternatively, planetary rings may consist of primordial material that never accumulated to form a satellite.

9.3 The moons of Saturn

(a) Titan – moon of mystery

Saturn's largest satellite, Titan, is slightly bigger than the planet Mercury and slightly smaller than that of Jupiter's largest satellite, Ganymede. Titan has an average density of 1.89 g/cm^3, suggesting that it is approximately half rock (3 g/cm^3) and half ice (1 g/cm^3).

Titan has a substantial atmosphere! It consists mainly of nitrogen (82 to 99 percent) with small amounts of methane, the natural gas we often use for cooking and heating. At the start of the 20th century, the Spanish astronomer, Comas Sola, reported faint variable markings which he attributed to clouds on Titan. In 1944, Gerard Kuiper discovered signs of methane in the spectrum of Titan's atmosphere. The presence of nitrogen was firmly established in 1980, when the ultraviolet detectors aboard Voyager 1 showed that nitrogen molecules account for the bulk of Titan's atmosphere.

Visible light cannot penetrate Titan's atmosphere, for it is covered by an

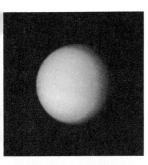

Figure 9.10. Titan's smog. The surface of Titan is hidden from view by a hazy layer of smog (top). When illuminated from behind, the smog forms a crescent several hundred kilometers above the satellite's surface (bottom). (Courtesy of JPL and NASA.)

obscuring veil of orange smog (Fig. 9.10) produced by photochemical reactions. Ultraviolet sunlight breaks methane and nitrogen molecules apart. Some of these fragments then recombine to create the smog, and in Titan's dry cold atmosphere, the smog builds to an impenetrable haze. (On Earth, smog also forms by the action of sunlight on hydrocarbon molecules in the air.)

But why does Titan have an atmosphere, when the planet Mercury and the larger satellite Ganymede do not? This paradox is discussed in Focus 9D, Why does Titan have an atmosphere?

At Titan's surface, the atmospheric pressure is about one and one-half times the surface pressure at sea level on Earth (Fig. 9.11). Titan's atmosphere is also dominated by nitrogen, like the Earth's. Nevertheless, there are many important – and fortunate, for us – differences. The Earth's air contains oxygen, and its oceans are made of liquid water; but there is no free oxygen in Titan's atmosphere and no liquid water covering its surface. Water would be frozen into Titan's cold surface as ice.

The surface temperature on Titan is only 93 degrees Kelvin. This temperature is interesting because it means that methane can exist in three forms – as a gas, liquid, or solid – depending on local conditions. Thus, methane on Titan may play the role of water on Earth. It can condense, forming lakes or oceans, and it can hang in clouds of methane that float above oceans. We do not know that there are oceans on Titan, for we cannot see through its smog, but chemists speculate that liquid hydrocarbons might be present. Calculations suggest that Titan's ocean – which may be a kilometer deep – might contain about 75 percent ethane and 25 percent methane.

Heavier organic substances, such as acetylene and smog particles, will fall through the atmosphere and sink into the ocean. Organic sludge may have accumulated at the bottom of the sea.

Figure 9.11. Titan's atmosphere. A study of the bending and fading of radio signals when the Voyager 1 spacecraft passed behind Titan's atmosphere led to this plot of the temperature and pressure in its atmosphere. Clouds of methane might float above an ocean of liquid ethane, while organic compounds and smog particles rain down to form a sludge on the solid surface.

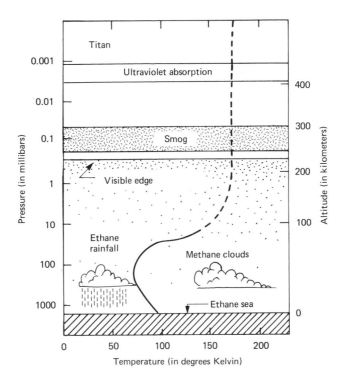

Focus 9D Why does Titan have an atmosphere?

Titan is the only moon in our solar system with a substantial atmosphere. Why does Titan have an atmosphere when Mercury and Ganymede, which are larger, do not? The ability of a planet or satellite to retain an atmosphere is determined primarily by the body's mass and temperature, and also by the composition of the atmosphere. Because Titan, Ganymede and Mercury have nearly the same mass, differences in temperature must account for their atmospheric differences. Mercury is so hot that even the heaviest molecules move fast enough to escape the planet's gravity. In contrast, the temperature on Titan is so low that only the lighter molecules, like hydrogen, can escape. Heavier molecules like methane and nitrogen are retained in Titan's atmosphere.

But Ganymede is now sufficiently cold and massive to retain a Titan-like atmosphere. The difference between Titan and Ganymede is most likely a consequence of the temperature at the time of their birth. Titan was born in the remote colder regions of the solar system, and nearby Saturn never became as hot as Jupiter did during its birth. The low temperatures permitted ammonia, methane and water ice to form on Titan's surface when it was born, and these ices probably sublimated to form a primeval atmosphere of ammonia and methane, while water remained locked into its surface as ice.

On Ganymede, water was probably the only ice that formed in the slightly warmer climate. If there was no ammonia or methane ice on Ganymede, and if the temperatures never became high enough for its water ice to sublimate, Ganymede would be left without any atmosphere.

(b) Saturn's large moons

In order of decreasing orbital distance, Saturn's largest moons are: Phoebe, Iapeteus, Hyperion, Titan, Rhea, Dione, Tethys, Enceladus and Mimas (see Fig. 9.12 and Table 9.2). Most of them are named after the Titans, who were the children and grandchildren of Gaea, the Earth goddess, who was impregnated with drops of Uranus' blood.

Those with known densities, excepting Titan, have densities between 1.1 and 1.4 g/cm^3, which suggests that they are composed mainly of ice. The surfaces of Saturn's icy moons are so cold that the ice is as rigid as steel, allowing their surfaces to preserve ancient impact craters.

Although most of Saturn's satellites revolve around the planet in pro-grade orbits, its outermost satellite Phoebe moves around Saturn in the opposite, retrograde direction. Phoebe is also very dark, in sharp contrast to Saturn's other bright, icy moons. These peculiarities suggest that Phoebe is a former asteroid that fell into the planet's gravitational embrace.

Curiously, early astronomers could observe Iapetus on only one side of Saturn. The tiny moon seemed to disappear when its orbit carried it to the other side of the planet. The reason for this strange behavior is that Iapetus is a divided world; half its surface is as bright as ice, and the other is as dark as coal. Iapetus keeps one side toward the planet, and as it revolves around Saturn the bright and dark parts are successively turned toward the Earth. When the dark half is pointed toward the Earth, the moon becomes very difficult to observe. The black substance appears to be organic tar,

Table 9.2. *Properties of Saturn's large icy moons*

Name	Mean distance from planet (radii[a])	Orbital period (days)	Radius (kilometers)	Mass[b] (10^{24} g)	Density (g/cm^3)
Mimas	3.08	0.942	196	0.04	1.4
Enceladus	3.95	1.37	255	0.08	1.2
Tethys	4.88	1.888	530	0.76	1.21
Dione	6.26	2.737	560	1.05	1.43
Rhea	8.73	4.518	765	2.5	1.34
Iapetus	59.0	79.33	730	1.9	1.16

[a]Saturn's radius is 60 330 kilometers, nearly ten Earth radii.
[b]Our Moon's mass is 73.5 in these units.

Figure 9.12. Saturnian satellites. Six of Saturn's large icy moons are shown in this one-minute exposure made with the US Naval Observatory's 26-inch (66-centimeter) refractor. From left to right, the satellites are Titan, Dione, Enceladus, Tethys, Mimas and Rhea (on the other side). The faint image below the planet is that of a star. A partially transparent metallic film was used to weaken the light from Saturn and its rings. (Courtesy of Dan Pascu.)

E

N

although there is no convincing explanation for its uneven distribution on the satellite.

Hyperion has an irregular, flattened shape, and it moves in an eccentric orbit. This combination produces a chaotic rotation – subject to fits and starts – about its own center of mass. This random speeding up and slowing down is explained by the same theory that accounts for the chaotic orbits of some asteroids.

Rhea, like Iapetus, is a little more than half the size of our Moon. Rhea's surface is battered with impact craters, as are the surfaces of Tethys and Dione (Fig. 9.13). A huge crack stretches three-quarters of the way around Tethys, and liquid seems to have been released from Dione's interior, flowing out to produce frozen, wispy marks.

Figure 9.13. Two icy worlds. Cracked Tethys (above) and wispy Dione (below) are two satellites with comparable sizes. They both have radii of slightly more than 500 kilometers. (Courtesy of JPL and NASA.)

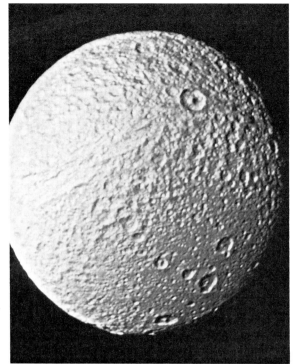

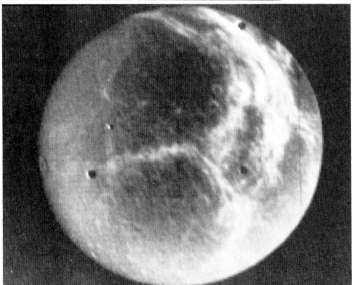

Figure 9.14. Mimas. The innermost of the large, icy satellites of Saturn has a radius of 196 kilometers. Its face is dominated by the giant crater, Herschel. The impact that produced this crater almost broke the satellite into pieces. (Courtesy of JPL and NASA.)

The surface of Mimas is saturated with impact craters, and is dominated by a gigantic crater, named Herschel (Fig. 9.14). The enormous impact that created Herschel must have nearly broken Mimas into pieces.

Enceladus has a bright, smooth icy surface that contains cracks, grooves and craters (Fig. 9.15). These diverse features suggest recurrent volcanic activity, although the erupting liquid would be primarily water, not rocky lava. Enceladus is caught in a gravitational tug-of-war between Saturn and the satellite, Dione, whose orbital period is twice that of Enceladus. Dione's repeated gravitational tug produces Enceladus' eccentric orbit, and causes recurrent tidal flexing that may heat the moon's interior. However, present calculations suggest that the heating is not sufficient to melt the interior, so the source of Enceladus' signs of activity remains a mystery.

It is also possible that some of the craters were produced by meteoric impacts. If such impacts released liquid from the interior, the violent escape and freezing of this liquid might have produced fine particles to feed Saturn's E-ring.

(c) The smaller satellites of Saturn

Spacecraft discoveries have extended the list of known Saturnian satellites to 17. All of the newly discovered satellites reside within the inner parts of its satellite system. They are all bright objects, probably composed of ice, and they all have orbits that are remarkable in one way or another.

The innermost known satellite of Saturn is Atlas. It skirts the outer edge of the bright A ring. The shepherd satellites, called the inner and outer F-ring shepherds, chase each other around the narrow F-ring and confine it as though they were two gravitational sheepdogs herding sheep into a narrow path (Fig. 9.16). Each shepherd tends one edge of the ring. The faster-moving inside satellite gravitationally pulls the inner F ring particles forward as it passes, causing them to accelerate and spiral outward. The slower-moving outer shepherd exerts a net backwards force on the outer ring particles, causing them to move inwards. The result is a very narrow ring.

Figure 9.15. Enceladus. Enceladus has an icy surface that reflects almost 100 percent of the incident sunlight, making it the most reflective object in the solar system. When viewed close up, its remarkable surface appears cracked and grooved, with smooth plains and deformed craters. (Courtesy of NASA headquarters.)

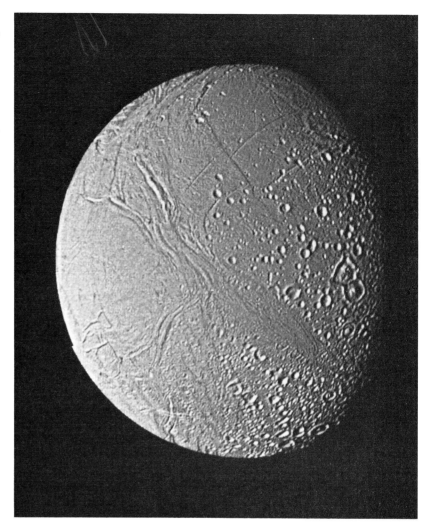

Figure 9.16. Saturn's shepherd satellites. Saturn's narrow F ring is confined by two shepherd satellites. The outer shepherd gravitationally deflects ring particles inwards, and the inner shepherd deflects particles outwards. (Courtesy of JPL and NASA.)

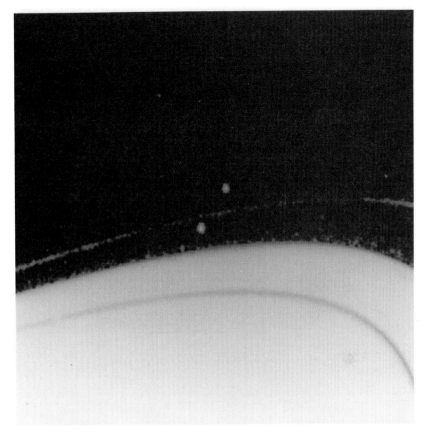

Saturn's two co-orbital satellites are even more bizarre. Janus and Epimetheus move in two nearly identical orbits. The satellite on the inner orbit moves slightly faster, overtaking the outer satellite every four years. But the bodies' diameters are greater than the distance between their orbital paths, so they cannot pass without some fancy pirouetting. They avoid a collision at the last moment by gravitationally exchanging energy and switching orbits. The inner one is pulled by the outer one and raised into the outer orbit, and *vice versa*. They then move apart, only to repeat this *pas-de-deux* four years later, and exchange again.

Three so-called Lagrangian satellites share the orbits of Saturn's larger satellites Tethys and Dione. The satellite Tethys shares its orbit with two small companions, one about 60 degrees ahead and the other about 60 degrees behind. These two positions are regions of gravitational stability specified by Jean Louis Lagrange in the nineteenth century. (See Focus 7A, Trojan satellites and Lagrangian points, in Chapter 7.) One additional miniature moon shares Dione's orbit, leading it by 60 degrees.

This concludes our survey of Saturn, the most distant planet known to the ancients (Fig. 9.17). We will now travel out beyond this enchanting world to the next wanderer, Uranus.

Figure 9.17. Leaving Saturn. When the Voyager 1 spacecraft sped past Saturn, it looked back to take this picture of the ringed planet from a perspective that cannot be enjoyed from Earth. (Courtesy of JPL and NASA.)

Focus 9E Saturn – summary

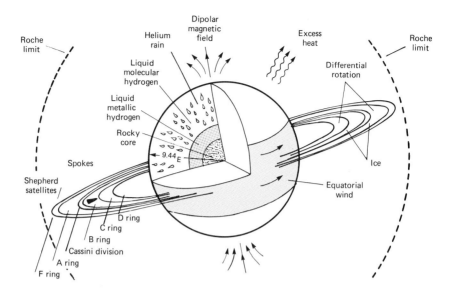

Mass: 5.68×10^{29} grams $= 95.18\ M_{\text{E}}$ (Earth $= 1$)
Radius: $60\,330$ kilometers $= 9.46\ R_{\text{E}}$ (Earth $= 1$)
Mean density: 0.71 g/cm^3
Rotational period: 10 hours 39 minutes 22 seconds
Orbital period: 29.46 years
Mean distance from Sun: 9.54 A.U.
Number of known satellites $= 17$
Magnetic field strength at cloud tops $= 0.2$ gauss

Uranus and its rings. Thin, spidery rings encircle the methane-rich atmosphere of Uranus in this artist's rendition. (Courtesy of NASA.)

10 Frozen worlds: Uranus, Neptune and Pluto

The known size of the solar system doubled when Uranus was discovered in 1781.

Uranus is tipped on its side and rotates in the retrograde direction.

Why is the Uranian ring system mostly empty space?

Neptune's stormy atmosphere may be driven by internal heat, and there may be icy volcanoes on its satellite Triton.

The planet Pluto could be an escaped satellite of Neptune.

Pluto is an icy planet with an oversized moon.

There is a growing suspicion that other unknown worlds lurk in the outer darkness of the solar system.

10.1 The first new world, Uranus

(a) An unusual comet

During the night of March 13, 1781, a professional musician and obscure amateur astronomer, William Herschel, discovered the planet Uranus. At first, he did not realize what he had found – he mistook it for a comet.

Herschel had come across an unusual faint "star" while surveying the heavens with his home-made, 15-centimeter (6-inch) reflector. When he increased the magnification of his telescope, the unusual object increased in size, and Herschel had the impression of seeing its body, while the neighboring stars still appeared as undefined points of light.

Another clue to the nature of the object was provided during the next few nights when he found that it was slowly moving across the background of distant stars. The object was clearly not a star; it belonged to our solar system, and the most natural explanation was a new comet. Herschel presented his discovery to England's Royal Society in a paper with the unexciting title "Account of a Comet."

(b) The new planet, Uranus

The object was soon lost in sunlight, but it was picked up a few months later. Its orbit was nearly circular, at twice the distance of Saturn, and this proved the object was no ordinary comet. It was quickly recognized as a new planet, and Herschel became world famous – almost overnight – as the first man in recorded history to have discovered a new planet. After some controversy, the new planet was named Uranus, after the Greek god of the heavens. (See Focus 10A, Naming a new world.) As we move outward from planet to planet, we also move upward among the generations of the gods: Mars' father is Jupiter, and Jupiter's is Saturn, while Saturn's is Uranus. Saturn's mother and subsequent mate was Gaea (Earth) who was born of Chaos.

In retrospect, orbit calculations proved that Uranus had been detected,

Focus 10A Naming a new world

This portrait shows Sir William Herschel thirteen years after his discovery of the first new planet in recorded history. He is holding a drawing with a Latin designation of the new planet as the "Georgian Planet," a name proposed by Herschel in honor of King George III, England's reigning monarch and a patron of the sciences. Eventually, the classicists prevailed, and the planet was named for Uranus, first ruler of Olympus, father of Titans and the grandfather of Jupiter.

mistaken for a star, on no fewer than 22 occasions during the century that preceded the realization that it was a planet. It is actually recorded as a star on several charts.

Uranus orbits the Sun at about 19 times the Earth's distance, or twice Saturn's distance from the Sun. As predicted by Kepler's law, Uranus requires 84 years to complete a round trip. Thus, Uranus has only completed two circuits around the Sun since it was discovered. The day on Uranus is only 17.24 Earth-hours, so the Uranian year is more than 45 000 Uranian days long.

(c) Uranus' atmosphere and clouds

Uranus' atmosphere contains about 63 percent hydrogen and 23 percent helium, and heavier methane gas may account for as much as 14 percent of the atmosphere (Fig. 10.1). The methane gives Uranus a bluish-green color.

When the Voyager 2 spacecraft sped past Uranus on January 24, 1986, several streaks were detected in its otherwise featureless cloud layer. Careful charting of the motion of these streaks showed that the clouds move around the planet with periods ranging from 14.3 to 16.8 hours. On the other hand, charting electrical storms on Uranus indicates that the magnetosphere rotates with the longer period of 17.24 hours. The magnetosphere is probably connected to the body of the planet, revealing the motion of the deep interior, so the implication of the shorter rotation periods for the clouds is that the clouds move ahead, in the direction of rotation. This disparity is a cause of dismay to most meteorologists.

(d) Tipped planet and tilted magnetic field

Uranus is a sideways world with a spin axis that lies almost in the plane of its orbit. In fact, the planet's equatorial plane is inclined some 98 degrees from its orbital plane so it spins backwards compared to the other planets (Fig. 10.2). Because the planet rotates on its side, with one pole now facing

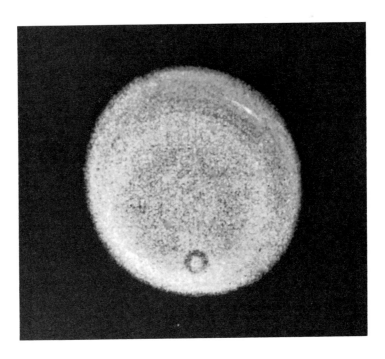

Figure 10.1. Uranus' methane atmosphere. A high-altitude, reddish brown haze obscures the south pole of Uranus. Because the methane-rich atmosphere absorbs red light, the planet appears blue-green. (Courtesy of NASA.)

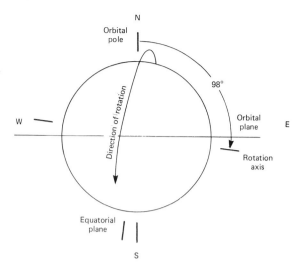

Figure 10.2. Tipped Uranus. Uranus is tipped on its side. The angle between the pole of its orbit and its axis of rotation is 98 degrees. As a result, the planet rotates in the retrograde direction. That is, it rotates from east to west, or in the clockwise direction as seen from the north.

Figure 10.3. Uranian magnetosphere. Electrically charged particles are trapped within Uranus' magnetic field. The planet's rotation axis, white arrow, lies roughly in the ecliptic, dashed line, the plane in which all the planets in the solar system orbit the Sun. But the axis of the magnetic field is skewed 55 degrees from the rotation axis. So as the planet rotates, the magnetosphere wobbles in space. The solar wind creates a bow shock on the sunlit side and draws the magnetosphere out into a long tail on the opposite side. (Courtesy of Rob Wood.)

the Sun, its moons form a bull's eye pattern revolving about the planet like Ferris wheels.

And the magnetic field is also tilted at an unusual angle. The north magnetic pole of Uranus is tipped 55 degrees from the rotational pole (Fig. 10.3). This would make compass navigation a rather awkward process on Uranus, as well as providing a mystery concerning the interior of the planet. Theoreticians expect a closer alignment between rotation and magnetism. (On the Earth the disparity is only 12 degrees.) Voyager 2 measurements also showed the magnetic field of Uranus to be about 50 times stronger than the Earth's. It is off-center by 8000 kilometers, as well as being tilted.

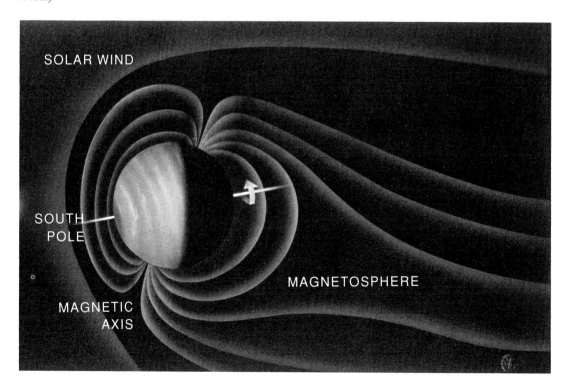

Because the rotational poles of Uranus lie near its orbital plane, the planet also has very strange seasons. The north pole is sunlit for intervals of 42 years. During this northern summer, the polar region receives more heat than the equatorial regions, while the south pole remains in darkness. Then, the north pole enters winter, and the south pole is bathed in sunlight for the next 42 years. Astronomers had expected this unequal heating to lead to polar warming, but the planet's outer atmosphere is remarkably uniform in temperature. Atmospheric circulation evidently distributes the heat uniformly.

(e) Rings around Uranus

Astronomers have had a history of happy accidents concerning Uranus, starting with Herschel's initial discovery. Another lucky incident occurred on March 10, 1977, when the planet was scheduled to pass in front of a faint star. Because of uncertainties in the predicted time of the star's disappearance, one telescope was set into action about 45 minutes early. Soon after the recording began, the starlight abruptly dimmed but then it almost immediately returned to normal, producing a quick, brief dip in the recorded signal. At first the dip was attributed to a cloud passing in front of the telescope or to an unexpected change in the telescope's orientation. But more dips occurred before the star disappeared behind the planet, and the pattern was repeated after the star had re-emerged. This symmetry indicated that Uranus was surrounded by narrow rings, and eventually the count rose to nine rings! Some were subsequently confirmed in telescopic observations from Earth, and the Voyager 2 spacecraft added two narrow rings and a broad one that was first seen in back-scattered light. (See Fig. 10.4 and Fig. 10.5.)

The rings are extremely difficult to see from Earth, because their

Figure 10.4. Beneath the rings of Uranus. This backlit image shows a multitude of Uranian rings and resembles Voyager views of Saturn's system (see Fig. 9.5). Here the spacecraft was looking back toward the Sun, which made the dust between the rings appear very bright. The bright dust is largely absent in the rings themselves. (Courtesy of NASA and JPL.)

particles are very dark – quite unlike the particles found in Saturn's rings. In fact, the particles of Uranus' rings reflect only about 2 percent of the sunlight falling on them, making them as dark as charcoal, or the blackest carbon-enriched meteorites and asteroids known. Uranus' rings may be fragments of a dark asteroid that once wandered too close to the planet and was torn into pieces. Or they may have been darkened by the action of cosmic rays on their chemical constituents.

The rings are also remarkable for their slenderness, judged from the brief interval that they obstruct starlight. Most of them are less than 10 kilometers in width.

Before the flyby of Voyager 2, it had been speculated that the slenderness of Uranus' rings might be caused by small shepherding moonlets, like those near Saturn. A shepherd satellite on each side of a ring could constrain its edges and prevent the ring from spreading. Relatively small, kilometer-sized satellites are large and massive enough to confine the observed rings, and these satellites could not be seen from the Earth. Alternatively, each ring might be confined and fed by a small moon hidden within it.

As expected, small satellites were found in abundance by Voyager 2, bringing the grand total of moons and moonlets to 15 (Fig. 10.6). Two of them confine the particles of one ring to a narrow track, just as two shepherd satellites control the F-ring of Saturn. But it was the closeup pictures of the four largest moons that really captivated astronomers.

Figure 10.5. Uranus' epsilon ring. These images of the epsilon ring were created from Voyager 2 observations of the light coming from the star Sigma Sagittarii as it passed behind the ring. Areas containing more material are lighter and areas containing less are darker. The widths of these sections are 31 kilometers and 22 kilometers. (Courtesy of JPL and NASA.)

Table 10.1. *Uranus' large moons*

Name	Mean distance from planet (radii[a])	Orbital period (hours)	Radius (kilometers)	Mass (10^{23} grams)	Density (g/cm^3)
Miranda	3.3	34	242	1	1.3
Ariel	5.0	60	580	14	1.6
Umbriel	7.4	100	595	13	1.4
Titania	10.3	209	805	35	1.6
Oberon	22.5	323	775	29	1.5

[a]The radius of Uranus is 25 900 kilometers.

Figure 10.6. Satellites and rings of Uranus. This diagram shows the orbits of Uranus' five largest moons, as well as those of its rings and the ten small satellites discovered by the Voyager 2 spacecraft.

(f) Satellite sculpture

The five largest satellites (see Table 10.1) are named for characters in literature. Oberon and Titania are named for the king and queen of the fairies in Shakespeare's *A Midsummer Night's Dream*. Inside their orbits is Umbriel, named after a "dusky, melancholy sprite" in Alexander Pope's *Rape of a Lock*. Closer to the planet is Ariel, described by Shakespeare as "an airy spirit" in *The Tempest*. Closer yet is a moon called Miranda, for Prospero's daughter in *The Tempest*. It was she who proclaimed,

> O, Wonder!
> How many goodly creatures are there here!
> How beautious mankind is!
> O, brave new world, That has such people in't.
>
> *The Tempest*, V, i, line 183

Because they are small, the moons of Uranus were expected to be heavily cratered iceballs, devoid of any signs of internal activity. But Voyager 2 surprised nearly everyone. The five previously known moons have, with the exception of Umbriel, surfaces that have been warped and twisted by geological upheavals.

Oberon, for example, has a mountain at least 4 kilometers high. The moon is pockmarked with craters whose dark, patchy floors suggest that dirty water has seeped through cracks in the crust. Titania's surface is cut by rift valleys, up to 1600 kilometers long, that may have been formed when internal water froze and expanded, shattering the overlying crust (Fig. 10.7).

Ariel is criss-crossed with huge canyon systems, perhaps as a result of repeated expansion and contraction (Fig. 10.8). But its bright surface reveals few craters, and this suggests that it has been resurfaced by icy extrusions. The ice may have welled up through long cracks in the surface resembling the mid-ocean rifts on the Earth. But, unlike terrestrial volcanoes that explosively emit liquid lava, Ariel's volcanoes oozed a solid icy mixture that crept outward like a glacier.

Miranda has the most amazing surface of all (Fig. 10.9). Icy volcanoes also seem to have been active on its surface, which is marked with a bizarre variety of grooves, valleys, mountains and cliffs, as though it were the work of a combination of faultings and eruptions. Some astronomers have argued that Miranda was once shattered by collision with an asteroid but that it managed to pull itself together again into a single body. Another possibility is that Miranda froze while it was still in an embryonic stage of differentiation, when rock masses were sinking to the interior and lighter chunks of ice were rising to the surface. In any case, Miranda has all the earmarks of a "brave new world."

After transmitting a fascinating sequence of images and radio signals, Voyager 2 left Uranus and headed for its rendezvous with Neptune.

10.2 Neptune – the twin of Uranus

(a) The discovery of Neptune

Neptune's discovery was no accident, in contrast to that of Uranus. It was a direct consequence of precise mathematical calculations of Uranus' motion.

Astronomers calculated predicted positions of Uranus using Newton's law of gravitation. Then they compared their predictions with recorded observations and found the two sets of positions disagreed. A large, unknown planet, located far beyond Uranus, was evidently producing a gravitational tug on Uranus. Two astronomers took on the challenge of locating that planet by a mathematical analysis of the wanderings of Uranus.

Figure 10.7. Oberon and Titania. The outermost moon of Uranus, Oberon (left), displays impact craters surrounded by bright rays of ejected material. Huge rifts of about 800 kilometers in length cross Titania (right). (Courtesy of NASA.)

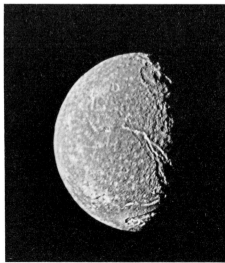

John Couch Adams' speculations about a remote planet began in 1841 when he was 22 years old. In 1845, Adams presented his results to two prominent astronomers, James Challis and George Biddell Airy, England's Astronomer Royal. Although he reported the location of the unknown world to both Challis and Airy, neither of these astronomers felt compelled to look for it.

In November 1845, the young French astronomer Urbain Jean Joseph Leverrier, published his own investigation of Uranus' motion, showing that an unknown planet was influencing Uranus (Fig. 10.10). When his solution for the location of the planet reached England, Airy compared Leverrier's work with Adams' solution, which Airy had set aside eight months earlier. The two solutions were almost identical!

On August 31, 1846, Leverrier presented a revised location to the French Academy of Sciences. Ironically, Leverrier was unable to convince his French colleagues to search for the suspected planet. He then wrote to Johann Galle at the Berlin Observatory, urging him to look for it. Within a night of receiving Leverrier's letter on September 23, 1846, Galle and his student Heinrich D'Arrest located the planet about one degree from Adams' and Leverrier's predicted locations. It was subsequently named Neptune after the Roman god of the sea. In hindsight, the name is

Figure 10.8. Ariel. This Uranian moon is cut by a network of branching, smooth-floored valleys, visible along the terminator at the right. The valley floors exhibit winding grooves, similar in appearance to terrestrial glaciers. (Courtesy of NASA and JPL.)

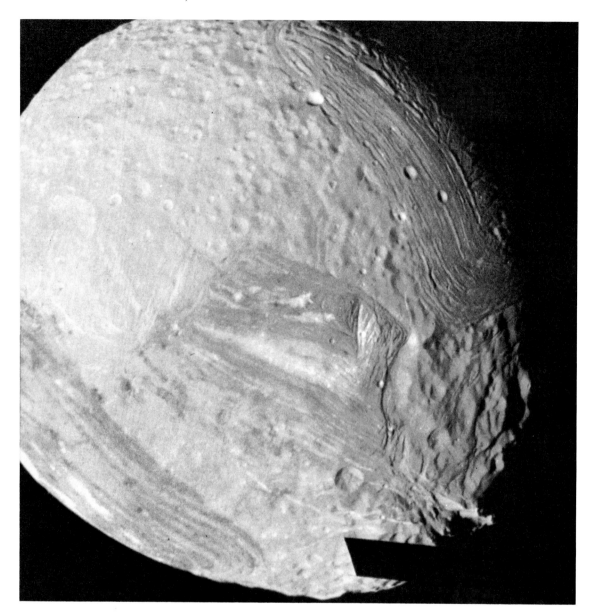

Figure 10.9. Miranda. Miranda is a bizarre collection of terrains. At the center is a pattern of light and dark ground nicknamed "the chevron." The light and dark lines at the lower left are jagged cliffs. Along the terminator at the upper right is a pattern of ridges and grooves. These three major regions are thought to have formed by the same process in which the chevron was created first. The ridges and grooves were then formed, followed by the jagged cliffs. At the lower right, the crust is broken by a cliff almost 8 kilometers high. (Courtesy of NASA and JPL.)

Table 10.2. *Some comparisons of Uranus and Neptune*

	Uranus	Neptune
Mass (10^{29} g)	0.868	1.024
Radius (kilometers)	25 359	24 760
Density (g/cm^3)	1.2	1.7
Distance from Sun (A.U.)	19.2	30.1
Orbital period (years)	84	165
Temperature at cloud tops	58 K	56 K
Length of day	17 h 14 min	16 h 03 min

appropriate, for current models suggest that a deep sea of liquid water makes up the bulk of the planet's interior.

Not only had the discovery of Neptune increased the radius of the known solar system by a factor of 1.6, it was also acclaimed as the ultimate triumph of Newtonian science, having resulted from mathematical calculations based on Newton's theory of gravitation. If proof were needed, this achievement certified the validity of Newton's theory.

(b) The planet Neptune

Neptune requires 165 years to revolve once around the Sun. It has not made a full orbit since it was discovered in 1846. Neptune's mass is 17 times the mass of the Earth; its mean density is about 1.7 times that of water. Both Neptune and Uranus have radii that are about four times that of the Earth. In fact, the mass, radius and mean density of Uranus and Neptune are so similar that they are often called planetary twins (Table 10.2).

The temperature of Neptune's outer atmosphere is 60 degrees Kelvin, but the temperature that would result from sunlight alone is 46 degrees Kelvin. This difference implies that heat is flowing from the interior of Neptune, as it does from Jupiter and Saturn. Uranus, on the other hand, shows no signs of such interior heating, and this may explain why its atmosphere is relatively bland and inactive, in contrast with the dynamic, stormy atmospheres found on the other giant planets.

The first close-up look at the stormy patterns of Neptune's dimly lit atmosphere were provided in 1989 by the Voyager 2 spacecraft. (At the time, Neptune was at the edge of the known solar system, outdistancing Pluto, which periodically swings inward and outward in an unusually elliptical orbit.) The largest storm system seen on Neptune is as broad as the Earth (Fig. 10.11). It is called the Great Dark Spot because it resembles the Great Red Spot of Jupiter. Both storms are found in the planetary tropics – at about one-quarter of the way from the equator to the pole – and both rotate counterclockwise. The main difference is that Jupiter's spot lies above the clouds while Neptune's seems to form a deep well in the atmosphere.

Like many other planets, Neptune acts as a radio transmitter, and variations in the intensity of the signals indicate that the day is 16 hours 3 minutes of earthly time. These variations are produced by rotation of the planet's magnetic field, which guides great streams of high speed electrons back and forth above the planet's clouds. The field is tilted at an angle of 50 degrees from the planet's axis, and in this respect the planet is nearly a twin of Uranus, whose field is tilted 59 degrees. (On Earth, the tilt is only 12 degrees.) Neptune's field is, however, much weaker than that of the other giant planets. In fact, it is only about as strong as the Earth's magnetic field.

Figure 10.10. Leverrier. Urbain Jean Joseph Leverrier, the young French astronomer who predicted the position of the then unknown planet Neptune using irregularities in the motion of Uranus. Leverrier and the English astronomer, John Couch Adams, share the credit for the discovery of Neptune. They made nearly identical predictions at about the same time. (Adapted from Camille Flammarion's Astronomie Populaire *(Flammarion et Cie, Paris, 1880).)*

Figure 10.11. Neptune's dynamic atmosphere. After a journey of 12 years and 4.4 billion miles, the Voyager 2 spacecraft was so far away that its radio signals, travelling at the speed of light, took more than 4 hours to reach Earth. Its narrow-angle camera captured this view of Neptune's Great Dark Spot (center) with its bright satellite cloud. The spot is as large as the Earth, and about half as large as the Great Red Spot of Jupiter. Also seen for the first time were the smaller bright cloud in the lower left, dubbed "Scooter" because it moves so quickly, and a lesser dark spot with a small bright cloud in its center. The atmosphere is swirled by winds up to 400 miles per hour. But the faint sunlight at Neptune's great distance cannot provide the energy of such winds; they are probably energized by heat from the interior of the planet. Voyager 2's images were sent by a 20-watt transmitter, one-tenth as powerful as radios used by amateurs to communicate thousands, rather than billions, of miles. (Courtesy of JPL and NASA.)

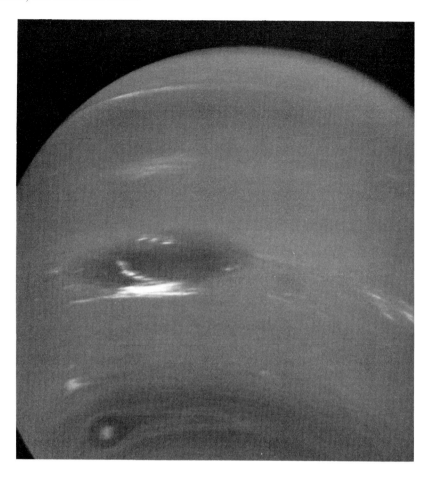

Its source is traced to currents inside the planet, perhaps driven by internal heat sources.

(c) Neptune's strange satellites

Before the arrival of Voyager 2, Neptune had two known satellites, Triton (a sea god, son of Poseidon) and Nereid (a sea nymph). Voyager's cameras found six more that had been hidden from our view by the light of the planet. All the new moons are small and irregular (a few hundred kilometers across) and have surfaces that are nearly as black as soot.

Triton and Nereid occupy unusual orbits that are highly inclined to Neptune's equatorial plane – Triton actually moves in the backward direction – and Nereid's orbit is more elongated, or eccentric, than that of any other natural satellite in the solar system. Tidal interaction between Neptune and Triton is gradually drawing the satellite inward. Some time in the distant future (100 million to 10 billion years) Triton will pass inside the Roche limit and will be gravitationally shattered into rubble, which will spread around the planet and add a new ring to the solar system.

The exceptional orbits of Neptune's satellites might be the result of a gravitational tug by a passing object, and such an event has been suggested to account for Pluto's origin. Another possibility is that Triton was once in an independent orbit about the Sun but collided with a smaller Neptunian moon and was captured by Neptune's gravity.

Triton, with a radius of 1400 kilometers, is slightly smaller than our Moon. Unlike our Moon, it has an atmosphere – a thin one of nitrogen and

Figure 10.12. Triton. In this composite photograph of Neptune's largest moon, the south polar cap is at the bottom of the picture. It consists of bright, highly-reflective methane and nitrogen ices that have been colored pink by the action of energetic radiation. Surrounding the ragged edge of the polar cap is a region whose surface features resemble those in the darker region to the north. At the time of this flyby, Triton was the coldest measured object in the solar system, and it appears to be the sight of icy volcanism. (Courtesy of JPL and NASA.)

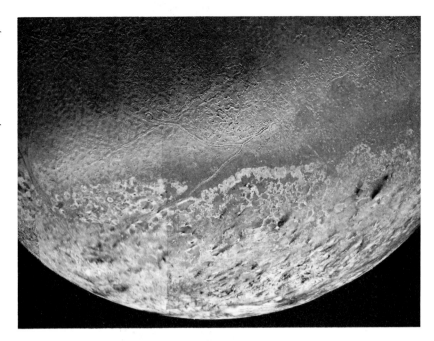

a bit of methane – that has a pressure only 1/100 000 that of the Earth's at sea level.

Astronomers once speculated that Triton's surface might be warm enough to sustain liquid nitrogen. (On Earth, of course, this would be considered extremely cold.) But Voyager 2 showed that Triton is too cold even for that. With a temperature of 50 degrees Kelvin (minus 369 degrees Fahrenheit) it is one of the coldest measured objects in the solar system. (Only Pluto and comets and asteroids at greater distances from the Sun are expected to be colder than Triton.) Its glazed surface of water, methane, and nitrogen ices is one of the shiniest in the solar system, but there is not much sunlight at that distance.

Voyager 2 had no difficulty seeing through Triton's thin atmosphere, and it revealed a jumbled terrain of pinkish gray, with smooth icy expanses occasionally marred by what appear to be meteor craters (Fig. 10.12). The color may be due to chemical reactions stimulated by high energy atmospheric particles and radiation.

Vast frozen basins on Triton appear to have been produced by the oozing of icy extrusions flowing out from the warmer interior through cracks in a layer of ice, perhaps resembling those of Ariel, moon of Uranus. This resurfacing would have erased ancient scars and it might provide a temporary seal over the interior. Streaks of dark material in the midst of bright icy features suggest volcanic activity, propelled by the pressure of liquid and vapor in the somewhat warmer interior. Where the seal is broken, an eruption would vent through the surface and leave a dark feature. (Halley's comet showed jets that had vented through portions of the surface, and this may be a common feature among active comets.)

(d) Rings around Neptune

On rare occasions, astronomers have been able to watch Neptune pass nearly in front of a star. With sensitive light-gathering equipment mounted at different parts of the Earth, they find the star blinks off and on, even when Neptune does not directly occult the star. They concluded that Neptune was surrounded by rings of dark matter orbiting far above its atmosphere.

But, mysteriously, these brief winks were not always seen, and they often appeared on only one side of the planet, so the hypothetical rings became shortened, in the minds of the astronomers, to ring-arcs that only reached part way around the planet. Chance would dictate which astronomers would detect the obscuration, but the idea of incomplete rings was difficult to understand.

Voyager 2 cleared up the mystery – or at least clarified the problem. Its cameras found 3 rings: an inner ring with a radius of 42 000 kilometers, a middle ring with a radius of 53 000 kilometers, and an outer ring with a radius of 63 000 kilometers (Fig. 10.13). The inner ring is faintest and it may actually extend all the way down to the top of Neptune's atmosphere. The inner and middle rings completely surround the planet, so they do not accord with the ring-arcs detected from Earth.

The outer ring, although extending all the way around, has three bright features, each about 6 to 8 degrees long, where ring material has accumulated. These features are responsible for the elusive occultations, and these clumpy but complete rings – which are easier to understand than incomplete rings – appear capable of explaining all the verified winks observed from Earth. The clumps are probably produced by shepherd moons (acting somewhat like those of Saturn) that accumulate the ring fragments here and there.

All in all, Voyager 2's journey past Neptune was a brilliant success for the surrogate eyes of a spacecraft in this first golden age of space exploration (Fig. 10.14).

(e) Inside Uranus and Neptune

Uranus and Neptune have substantial quantities of methane in their atmospheres. They are also more dense than their giant neighbors, Jupiter and Saturn, indicating that they cannot be made solely of the lightest elements, hydrogen and helium. On the other hand, Uranus and Neptune are less dense than the Earth, so they cannot be entirely composed of rocks. The bulk of their interior is probably composed of a deep ocean of water. According to one model, this liquid mantle surrounds a rocky core, while a thick gaseous atmosphere hides the water from view (Fig. 10.15).

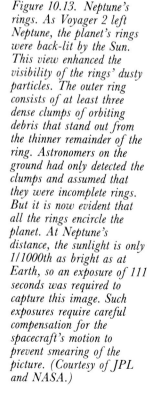

Figure 10.13. Neptune's rings. As Voyager 2 left Neptune, the planet's rings were back-lit by the Sun. This view enhanced the visibility of the rings' dusty particles. The outer ring consists of at least three dense clumps of orbiting debris that stand out from the thinner remainder of the ring. Astronomers on the ground had only detected the clumps and assumed that they were incomplete rings. But it is now evident that all the rings encircle the planet. At Neptune's distance, the sunlight is only 1/1000th as bright as at Earth, so an exposure of 111 seconds was required to capture this image. Such exposures require careful compensation for the spacecraft's motion to prevent smearing of the picture. (Courtesy of JPL and NASA.)

*Figure 10.14. Farewell to
the planetary system. In
parting, Voyager 2 provided
this last picture show, with
Neptune's rim in the
foreground and Triton
appearing as a thin crescent
in the distance. The hardy
spacecraft, launched in 1977,
visited the outer giant
planets, using the gravity of
each to sling-shot the
spacecraft on its journey to
the next. Its camera recorded
the swirling face of Jupiter
in July 1979, the myriad
rings of Saturn in August
1981, pale Uranus in
January 1986, and stormy
Neptune in August 1989
before heading out of the
solar system. All the planets
except Pluto have now been
visited by spacecraft and
recorded in detail far beyond
what had been seen from
Earth – a glorious
achievement of space
engineering. (Courtesy of
JPL and NASA.)*

*Figure 10.15. Inside
Uranus and Neptune. The
interiors of Uranus and
Neptune are described by a
three-layer model consisting
of a central rocky core, a
mantle, and an outer gaseous
atmosphere of hydrogen,
helium and methane. At the
high temperatures and
pressures within these
planets, the water, methane
and ammonia are in a liquid
state.*

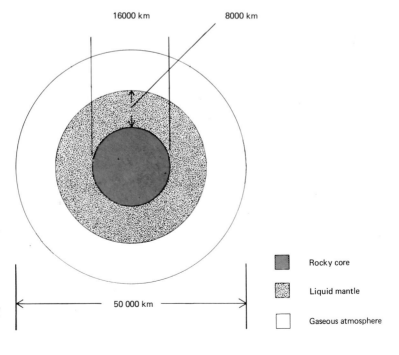

10.3 Pluto – an icy planet with an oversized moon

(a) The discovery of Pluto

Astronomers expected that irregularities in Neptune's motion would lead to the discovery of another remote, unknown planet; but because of Neptune's long 165-year orbit there were insufficient observations. Detection of an unknown planet therefore had to be based upon perturbations in Uranus' motion, after accounting for the gravitational effects of Neptune.

The first such prediction was made in 1909 when William Pickering argued that both Neptune and a remote Planet O were producing gravitational tugs on Uranus. The next attempt was made by Percival Lowell in 1915. He called his unknown object Planet X.

Astronomers searched unsuccessfully for Planets O and X in their predicted locations for more than two decades. Then, on February 18, 1930, the faint, star-like images of a previously unknown planet were discovered by young Clyde Tombaugh at the Lowell Observatory (Fig. 10.16). The new planet was named Pluto, for the god of the underworld.

As it turned out, Pluto was not found because it was correctly predicted! Recent studies of its companion, Charon, prove that Pluto's mass is too small to account for the apparent perturbations of Uranus' motion. Thus, the predicted position had nothing to do with Pluto itself and the discovery of the planet was entirely accidental. The discovery was the result of a meticulous and systematic search that was guided by an incorrect prediction which merely happened to point in the general direction of Pluto.

(b) An escaped satellite of Neptune?

Pluto's highly elongated orbit ranges from 29.7 to 49.3 A.U. from the Sun. In 1979, Pluto actually passed inside Neptune's orbit, so Neptune will remain the most distant known planet until 1999 when Pluto will pass outside its orbit once more. The cross-over of the orbits of Pluto and Neptune led to speculations that Pluto might be an escaped satellite of Neptune. If so, this might account for the odd orbits of Neptune's moons. The idea that Pluto might be an escaped satellite of Neptune has become less likely with the discovery that Pluto itself has a satellite. Planetary physicist William McKinnon argues that Pluto and Charon are too small and too loosely bound to each other for such a scenario to be likely.

Figure 10.16. Discovery of Pluto. A process called blinking provided the basis for Clyde Tombaugh's discovery of Pluto. Once three photographs had been taken at intervals of several days, they were set in pairs in a blink comparator that would show the apparent motion of a planet, asteroid or comet against the background stars. After months of painstaking work, Tombaugh finally placed this pair of plates in the blink comparator. Because of its slow apparent motion across the sky, the planet images were only separated by 3.5 millimeters on the two photographs (see arrows). (A Lowell Observatory photograph.)

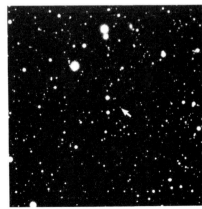

January 23, 1930   January 29, 1930

(c) A tiny world of ice and rock

Pluto has a diameter of about 2284 kilometers, which makes it smaller than our Moon (3476 kilometers). Infrared measurements show it to be the coldest planet, with a surface temperature of about 40 degrees Kelvin when farthest from the Sun, although Pluto warms to about 70 degrees Kelvin when it comes inside Neptune's orbit. Pluto's density (determined from its mass and size) is estimated to lie between 1.84 and 2.14 g/cm^3, suggesting that it may be quite rocky. It is not just a ball of ice; it may contain 70 percent rock, by mass.

Pluto's surface is covered with frozen methane. Some of this ice probably sublimates to produce the very thin atmosphere of gaseous methane (with a pressure of 5 to 20 millionths of the Earth's sea-level pressure). Although Pluto's weak gravity is insufficient to hold an atmosphere for long, the surface ice may replenish the gas that escapes into space.

(d) Pluto's companion, Charon

The discovery of Pluto's companion was an accidental by-product of observations made for another purpose. In 1978, astronomers at the United States Naval Observatory were obtaining a series of photographs to improve the accuracy of Pluto's orbit, when several of the images appeared slightly elongated. The elongation seemed to disappear every few days, and careful examination showed that Pluto has a companion that orbits every six days at a distance of 19 130 kilometers (Fig. 10.17). Planet and moon are more nearly alike in size than any other pair in our solar system. They are more appropriately called a double planet.

Charon was the name of the boatman who ferried new arrivals across the river Styx at the entrance to Pluto's underworld, Hades. Penniless ghosts are said to have waited endlessly because Charon gave no free rides.

The announcement of this remarkable doubling was a happy surprise, because it permitted determining the mass of Pluto. Charon orbits Pluto once every 6.38718 Earth-days. For comparison, a satellite that close to Earth would orbit in 7 hours. Charon's leisurely pace is a result of Pluto's small mass – only 1/440 times the mass of the Earth.

This small mass is only about one-sixth the mass of our Moon and is far too small to have influenced the past motions of Uranus and Neptune. In fact, the Earth exerts a larger gravitational influence on those planets than Pluto does.

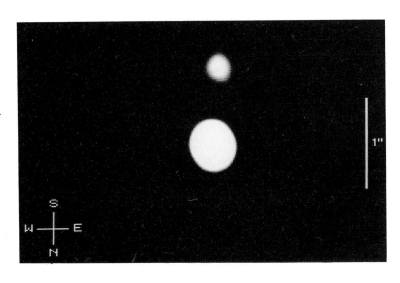

Figure 10.17. A double planet. Pluto and Charon appear distinctly when they are viewed by a special technique that removes the blurring effect of the Earth's atmosphere. This composite picture shows the true separation of the two bodies, and their relative sizes. They are so similar that Pluto and Charon can be characterized as a double planet. (Courtesy of G. Weigelt, Physikalisches Institut der Universitat Erlangen-Nurnberg.)

Table 10.3. *Parameters of the Pluto–Charon System*

Semimajor axis	19 130 km
Orbital period	6.38718 days
Mass of system	1.36×10^{25} grams
	(0.0023 Earth masses)
Pluto's radius	1123 km ($\pm$ 20 km)
Charon's radius	560 km ($\pm$ 20 km)
Mean density of system	1.99 g/cm^3 ($\pm$ 0.09 g/cm^3)
Pluto's density	1.84 to 2.14 g/cm^3
Charon's density	1 to 3 g/cm^3
Pluto's mass	1.25×10^{25} g
	(0.0021 Earth masses)
Orbit tilt	120° (retrograde)

Pluto and Charon are locked in a gravitational dance so Charon keeps the same face to its partner, as does our Moon. But unlike the Earth, Pluto also keeps the same face toward Charon.

Astronomers were offered a rare opportunity to observe Pluto and Charon's pirouette edge-on between 1985 and 1991, an event that only occurs twice in a Plutonian year – that is, every 124 Earth-years. When Charon's orbit narrows to a straight line as seen from Earth, we see a series of mutual eclipses as Charon and Pluto alternately pass directly in front of one another. Timing the starts and ends of such occultations and measuring the amount by which the light decreases permitted an accurate mapping of the outlines of this intriguing pair. (See Table 10.3.)

(e) Are there unknown planets beyond Pluto?

It seems likely that all the sizeable planets inside the orbit of Pluto have been discovered. The possibility of planets beyond Pluto is an open question. Pluto is now known to be too small to have caused the unexplained waverings in the motions of Uranus and Neptune. In fact, the early-recognized difference between the observed and predicted properties of Pluto had led to a continuation of the Lowell Observatory search for a trans-Neptunian planet, but after examining 90 million star images, Clyde Tombaugh felt he could rule out the presence of a Neptune-sized body in the ecliptic within a range of 270 times the Earth's distance from the Sun. Scientists at the United States Naval Observatory are nevertheless searching for a presently unknown Planet X in the extreme southern sky out of the ecliptic and not visible from mid-northern latitudes.

Several astronomers have proposed that there may be a swarm of small planets in the outer reaches of the solar system. Perhaps they will be found by telescopes raised above the atmosphere that blurs our vision. Such planets would be in the realm of the comets, to which we now turn.

Focus 10B Uranus – Summary

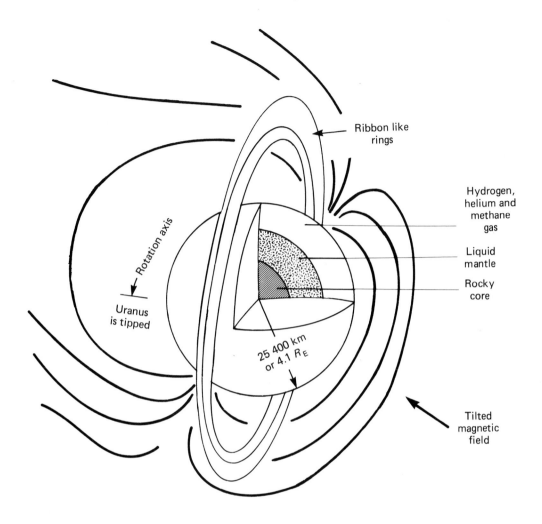

Mass: 8.72×10^{28} grams = 14.54 M_E (Earth = 1)
Radius: 26 150 kilometers = 4.10 R_E (Earth = 1)
Mean density: 1.21 g/cm^3
Rotational period: 17.24 hours
Orbital period: 84 years
Mean distance from Sun: 19.2 A.U.
Number of known satellites = 15
Magnetic field strength at cloud tops = 0.1 to 1.1 gauss

The black heart of Comet Halley. The coal-black nucleus of Comet Halley is silhouetted against bright jets that stream sunward (right) from at least three active regions. In this projection, the nucleus measures 14.9 by 8.2 kilometers. This is a composite of 60 images taken by the Halley Multicolor Camera aboard the European Space Agency's Giotto spacecraft. (Courtesy of Harold Reitsema of the Ball Aerospace Corporation and Horst Uwe Keller.)

11 Comets: icy wanderers

The sudden apparition, changing shapes, and unpredictable movements of comets have puzzled humanity for centuries.

Halley's comet has returned to fascinate and frighten the world for more than 2000 years.

The nucleus of Halley's comet is about the size of Manhattan, and blacker than coal.

Jupiter's immense gravity can propel a comet into a planet-like orbit, or hurl it into the interstellar void.

When a bright comet nears the Sun, it turns on its celestial fountain, spurting out about a million tons of water each day.

Visible comets are in their death throes, but they carry the residues of creation in their ice and dust.

11.1 Cometary motions

(a) Mysterious apparitions

Every decade or so, an unusually bright comet will blaze forth in the night sky becoming visible to the naked eye and sporting a graceful tail resembling long hair blowing in the wind. In fact, the word comet is derived from the Greek name *aster kometes*, meaning long-haired star (Fig. 11.1).

Unlike the planets, which are confined to the ecliptic, comets can appear almost anywhere in the sky, remain visible for a few days, weeks or months and then vanish into the darkness. (Astronomers call this period of visibility an "apparition.") The enigmatic comets also travel far outside the paths of

Figure 11.1. The Great Comet of 1577. This drawing by a Turkish astronomer appeared in the book Tarcuma-i Cifr al-Cami *by Mohammed b. Kamaladdin written in the 16th century. The yellow Moon, stars and comet are shown against a light blue sky. (Courtesy of Erol Pakin, Director – Istanbul Universitesi Rektorlugu.)*

the planets while moving in every possible direction and changing in shape from night to night.

By their unexpected arrivals, comets seemed to violate the elegant order of the heavens, and to presage changes in the order of things on Earth – wars and the death of rulers. To ancient and medieval minds, the apparition of a bright comet was an appalling omen of evil and the harbinger of epidemics and other disasters (Fig. 11.2). A famous example was the Norman conquest of England in 1066, which was coincident with the appearance of Halley's comet. This comet was also seen in 1456 when the Turks conquered Constantinople.

The idea that comets foretell disaster has persisted to modern times. For instance, in 1910 there were speculations that Halley's comet would impregnate the air with poisonous vapors and wipe out life on Earth; but there were no noticeable effects on humans or other living things when the Earth passed near the comet's tail.

During the middle ages, the sudden apparitions, irregular shapes and unusual motions of comets were explained by assuming that they were fiery exhalations in the Earth's atmosphere. The idea of a terrestrial origin for comets was challenged by Tycho Brahe when he showed that the Great Comet of 1577 travelled outside the terrestrial atmosphere. Tycho compared his observations of the comet, made at the island of Hveen (near Copenhagen), with those made by another astronomer in Prague at about the same time. Both astronomers observed the comet at the same position in the night sky, although the Moon showed a measurable shift in position with respect to the background stars. This meant that the comet was far beyond the Moon.

Tycho Brahe's conclusion that comets are distant objects did not, however, lead to an immediate unravelling of their mysteries.

(b) Sharply veering ways

The first mystery to be solved was the nature of cometary motions. In his *Principia*, published in 1687, Isaac Newton showed that the comets are subject to the Sun's gravity and that they move in orbits that are controlled by the Sun and described by his gravitational theory. Like the planets, they follow elliptical paths with the Sun at one focus (see Chapter 1).

Unlike the planets, the comets are seen for only a brief interval because their orbits are highly elongated and they only become visible when they pass near the Sun and the Sun's heat causes them to emit bright gas and dust. When far from the Sun, they remain cold and inert.

(c) The return of Halley's comet

In the late 17th century, Edmond Halley (Fig. 11.3) computed the orbits of many of the previous comets and he found that the retrograde orbit of the comet of 1682 was similar to those of comets observed in 1607 (by Kepler) and in 1531 (by Petrus Apianus). All three comets moved in retrograde orbits with a similar orientation. Halley also knew that the Great Comet of 1456 had travelled in the retrograde direction, and he concluded that all four comets were returns of the same comet in a closed elliptical orbit with a period of about 76 years (Fig. 11.4). Halley predicted its return, noting that he would not live to see it, and the comet was re-discovered on Christmas night in 1758 with a homemade telescope by a German farmer and amateur astronomer named George Palitzch. Halley's achievement was acknowledged, albeit posthumously, by naming it Halley's comet. Since that time, the name of the discoverer (or discoverers) is given to each new comet.

Halley's is the most famous of the comets because it was the first to arrive on schedule. Its fame is deserved on other counts as well, because Halley's

Figure 11.2. The Eve of the Deluge. People believed for centuries that comets presaged wars, death and other disasters. Here the arrival of a comet foretells the great flood at the time of Noah. The artist, John Martin, may have been influenced by the 1835–36 apparition of Halley's comet, for he finished this painting a few years later in 1840. (Collection of Her Majesty the Queen.)

Figure 11.3. Edmond
Halley. In this portrait,
Edmond Halley, at age 30,
is holding a drawing of the
orbit of a comet in the
vicinity of the Sun. The
figure heading notes that
Halley was eventually
elected secretary of the Royal
Society of London and
appointed Savilian Professor
of Geometry at Oxford,
although he left the
University as a student
without a degree. (Painting
by Thomas Murray,
Courtesy of the Royal
Society of London.)

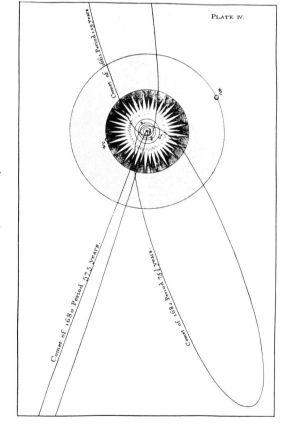

Figure 11.4. Comet orbits.
Of the three comets shown,
the Comet of 1680 is the one
Newton had first computed
the orbit for, and the Comet
of 1682 is the comet Halley
had predicted would return –
less than a decade after the
publication of this figure in
1750. At the center are the
planets Mercury, Venus,
Earth, and Mars, each
represented by its
astronomical symbol, and
then, just outside the Sun's
rays, the orbits of Jupiter
and Saturn. (From Thomas
Wright's An Original
Theory or New
Hypothesis of the
Universe, London, 1750.)

comet displays a complete range of cometary fireworks including an exceptionally long tail, a bright head, and jets, rays, streamers and halos. Moreover, it has been observed for more than 2000 years, a longer period of time than any other comet (Fig. 11.5). The earliest apparition established with confidence from Chinese chronicles dates back to 240 BC; since then, all its perihelion passages have been retraced in the records of oriental astronomers (see Table 11.1).

In 1910, Halley's comet moved away from the Sun into the outer darkness, arriving in 1948 at the remotest part of its orbit at 35 A.U. from

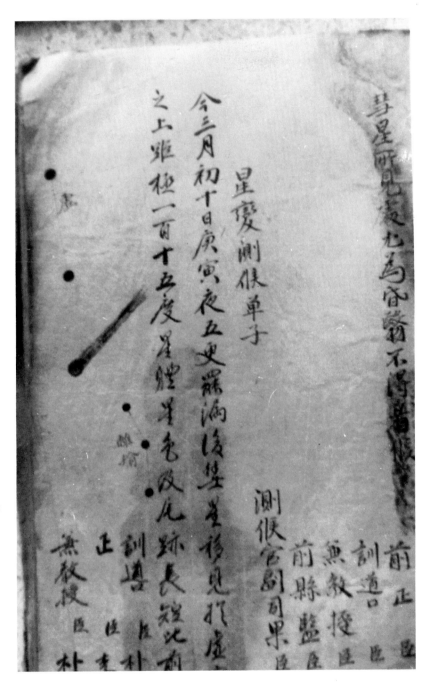

Figure 11.5. Halley's comet in AD 1759. A Korean record of Halley's comet during its first predicted return in AD 1759. The Korean astronomers have been recording the appearance of comets and other unusual celestial objects for more than 3000 years. (Courtesy of Il-Seong Na, Yonsei University, Seoul.)

Table 11.1. *Thirty-two perihelion passages of Halley's comet (all but the last two have been recorded)*

240 BC	May 25
164	November 13
87	August 6
12 BC	October 11
AD 66	January 26
141	March 22
218	May 18
295	April 20
374	February 16
451	June 28
530	September 27
607	March 15
684	October 3
760	May 21
837	February 28
912	July 19
989	September 6
1066	March 21
1145	April 19
1222	September 29
1301	October 26
1378	November 11
1456	June 10
1531	August 26
1607	October 28
1682	September 15
1759	March 13
1835	November 16
1910	April 20
1986	February 9
2061[a]	July 28
AD 2134[a]	March 27

[a]Predicted dates.

the Sun. The comet then turned the direction of its course, and began falling back towards the heart of the solar system with ever-increasing speed. The first glimpse of its return was made in 1982, when the comet was more distant than the planet Saturn. It reached perihelion, or its closest distance from the Sun, on Sunday, February 9, 1986. Halley's comet and the Earth were then on opposite sides of the Sun, so this was among the least favorable apparitions for naked eye observing in 2000 years. Nevertheless, the apparition was the most thoroughly studied in the history of cometary research (Fig. 11.6).

An international flotilla of spacecraft flew by the comet during its post-perihelion crossing of the ecliptic (the plane of the Earth's orbit) in March 1986. The first to arrive were two Soviet spacecraft, named Vega 1 and 2, which dove through the comet's atmosphere, conducting a comprehensive

Figure 11.6. The return of Comet Halley. Rays, streamers and kinks can be seen in the ion tail of Comet Halley during its 1986 return to the inner solar system. The broad, fan-shaped dust tail can also be seen. The radio galaxy known as Centaurus A, or NGC 5128, can be seen in the bottom left corner. It is about ten trillion times further away from the Earth than the comet is. Photograph taken by Arturo Gomez on April 15, 1986 with the Curtis Schmidt telescope at Cerro Tololo. (Courtesy of the National Optical Astronomy Observatories.)

study of the gas, dust, charged particles and magnetic fields in the vicinity of the comet.

The European spacecraft Giotto was directed to fly even closer to Halley's comet, and it revealed a dark nucleus and provided a detailed record of the composition, physical processes, and chemical reactions in the comet's atmosphere. The Japanese spacecraft was equipped to detect the ultraviolet light of the comet's hydrogen cloud, and it also observed cometary growth and decay from a safe distance.

After these visitations, Halley's comet headed for the cold reaches of space beyond Saturn, to return to the Sun's neighborhood in 2061.

11.2 Where do comets come from?

(a) Vagabonds in space
Before describing the results of recent research, we step back and ask where the comets have come from. Most of the comets (84 percent) with known orbits have periods of 200 years or more. These are called the long-period

comets to distinguish them from the short-period comets, whose periods are between 3 and 200 years. The two classes are quite distinct and they differ in many other respects as well.

The long-period comets come into the planetary realm at every possible angle – their orbits are inclined at all angles to the ecliptic. Roughly half of them orbit the Sun in the retrograde direction, opposite to the motion of the planets. The long-period comets are true vagabonds in space! (See Focus 11A, Discovering comets.)

It may take millions of years for a long-period comet to move from the depths of space into Earth's neighborhood, where it becomes visible. As the comet falls towards the Sun, it moves faster and faster, and when it enters the inner part of the solar system, the Sun's heat produces an expanding cometary atmosphere. Then the comet whips around the Sun in a few weeks and heads outward, fading into darkness on an elongated trajectory that carries it in more or less the same direction that it came from. Thus, most comets only light up and become visible for a brief fleeting interval during their long journey through space. Each passage strips off a portion of the comet's body.

(b) Playthings of the planets

Every year several new comets travel into the realm of the planets and light up during their first swing around the Sun. They are called new comets because they have entered the inner part of the solar system for the first time. These new comets have probably been moving in dark outer space for billions of years, but their orbits have never before brought them close to the Sun's heat.

When new comets enter the planetary realm, they can be tossed about by the massive planets, particularly by Jupiter, which acts like a cosmic street cleaner, throwing many of the new comets entirely out of the solar system. Others proceed in elliptical orbits.

A few of these survivors are slowed down enough to be gravitationally confined within the inner part of the solar system (Fig. 11.7). Such comets are called short-period comets and most of them have periods of less than 20 years. But close encounters with Jupiter are rare, and they sometimes lead to catastrophe – the comet crashes into one of the giant planets or is tossed out of the solar system altogether. Only a small fraction will be captured into a short-period orbit.

The large majority of short-period comets move in the same prograde direction as the planets, and they are usually confined within 20 degrees of the ecliptic. If a comet is to join the realm of the planets, it must conform somewhat to the pattern of planetary motion. Encke's comet is a short-period comet that revolves about the Sun with a 3.3 year orbital period, moving from just within Mercury's orbit to out beyond the orbit of Mars. Each time a short-period comet passes through perihelion, the Sun's heat vaporizes its outer layers – about a meter's worth per orbit of Encke's comet. These comets may eventually vaporize completely away. Alternatively, their gas and dust may be preferentially removed, leaving a black rocky corpse behind. The discovery of some asteroids in cometary orbits suggests that some asteroids might be the burned-out cores of comets that have outgassed all of their volatile substances.

(c) The comet cloud

All of the comets that periodically return to the vicinity of the Sun are caught in a life of continual decay as they evaporate and blow themselves away during repeated passes by the Sun. In fact, it is estimated that all of the known periodic comets will either be ejected into distant orbits or

Focus 11A Discovering comets

In recorded history, fewer than one thousand comets have been discovered, and fewer than one hundred have been seen in more than one passage by the Sun and Earth. New comets are often discovered by amateur astronomers who diligently search for them with small telescopes. Comets are also occasionally found by professional astronomers who accidentally come across one while using a large telescope for another purpose.

This Lick Observatory photograph shows the beautiful tail of Comet Ikeya-Seki (1965 VIII) that stretched across 120 million kilometers, or nearly the distance between the Earth and Sun. It was discovered by two Japanese comet hunters, Kaoru Ikeya and Tsutomu Seki.

The Infra-Red Astronomy Satellite (IRAS) has found more new comets in a shorter time than any observer in history. Because of its extreme sensitivity to the infrared radiation from warm dust, IRAS discovered 6 comets in 1983, one of which skimmed past the Earth at a distance of only 5 million kilometers, closer than any other comet since Lexell's in 1770.

Figure 11.7. Capturing comets. A planet's gravity may transfer a comet from an extremely elongated orbit (dashed lines) into a shortened elliptical orbit (solid line with arrows). In this case, the comet is said to be captured by the planet. The most massive planet, Jupiter, has captured a family of comets in this manner. The majority of short-period comets (about 70 percent) belong to Jupiter's comet family. These comets usually have orbital periods that are less than that of Jupiter (or less than 12 years). They all have orbits that pass near the Sun and reach out to Jupiter's orbit (shown as a partial circle at 5.2 A.U.)

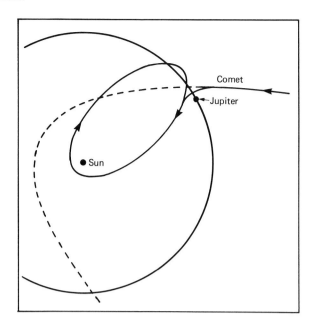

become exhausted, fall apart, and vanish from sight in less than a million years. This is a relatively short lifetime by cosmic standards, for the solar system is about 4.6 billion years old. Unless comets were recently born, which is very unlikely, some source must be furnishing the planetary realm with new comets.

Where could these new comets come from? In order to determine the birth-place of comets, we must determine their original orbits before entering the realm of the planets. These original orbits can only be established for comets that are entering the inner part of the solar system for the first time.

These new comets appear to have approached the Sun from enormous distances of 50 000 A.U. or more (Fig. 11.8). Because their enormously elongated orbits are inclined at all angles to the ecliptic, the new comets must have come from a remote spherically-shaped reservoir called the comet cloud. No one has seen this comet cloud, but the trajectories of new comets indicate that they must have come from such a region.

The comet cloud probably extends almost one-quarter of the way to the nearest star (Alpha Centauri at 250 000 A.U.). Even at these enormous distances, the Sun's gravity is usually powerful enough to hold the unseen comets in gigantic elliptical orbits with the Sun at one focus. At greater

Figure 11.8. The comet cloud. About 200 billion comets hibernate in the outer fringes of the solar system up to distances of perhaps 50 000 A.U. from the Sun. This illustration shows a cross-section through the cloud. By comparison, the distance to the nearest star, Alpha Centauri, is 268 000 A.U., while Neptune orbits the Sun at a mere 30 A.U. The planetary realm therefore appears as an insignificant dot when compared to the comet cloud, and has to be magnified by a factor of 1000 in order to be seen.

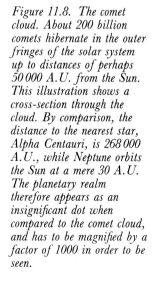

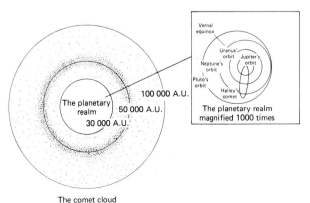

The comet cloud

distances are the stars of our neighborhood – each is imagined to have its own retinue of comets (Fig. 11.9).

The comet cloud marks the outer boundary of the solar system, where the sunlight is so diminished that everything is frozen. Comets therefore spend most of their lives hibernating in the deep freeze of outer space where they remain invisible.

But how do comets fall from the comet cloud into the heart of the solar system? At their great distances, the comets circulate about the Sun at

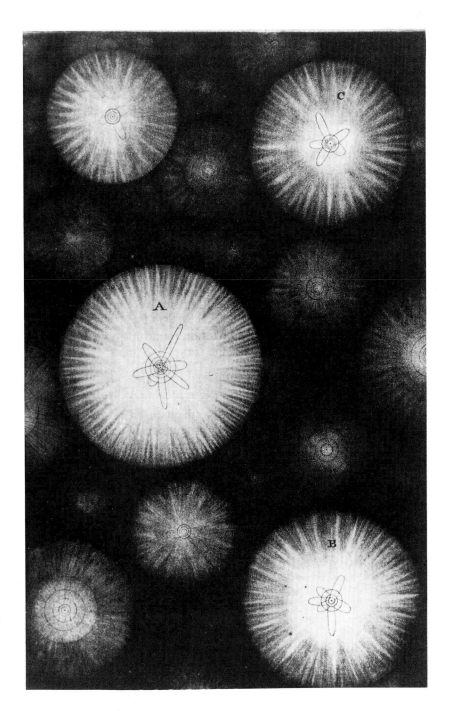

Figure 11.9. Cometary rosettes. Thomas Wright's vision of innumerable stars surrounded by cometary orbital rosettes. From his An Original Theory or New Hypothesis of the Universe, *London, 1750.*

leisurely rates of about 40 meters per second, and they require millions of years to complete a loop. Occasionally, a passing star or a giant cloud of interstellar molecules will impart a slight gravitational tug on some of the comets that happened to be in the right place and moving with the right velocity.

More important, perhaps, is the action of the Milky Way. A careful study of the orbits of new comets shows that they do not come from completely random directions in the sky. They show a slight tendency to avoid the plane of the Milky Way and its poles. This tendency indicates that they are influenced by the tidal force of the Milky Way – a force acting perpendicular to the disc of stars that forms our Milky Way. As the Sun travels about the center of our Galaxy, it moves up and down through the disc every hundred million years or so and each trip leads to a slight disturbance of the comet cloud.

As time goes on, the accumulated effect of these tugs will send a trickle of comets in toward the Sun – or outward to interstellar space. If the several hundred comets observed during recorded history have been shuffled into view by the perturbing action of the galactic tides and nearby stars, then the comet cloud may contain at least one hundred billion comets. There may even be a trillion of them! This large population of unseen comets can sustain the visible ones and persist without serious depletion for billions of years.

Thus, we conclude that there may be a trillion unseen wanderers in our neighborhood, reaching one-quarter of the way to the nearest star. Long ago, Seneca (4 BC to AD 65) expressed the question:

How many bodies besides these comets move in secret, never rising before the eyes of men! For god has not made all things for man.

(*Natural Questions*, Book 7, Comets)

How much matter is contained in a trillion comets? If an average comet is one kilometer in radius and has the density of water, the total mass is roughly the same as that of a single Earth.

(d) From dust to dust

Astronomers generally agree that there is an invisible comet cloud, but there is controversy over how it originated. Some astronomers argue that there may not have been enough material to create comets in the outer fringes of the primeval solar nebula. They could have been created closer in, near the orbits of the two icy planets, Uranus and Neptune, and then gravitationally ejected to the outer precincts of the solar system. Alternatively, the comets might have formed within cold, dense fragments of the primeval nebula as it collapsed to form the Sun and planets.

(e) Time capsules

Regardless of their exact place of origin, the comets were born cold and have stayed cold for billions of years. They have almost certainly spent their lives hibernating in cold storage, remaining virtually unaltered since their birth. The comets we see today are therefore time capsules that may date back to the origin of the solar system.

Some particularly speculative astronomers have endowed comets with life- and death-giving properties, arguing that they might spread disease or breathe life into dead planets. They speculate that comets might inject deep-frozen viruses into the Earth's air, bringing about the sudden, widespread appearance of ancient plagues and modern flu. One fertile comet is even supposed to have impregnated the Earth with the seeds of life. Very few astronomers take this idea seriously.

Table 11.2. *Structural features of a comet*

Feature	Size	Composition	Appearance
Nucleus	0.1 to 100 km	Dust or solid particles, ice	Very dark
Coma	Up to 0.01 A.U.	Neutral molecules and dust particles	Slightly yellow
Ion tail	Up to 1 A.U.	Ionized molecules	Blue, straight
Dust tail	Up to 0.1 A.U.	Dust particles	Slightly yellow, usually curved
Hydrogen cloud	Up to 0.1 A.U.	Hydrogen atoms	Ultraviolet

11.3 The nature of comets

(a) Anatomy of a comet

When a comet emerges from the deep freeze of outer space and moves towards the Sun, the increased solar heat eventually causes its ices to sublimate and blow dust away with the escaping gas. The comet then becomes visible as an enormous moving patch of light. When a comet first lights up at several A.U. from the Sun, a glowing diffuse ball of light is developed. It is called the coma, the Latin word for hair. If the comet travels even closer to the Sun, it develops a gossamer tail; but comets that stay outside a distance of about 1.5 A.U. usually have no tail. The ghost-like tails are paler and more tenuous than the cometary heads.

The glowing coma, or head, is a spherical cloud containing roughly equal amounts of gas and dust. It is much larger than the Earth in size. A typical coma is between thirty and one hundred thousand kilometers across, whereas the Earth has a diameter of about ten thousand kilometers (see Table 11.2). A comet exerts very weak gravity, so the gas and dust that are blown out of the central regions easily escape to interplanetary space. They are continuously replenished by gas that has vaporized off the solid cometary nucleus.

Comets are enveloped by a vast cloud containing hydrogen atoms that emit ultraviolet radiation. Observations of this glow – invisible to the eye – indicate that the hydrogen halo can be ten million kilometers across, or about ten times bigger than the Sun (Fig. 11.10). The halo is probably the result of photo-dissociation of water released from the comet's nucleus.

The long, flowing tails sweep across the sky in regal splendor, attaining lengths of ten million and even one hundred million kilometers (or about 1 A.U.). Thus, the tails of comets can briefly become the largest structures in the solar system.

Some comets show two types of tails at the same time. There are the long, straight ion tails and the shorter, curved dust tails (Fig. 11.11). An individual comet may have a dust tail, an ion tail, or both types of tails (see Fig. 11.12 and Focus 11B, Comet tails and solar forces).

Because it contains both electrons and ions, the ion tail is sometimes called a plasma tail. A plasma contains equal numbers of free electrons and ions, and it is often referred to as a fourth state of matter in addition to the solid, liquid and gaseous states.

As noticed by the ancients, stars shine clearly through comet tails. It is

Focus 11B Comet tails and solar forces

Cometary gas and dust are initially ejected in all directions from the comet, but primarily in the general direction of the Sun; solar forces push them into tails that flow away from the Sun. As a result, a comet tail usually points away from the Sun, and the comet travels head first when it approaches the Sun and tail first when moving away.

But what are the solar forces that blow gas and dust into comet tails? The powdery dust is blown out of the coma by the pressure of the Sun's light. When sunlight bounces off the dust particles, it gives them a little outwards push, called radiation pressure, and this forces them into the dust tails.

The ion tails stream faster and straighter than the dust tails. A wind of charged particles (electrons and protons) is continuously flowing away from the Sun's surface. It is this "solar wind" that accelerates the ions to high velocities and pushes them into the relatively straight ion tails. If the comets were not moving, their ion tails would point outward in the direction of the solar wind, and they would move at nearly the wind's speed – often 400 to 500 kilometers per second. Because comets move with orbital speeds of about 40 kilometers per second when they are near the Earth, their tails are slanted backward by about 5 degrees.

As illustrated in this figure, a magnetic field is entrained within the solar wind and wraps around the comet coma, producing a tail that confines the ions. The magnetic field of the solar wind is divided into sectors with

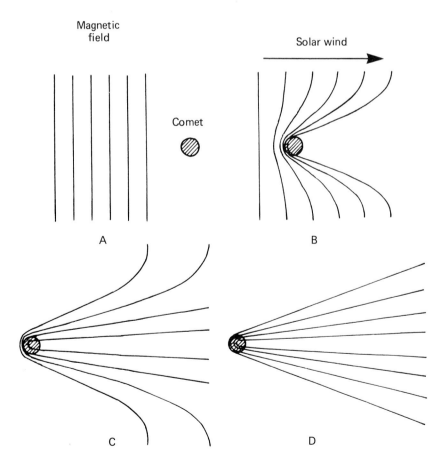

opposing field directions. When a comet crosses from one sector to another, the magnetic tail becomes pinched and the comet loses its tail, somewhat like a tadpole. But unlike a tadpole, the comet soon grows another.

Thus, the ion tails act like a wind sock and, in fact, the existence of the solar wind was hypothesized from observations of comets before the age of space exploration. Spacecraft have now confirmed these predictions, and have permitted measuring the ionized gas that is blown away from the Sun.

Figure 11.10. Hydrogen halo. A comparison of the visible image (left) of Comet Kohoutek with an ultraviolet image (right) on the same scale. Both pictures were taken from an Aerobee rocket on January 5, 1974. The ultraviolet image shows the gigantic halo of hydrogen, nearly 10 million kilometers in size, which is being fed by the cometary nucleus at the rate of 500 billion billion billion atoms of hydrogen every second. (Courtesy of Chet B. Opal, Naval Research Laboratory.)

therefore no wonder that the Earth has passed through many comet tails unscathed. Comets look awesome, but they contain so little matter that they are very close to being nothing at all.

A comet's coma and tail are continuously being blown away and renewed from the nucleus. The gas and dust in the coma leak into the tails and escape into interplanetary space, being completely replaced in a few days. An entire tail is completely renewed in a few weeks. All this material is lost to the comet forever, and must be continuously replenished from the comet's nucleus.

A comet's anatomy is not a static thing, for comets are always changing shape. The tail of a comet usually grows when the comet approaches the Sun, and shrinks when the comet moves away from the Sun. There is no such thing as a typical comet tail. They differ in shape, size, and structure. Some comets have multiple tails, some have only one tail, and others have no tail.

(b) A comet's nucleus
The tiny nucleus is buried inside the coma where it is hidden within the brilliant glare of fluorescing gases and reflected sunlight. As a result, no one had ever seen the bare surface of a comet's nucleus until spacecraft peered into Halley's core. They found the nucleus to be blacker than coal (Fig.

Figure 11.11. Comet tails. A drawing of Comet Donati (1858 VI) as viewed from Paris. As the comet accelerates about the Sun, its dust tail is swept into a broad arc that resembles a scimitar. In contrast, the two narrow ion tails are nearly straight. The dust tail is composed of particles that are typically one ten-thousandth of a centimeter across, and may consist of silicate materials. The ion tails are composed of molecules that have been ionized by the Sun's light and wind and they are swept rapidly outward. (Adapted from Amedee Guillemin's Le Ciel, Librairie *Hachette, Paris (1877).)*

11.13). The oblong chunk probably resembles icy snow and its blackness may be due to an admixture of minerals, organic compounds and metals. It is evidently surrounded by a crust that protects the underlying ice and snow. Where the crust has broken, localized emission of gas produces jets, and these jets are the source of material in the coma and tails.

How big is a comet's nucleus? The diameters of cometary nuclei may be inferred from the brightness of remote comets that have not yet formed a coma. The amount of reflected sunlight indicates sizes of between 100 meters and 10 kilometers for most comets. Spacecraft demonstrated that these sizes are of the right order of magnitude when they took close-up

Figure 11.12. Dust and ion tails. This photograph of Comet West (1976 VI) shows a broad, curved, pearly-hued dust tail. Because dust particles scatter sunlight, the dust tail has a slightly yellow color. The ion tail absorbs sunlight and re-emits it by the fluorescence process. Its visible radiation is dominated by the fluorescence of ionized carbon monoxide, CO^+, that gives the ion tails a blue color. This photograph was taken by Dennis DiCiccio at 4:30 a.m. on March 8, 1976 from Duxbury Beach, Massachusetts with a 2 minute exposure and at f/2. (Courtesy of Dennis DiCiccio, Sky and Telescope.)

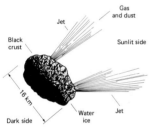

Figure 11.13. The nucleus of Halley's comet. Dust and gas geyser out of narrow jets from the sunlit side of a comet's nucleus (also see the frontispiece to this chapter). The gas is mainly water vapor sublimed from ice in the nucleus, while a significant constituent of the dust may be dark carbon-rich matter. The dark surface crust is blacker than coal, reflecting only about 4 percent of the incident light. The nucleus of Halley's comet was larger than anticipated, some 16 kilometers long and 8 kilometers across, or roughly the size of Manhattan.

photographs of the nucleus of Halley's comet. It is a peanut-shaped object 16 kilometers long by about 8 kilometers wide. It was darker than expected, and if this is typical, then comet nuclei may be somewhat larger than has been estimated in the past.

The Giotto spacecraft also gave us measurements of the gas and dust composition in Halley's coma. At the moment of measurement, the comet was spewing out 25 tons of water and 5 to 10 tons of dust every second – propelled by the vaporizing ice which absorbed the heat of sunlight. The proportions of various gases released by the icy nucleus were those to be expected if the comets are indeed primordial remnants of the outer regions of the early solar system. It seems likely that the comets carry mementos of creation within their ice and dust.

A comet's nucleus also rotates. Typical rotation periods are a few hours to a few days (see Fig. 11.14 and Focus 11C, Cometary fountains). Observations of Halley's comet, for example, indicate that it rotates around its longest axis once every 7.4 days, and that it wobbles about its shortest axis once every 2.2 days. (Most solid objects have three distinct axes, but motion about the intermediate axis is unstable.) As the nucleus rotates, new regions turn to face the Sun, heat up and become active, while others face away from the Sun and momentarily turn off their activity.

The cometary nuclei originate in the outer fringes of the solar system where the temperatures are below the freezing point of water vapor, carbon dioxide, and other gases. The cometary gases are therefore initially frozen, and according to the icy conglomerate model, a cometary nucleus is just a

Figure 11.14. Rotating comet. This photograph of Comet Halley shows jets of dust ejected from a rotating nucleus. The image was taken on January 6, 1986 by Stephen Larson and David Levy with the 1.5-meter (61-inch) Catalina reflector on Mount Lemmon. They used a CCD camera and a red filter to enhance the dust component of Halley's light. (Courtesy of Stephen Larson, Lunar and Planetary Laboratory, University of Arizona.)

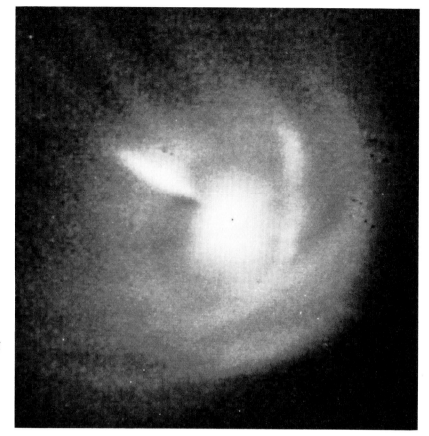

Figure 11.15. Icy comet and snow-covered ground. Comet Bennett (1970 II) photographed over the snow at Gornergrat, Switzerland by Claude Nicollier. (Courtesy of C. Nicollier.)

gigantic ball of snow crystals laced with darker dust (Fig. 11.14). According to this model, a comet resembles a dirty snowball, but it differs from the snowballs made by children on Earth, because the cometary snowball lacks compactness and strength. The pressure at the center of a comet's fragile nucleus is roughly comparable to that under a thick layer of loose bedclothes.

The icy conglomerate model can help explain comets that move erratically and seem to defy gravity (Fig. 11.16). These erratic motions are attributed to non-gravitational forces caused by jets of matter ejected from the spinning, icy nucleus (see Focus 11D, How to make a jet engine out of a dirty ball of ice).

The gas pressure in a coma is too small for liquids and boiling to occur. The solid material goes directly to gas, in a process called sublimation. Terrestrial frost vanishes in the morning sunlight by a similar process. Calculations suggest that the nucleus can lose a layer of only a few meters thick during each passage through perihelion. If the nucleus is a kilometer or more in diameter, it can withstand hundreds or thousands of approaches to the Sun before it sublimates away.

What are the gases that are frozen within a comet's nucleus? Judging from the material in the coma, the frozen material of the nucleus ought to be composed of compounds of the most abundant elements, hydrogen, carbon, nitrogen, and oxygen. These compounds include ices of water, carbon monoxide, carbon dioxide and hydrogen cyanide. A comet's hydrogen cloud is supplied by the release of about a million tons of water vapor a day. If a cometary nucleus provides such huge quantities of water and still

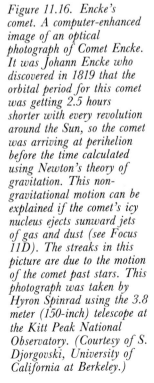

Figure 11.16. Encke's comet. A computer-enhanced image of an optical photograph of Comet Encke. It was Johann Encke who discovered in 1819 that the orbital period for this comet was getting 2.5 hours shorter with every revolution around the Sun, so the comet was arriving at perihelion before the time calculated using Newton's theory of gravitation. This non-gravitational motion can be explained if the comet's icy nucleus ejects sunward jets of gas and dust (see Focus 11D). The streaks in this picture are due to the motion of the comet past stars. This photograph was taken by Hyron Spinrad using the 3.8 meter (150-inch) telescope at the Kitt Peak National Observatory. (Courtesy of S. Djorgovski, University of California at Berkeley.)

Focus 11C Cometary fountains

The trajectories of the gas and dust that flow from a comet's nucleus resemble the paths of water droplets ejected by a rotating fountain. When the nucleus is activated by the Sun's heat, it ejects material towards the Sun. These solar-activated jets then bend away into the tail in much the same way that a water jet appears to curve back as the nozzle spins.

This fountain-like ejection of cometary particles is strikingly illustrated by this drawing of Comet Donati (1858 VI). These nested halos are due to the variable ejection of gas and dust. According to one hypothesis, the matter is being ejected from one part of the comet's surface that is periodically exposed to the Sun by the comet's 4.6-hour rotation. (Adapted from George P. Bond, *Annals of the Harvard College Observatory* **3**, 1 (1862).)

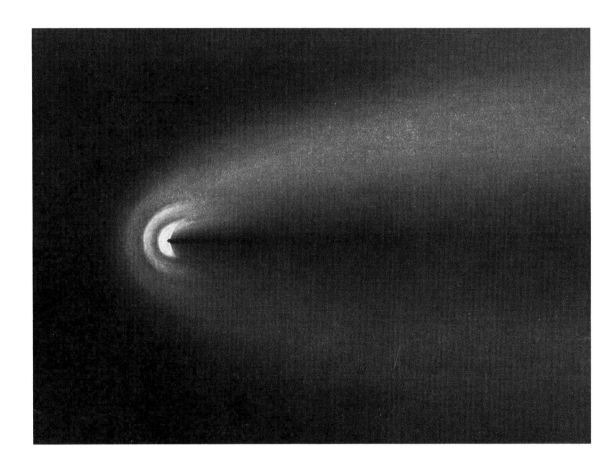

Focus 11D How to make a jet engine out of a dirty ball of ice

Many comets seem to defy gravity by arriving at perihelion before or after the time calculated using Newton's theory of gravitation. Some comets are speeding up while others are slowing down. These non-gravitational motions can be explained by sunward jets of matter ejected from a spinning, icy nucleus.

The exposed snow is sublimated on the comet's sunlit side, thereby releasing a jet of gas and dust into the comet's afternoon sky. For every action there is an equal and opposite reaction. The matter ejected from a comet's nucleus therefore propels the comet in the opposite direction. A similar effect explains the darting action of a small balloon when it is released, as well as the forward thrust of a rocket engine.

The comet's rotation will send the ejected mass slightly askew from the Sun's direction. If the nucleus is spinning in the direction opposite to its orbital motion around the Sun, the jets will produce a reaction that opposes the comet's motion. The comet will tend to fall toward the Sun into a shorter orbit and it will arrive early. On the other hand, if the direction of spin coincides with the orbital motion, the rocket action will push the comet away from the Sun into a longer orbit, and the comet will arrive later than expected. Halley's comet, for example, returns about 4 days later than it would under the influence of gravity alone.

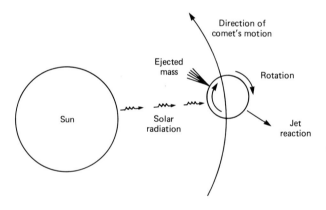

survives for hundreds of thousands of revolutions, then it ought to be mainly composed of water ice – perhaps in the form of snow.

Of course, the cometary coma is filled with all sorts of noxious substances. There is cyanogen, C_2N_2 (a poisonous, flammable and colorless gas), and hydrogen cyanide, HCN (or prussic acid). Nevertheless, these complex molecules are almost certainly only minor cometary ingredients, and it is water ice that seems to dominate the nucleus.

(c) What turns a comet on?

The distance at which an invisible nucleus turns on and creates a coma varies from comet to comet. The short-period comets turn on at several A.U. from the Sun, but new comets that are traversing the planetary system for the first time can turn on at greater distances. For instance, many first-time visitors to the solar neighborhood develop an unusually bright and extensive coma at distances of 5 A.U. or more. The icy conglomerate model helps explain this difference.

New comets turn on sooner because their outer layers more closely reflect primordial conditions in the solar system. Laboratory studies show that ices formed at the extreme low temperatures (15 degrees Kelvin) of the solar nebula will not be of the usual crystalline form of snowflakes. Instead, they are "amorphous" (from the Greek *without form*). When heated by passage near the Sun, this ancient amorphous ice is more volatile and produces more active jets than the ice of a periodic comet. The ancient ice within a meter of the surface of the comet has also been exposed to cosmic rays for billions of years, and this may alter its form and also lead to increased activity when it finally approaches the Sun.

Periodic comets, on the other hand, have made many passages close to the Sun, and their outer layers have been "cooked" and partially stripped off. Under the influence of solar heating, the atoms and molecules of gas in a comet's nucleus can percolate through the lattice of outer material, leaving behind an insulating crust, composed largely of dust. This layer is less prone to produce jets and it will hinder solar radiation from penetrating. This explains the limited loss of material from periodic comets that have been repeatedly exposed to the Sun.

The fact that comas of periodic comets appear near 3 A.U. from the Sun suggests that water ice dominates the nucleus, since the temperature of the Sun's radiation at 3 A.U. is approximately that required to vaporize water ice. Ices of other possible molecules begin to sublimate off the nucleus at much lower temperatures and greater distances from the Sun. The vaporizaton of these more volatile substances may initiate the production of gas and dust in the new comets. Volatile parent molecules like carbon monoxide, methane, carbon dioxide and ammonia are required to explain the presence of carbon molecules and nitrogen–hydrogen radicals in the comae of some periodic comets. These more volatile molecules are probably trapped within cavities in the water ice crystals, and as a result, all of the volatile substances are kept from being released until the short-period comet is about 3 A.U. from the Sun.

11.4 Cometary debris

(a) Decay and disintegration

The heads and tails of all the comets are continuously blown away as they seed space with gas and dust. Sooner or later the comet will be exhausted. It will either vaporize itself away into gas and dust, or it will fade into a black invisible corpse. In either event, comets are consumed by their own emissions. They are in the death throes of continual decay and disintegration.

The dust detectors on the Vega 1 spacecraft that visited Halley's comet in 1986 found several times as much dust as those on Vega 2 which arrived three days later, and this change was consistent with the bursts of cometary activity witnessed from Earth. The spacecraft that penetrated the coma during the 1986 flyby were damaged to various degrees (but not destroyed) by high-speed collisions with small dust particles that vaporized on impact. Most of the particles were smaller than those found in meteorites, and they are typical of grains that may exist in interstellar space, so the probes may, indeed, have detected the type of material from which the Sun and planets were formed.

The products of cometary decay can interact with the Earth in a variety of ways. Larger clumps of dust particles spread out along a comet's orbit, giving rise to meteors when the Earth chances to cross its path. Occasionally an exceptionally large cometary fragment enters the Earth's air, producing a cosmic explosion equivalent to a nuclear bomb (see Focus 11E, The Tunguska event).

(b) The Sun-grazing comets

Some comets plunge deep within the Sun's thin, hot, outer atmosphere, or corona. These sun-grazing comets pass within one million kilometers (0.01 A.U.) of the Sun's fiery surface.

They pay a heavy price for this trip. Some Sun-grazing comets break apart, probably because of the Sun's intense tidal forces. As an example, shortly after perihelion, the Great September Comet (1882 II) divided into four or more pieces stretched along nearly the same orbit like a string of pearls. Occasionally the demise is more spectacular, and the comet plunges directly into the solar atmosphere (see Fig. 11.17 and Fig. 11.18).

(c) Comet clones

A comet's nucleus sometimes splits into several pieces when it passes near the Sun (Fig. 11.19). The pieces of a split nucleus have too little mass to pull themselves together gravitationally. Once a nucleus splits, its pieces remain forever separated; the jets of escaping gas kick them away from each other and they continue to drift farther apart.

We now know of 21 comets that have split into two or more pieces, although many ancient comets may also have split apart, unobserved. Astronomers have observed more than 750 individual comets, and they estimate that about 7 percent of the comets break apart when they enter the planetary realm and approach the Sun.

(d) Showers of fire

When the Earth passes through the debris in a comet's path, we are treated to showers of fire that rain down through the atmosphere (Fig. 11.20). Particles in interplanetary space receive the designation meteor when they briefly dash into the atmosphere and light up. They are often called shooting stars (Fig. 11.21), and were immortalized by the ancients. For example,

> Oft you shall see the stars, when the wind is near,
> Shoot headlong from the sky, and through the night
> Leave in their wake long whitening seas of flame.
>
> (*Georgics*, Book 1)

Figure 11.17. Comet colliding with the Sun. In this photograph made from a satellite in 1979, a comet hurtles towards a fiery death on a kamikaze collision course with the Sun. The bright round object is an occulting disc of the coronagraph that was used to block out the Sun's bright glare. Six Sun-grazing comets were discovered with this satellite, leading to speculations that astronomers have missed many of these comets. (Courtesy of the US Naval Research Laboratory.)

Focus 11E The Tunguska event

On June 30, 1908, a giant blue-white ball of fire, said to have appeared brighter than the Sun, streaked across the daytime sky above the stony Tunguska river in Siberia. The Tunguska fireball then exploded with a violence equivalent to a nuclear bomb. The devastating blast felled trees like matchsticks. The sound of the explosion was heard thousands of kilometers away, and the skies over Europe glowed throughout the night for several days.

Curiously, no deep crater has been found at the site of the Tunguska event, and a clump of trees remains standing at the center of the devastation. Evidently, the exploding object never reached the ground! It instead blew up in the atmosphere.

The Tunguska event was similar to the aerial explosion of a nuclear bomb at Hiroshima. It also produced devastating results, but left no crater and did not affect trees directly beneath it.

Scientists have offered all sorts of imaginative explanations for the Tunguska explosion, including an encounter with a mini-black hole or anti-matter, and the explosion of an alien nuclear-powered spacecraft. A collision with an asteroid or a comet is a more plausible explanation.

According to some estimates, only 3000 comets have collided with the Earth since its formation. Size for size, asteroid impacts are about 50 times more likely than cometary ones.

Table 11.3. *Major naked-eye meteor showers, late 20th century*

Shower	Date of maximum	Hourly rate for single observer	Speed (km/s)	Duration (days)	Associated object
Quadrantids	Jan 3	40	41	1.1	?
Lyrids	Apr 22	15	48	2	1861 I
Eta Aquarids	May 4	20	65	3	Comet Halley
S. Delta Aquarids	July 28	20	41	7	?
Perseids	Aug 12	50	60	4.6	1862 III
Orionids	Oct 21	25	66	2	Comet Halley
S. Taurids	Nov 3	15	28	—	Comet Encke
Leonids	Nov 17	15	71	—	1866 I
Geminids	Dec 14	50	35	2.6	Phaethon[a]
Ursids	Dec 22	15	34	2	Comet Tuttle

[a]Phaethon may be a comet or an asteroid.

Figure 11.18. The Sun-grazing Comet Ikeya-Seki. In 1965 Comet Ikeya-Seki became brighter than the full Moon when it moved near the Sun. It is near perihelion in this daytime picture from Tokyo Observatory's Norikura solar station. The comet passed within 0.008 A.U. (483 000 kilometers) from the Sun, well within the outer atmosphere, or corona, of the Sun. The Sun's light is blocked by the disc of a coronagraph that shows the surrounding halo due to scattered photospheric light. (Courtesy of F. Moriyama, Tokyo Observatory.)

In addition to spewing off gas and dust, a comet blows away larger pieces of material ranging in size from sand grains to large pebbles. This debris continues to orbit the Sun and is spread along the comet's orbit, forming a continuous swarm of material that astronomers call a meteor stream (Fig. 11.22). When the Earth passes through a meteor stream, it intercepts millions of the orbiting particles, and as these particles race down towards the ground at speeds of 10 to 60 kilometers per second they rub against the air and produce the luminous trails of meteor showers. (See Table 11.3.) About one percent of the energy released during this process is converted to light, and a fragment the size of a tiny pebble is capable of creating a bright meteor trail.

The meteors associated with comets are fragile and they vaporize completely in flight. In fact, no meteor associated with a cometary orbit has ever reached the ground. There is a great difference between the fragile meteoric material associated with comets and the tough meteorites that survive the atmospheric flight to the ground. (See Chapter 7.)

(e) The zodiacal light

The zodiacal light is a luminous pyramid of light that appears brightest and widest in the direction of the Sun, and gradually becomes fainter and narrower with increasing distance (Fig. 11.23). Its name comes from the fact that this glowing pyramid stretches along the zodiac – the ecliptic, marking the paths of the Sun and planets around the sky. As the Earth rotates, the zodiacal light gradually sinks below the western horizon shortly after sunset, and then rises out of the eastern horizon a few hours before sunrise.

The zodiacal light is caused by sunlight scattered off a cloud of cometary dust that has been ejected into interplanetary space by short-period comets. The cloud is flattened and extends along the ecliptic within the orbit of Mars. The dust is circling in orbits that gradually sink toward the Sun, and it is constantly replenished by the short-period comets, at an average rate of 10 tons every second.

Figure 11.19. Nuclear splitting of Comet West. These photographs were taken (left to right) on March 8, 12, 14, 18 and 24, 1976, in yellow-green light using a 60 centimeter (23.6 inch) Cassegrain reflector. On March 18, the diameter of the four features was about 10 000 kilometers or 6210 miles. (Courtesy of C. Knuckles and S. Murrell, New Mexico State University Observatory.)

Figure 11.20. Leonid meteor shower. A woodcut showing the Leonid meteor shower on the night of November 12–13, 1833. The meteor trails seem to be raining down from the same radiant point in the constellation Leo. This is an effect of perspective. Meteors that appear to diverge from a point are actually moving on parallel paths, just as parallel railroad tracks seem to come from a point on the distant horizon. The meteoric particles were therefore travelling in parallel paths in space. (Courtesy of the American Museum of Natural History.)

Figure 11.21. The Shooting Stars. Two couples portray meteor showers or falling stars in this painting by Jean-François Millet (1847–48?). They soar through the skies, perhaps illustrating the transcendental nature of erotic love. (Courtesy of the National Museum of Wales, Cardiff.)

Figure 11.22. Radiant meteors. The apparent paths of shooting stars on November 27, 1872. Meteor showers are, in fact, named after the constellation in which their radiant appears. This meteor shower is called the Andromedids meteor shower because its radiant appears in the constellation of Andromeda. The shower occurs every November when the Earth intersects the debris that has been scattered along the orbit of Biela's comet. (Adapted from Amedee Guillemin's Le Ciel, *Librairie Hachette, Paris (1877).)*

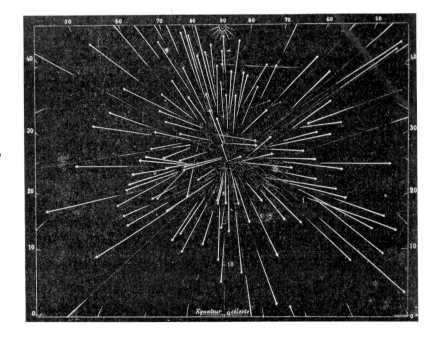

Figure 11.23. The zodiacal light. The zodiacal light observed from Japan in the mid 19th century. The light was once attributed to sunlight reflected from a cloud of material near the Sun; but it is now known that it is sunlight scattered by an interplanetary dust cloud that lies near the Earth. The zodiacal dust cloud is replenished by cometary dust. (Adapted from Amedee Guillemin's Le Ciel, *Librairie Hachette, Paris (1877).)*

(f) Cosmic dust

Some of the dust that has been ejected from the heads of comets even enters our own hair. The Earth is now sweeping up comet dust at the rate of about a million tons per year. This cosmic dust is everywhere.

Dust particles that pass through the air without vaporizing are called micrometeorites. They are so small that the air slows them down rather than burning them up. Hundreds of suspected micrometeorites have been collected from the stratosphere and examined in the terrestrial laboratory. On the average, their composition and elemental abundances closely match those of primitive meteorites (the carbonaceous chondrites) but with

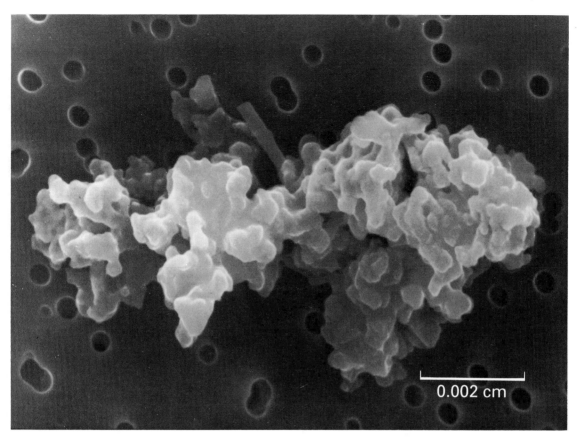

0.002 cm

Figure 11.24. Cosmic dust. This interplanetary dust particle is probably of cometary origin. It is a mere one-hundredth (0.01) of a centimeter long. This particle was collected at 20 kilometers altitude by a NASA U-2 aircraft, and then photographed with a magnification of 16 000 using a scanning electron microscope. The embedded rod-shaped crystals were probably formed in the primeval solar nebula, or perhaps in the pre-solar interstellar environment. (Courtesy of Donald E. Brownlee.)

more carbon. Micrometeorites also have the fragile, porous structure expected of cometary debris (Fig. 11.24).

Comet dust is probably too fragile to survive the mildest heating, so, if it survives at all, it cannot have been significantly altered from the moment of its creation. The delicate primordial dust is therefore thought to preserve a record of chemical conditions at the time of planet formation, and it may even contain the ashes of stars that existed before the Sun was born. This brings us to the birth of the Sun and planets and evidence for comet clouds around other stars.

Focus 11F The anatomy of a comet – summary

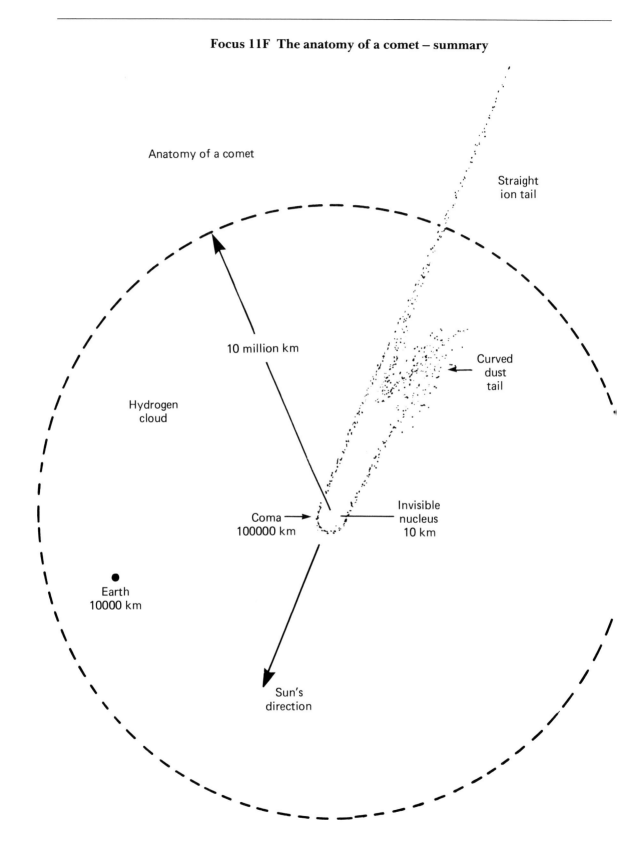

Anatomy of a comet

Straight
ion tail

10 million km

Hydrogen
cloud

Curved
dust
tail

Coma
100000 km

Invisible
nucleus
10 km

• Earth
10000 km

Sun's
direction

Comet nucleus

Mass: 10^{11} to 10^{21} grams = 10^{-17} to 10^{-7} M_E (Earth = 1)
Radius: 0.1 to 100 kilometers = 10^{-5} to 10^{-2} R_E (Earth = 1)
Mean density: less than 1 to 2 g/cm^3
Rotational period: 4 to 200 hours
Orbital period: 3.3 to more than 1 million years
Mean distance from Sun: 2 to 30 000 A.U.

Formation of the solar system. Unformed planets circle a nascent star, our Sun, before its nuclear fires burst forth. According to the nebular hypothesis, the Sun and planets were formed at the same time during the collapse of an interstellar cloud of gas and dust that is called the solar nebula. (Courtesy of Helmut K. Wimmer, Hayden Planetarium, American Museum of Natural History.)

12 Birth of the solar system

12.1 How old is the solar system?

One thing is certain about the age of the solar system – it cannot be greater than the age of the Universe, whose expansion from a flash of light began about 12 to 18 billion years ago. We shall start our story from the childhood of the Universe.

At first, the Universe was pure energy; there was no solid matter. Quickly this energy condensed to make energetic particles which, in turn, formed into electrons and the nuclei of hydrogen and helium. As the young Universe expanded and became rarified, it also cooled. The gas became clumpy and formed into swirling concentrations that gradually separated into isolated eddies that comprised primordial galaxies. Within these galaxies, smaller concentrations appeared, and these in turn fragmented, giving birth to clusters of stars.

The first stars that formed in such galaxies could not have been accompanied by a family of rocky planets like the Earth. There was nothing but hydrogen and helium at that time, and because there was no carbon, life as we know it could not have evolved on such planets.

The nuclei essential to life were created in the crucibles of the first generation of stars. These crucibles were of two types: (1) occasional spectacular explosions of supermassive young stars, called "supernova explosions," and (2) the interiors of more sedate stars that gradually converted hydrogen to helium (which has approximately the weight of four hydrogen atoms), helium to carbon, and so on up the chain of the nuclear species. It now appears that all the elements as heavy as carbon or heavier were formed inside the stars or during their explosions. The enriched stellar matter, first scattered into space by stellar winds or supernova explosions, was then recondensed into the next generation of stars.

Photographs of the star fields around us, and of galaxies like our own, show that star formation is a continuing process and successive new generations of stars are forming out of the interstellar gas and dust (Fig. 12.1). Our Sun must be a member of one of these later generations, because, in addition to hydrogen and helium, it contains about 2 percent by mass of heavy elements that could only have been produced in stars of the earlier generations. Theories for the lives of stars suggest that our Sun has been shining as it is (with a gradual and slight increase of brightness) for about 5 billion years. Thus, the solar system is not likely to be more than about 5 billion years old.

A more precise dating of the birth of the planets is possible by examining meteorites. The most primitive of these are the carbonaceous chondrites, which probably originated in the relatively cool precincts of the solar system during the earliest stages of planet formation. These relics have remained unaffected by the erosion that removed the primordial record from terrestrial rocks, and measurements of their radioactivity indicate that carbonaceous chondrites were formed about 4.6 billion years ago. This is probably also the age of the planets.

If we accept an age of 4 to 5 billion years for the solar system we conclude that the planets were formed at about the same time as the Sun. This provides an important clue to the early days of the planets. Additional clues are provided by the presently observed regularity of the system of the planets.

12.2 Patterns in the solar system

Some aspects of the regular pattern of the planets had been noted by the ancients:

1. The planets move in a narrow band around the sky: the band of the "zodiac." This pattern implies that the orbits all lie in nearly the same plane.

2. Viewed from the northern hemisphere of the Earth, the orbits of the planets all take them from right to left across the sky, corresponding to prograde or counter-clockwise revolution about the Sun.

Earth-based telescopes and spacecraft have revealed several other regularities.

3. The orbits are nearly circular.

4. The equator of the Sun's rotation nearly coincides with the plane of the planets' orbits.

5. The orbits of most of the satellites imitate the planets in being confined to the planet's equatorial plane and following prograde motion.

6. With the exception of Venus, Uranus, and probably Pluto, the Sun and the planets all rotate about their axes in the same prograde direction (Fig. 12.2).

7. The orbital radii of the planets follows a fairly regular pattern. In the inner solar system each planet's orbit has about 1.5 times the radius of its inward neighbor, and this ratio increases to roughly a factor of 2.0 in the outer solar system.

8. The angular momentum of the solar system is concentrated in the larger planets, and the Sun has less than one percent of the amount carried by the planets, even though it has more than 100 times the mass of the planets. We might say that the Sun has less than one-hundredth percent its fair share.

9. The most massive planets, Jupiter and Saturn, are in the middle zone, and the planets nearest to and farthest from the Sun, Mercury and Pluto, are the smallest of the nine.

10. The predominant ingredients of the planets (rock, ice, and gas) are not uniformly distributed (see Table 12.1). The dense rocky substances dominate the four planets nearest the Sun (Mercury,

Venus, Earth, Mars) while the lighter icy and gaseous substances dominate the outer, giant planets, Jupiter through Neptune. Furthermore, the giant planets all appear to have cores consisting of about 10 Earth-masses of rocky material, but they differ in the amount of icy material covering these cores.

Laplace appears to have been the first to try a mathematical analysis of the significance of such patterns. He computed the likelihood that the orbital arrangement of the planets could be accidental by the following argument. Imagine that a large number of solar systems, perhaps one

Figure 12.1. The Orion nebula. Interstellar gas becomes visible when it is close to very hot, massive stars which ionize it to fluorescence. Such stars are found in the bright central region of the Orion nebula (M 42 or NGC 1976) shown here. The dark, red dusty knots may be condensing to form future stars. This photograph was taken in red light during a 90 minute exposure on the 1.2 meter UK Schmidt Telescope. The original photographic plate has been copied using the unsharp masking technique for controlling contrast, thereby permitting fine detail to be resolved in both the bright central region and the fainter outlying filaments. (Royal Observatory Edinburgh © 1978, prepared by David F. Malin.)

3 light-years

Table 12.1. *Principal ingredients of the solar system*

Ingredient	Examples	Density (g/cm^3)	Melting temperature (degrees K)	Dominant location
Rock	Iron, Fe	7.9	2000	Terrestrial
	Silicon, Si	3.3		planets
	Magnesium, Mg			
Ice	Water, H$_2$O	1.0	273	Satellites,
	Ammonia, NH$_3$			comets,
	Methane, CH$_4$			Uranus, Neptune
Gas	Hydrogen, H	<1	14	Jovian
	Helium, He	<1	1	planets

million, each containing nine planets were thrown together haphazardly. In how many of these systems would we expect to find all the planets revolving about the Sun in the same direction? The direction of rotation of the Sun's surface is what we will call prograde. Each succeeding planet has two choices, prograde and retrograde, and we may consider them one at a time. The chance that the second will also be prograde is $\frac{1}{2}$, as each of the two possibilities is assumed to be equally likely. The chance that the second and the third will follow suit is $(\frac{1}{2})\,(\frac{1}{2}) = \frac{1}{4}$.

If there are 9 planets altogether, we will have 9 factors of $\frac{1}{2}$, so we can at once say that the probability that all will have the same direction of motion is

$$p = (\tfrac{1}{2})(\tfrac{1}{2})(\tfrac{1}{2})(\tfrac{1}{2})(\tfrac{1}{2})(\tfrac{1}{2})(\tfrac{1}{2})(\tfrac{1}{2})(\tfrac{1}{2}) = 0.002$$

This is the same as the chance of a tossed coin coming up heads 9 times in a row. If we were to include all the known moons – about 50 – this number would become so small that it is difficult to keep track of the zeros:

$$p = 0.000\,000\,000\,000\,000\,003\,5$$

Even if a million million million solar systems were made haphazardly and the planets and moons were thrown into randomly oriented orbits, only one of these solar systems would be expected to look like our own. And this is only part of the calculation. If we include the other regularities, such as the planetary spins and the orbits of the asteroids, the chances become small, indeed!

Faced with this dilemma, Laplace looked for a solution other than chance alignment. He realized that although Newton's laws could describe the present behavior of the solar system, they would not explain the remarkable arrangement. In the same way, even if we understand the working of our bedside clock, this will not explain how the clock was built or was set to the correct time. We must assume the action of an outside force to set the clock in motion – perhaps our own hand. We must have some additional information, and in the language of mathematics this information is called the "initial conditions." They describe the state of affairs before the moment when Newton's laws would describe the motion. For insights into the initial conditions, we must examine the processes by which the Sun and other stars are formed.

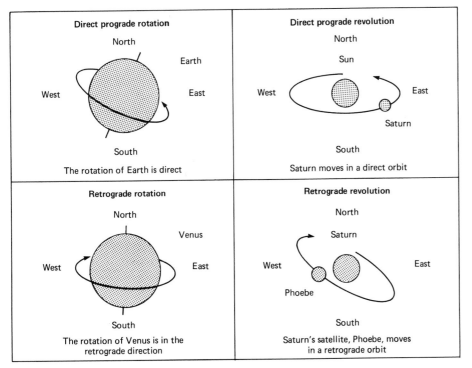

Figure 12.2. Prograde and retrograde motions. The planets all orbit the Sun in coplanar orbits, moving in the same prograde direction. The Sun and most of the planets also spin or rotate in the prograde direction. Moreover, most of the satellites orbit within their planet's equatorial plane in the same direction. This dominant direction of motion is counter-clockwise, or from west to east, when viewed from the north. There are nevertheless some exceptions to this astonishing dynamical regularity of the solar system. Venus and Uranus spin in the opposite retrograde direction, from east to west when viewed from the north. Saturn's outermost satellite and several of Jupiter's outer satellites move in retrograde orbits that are very eccentric and inclined.

12.3 Initiating star birth

The space between the stars may look empty to the eye, but it contains clouds of gas (largely hydrogen) and dust composed of heavier elements concentrated into tiny solid particles resembling smoke. Some of the clouds are debris left over from earlier stages of star formation; some are the ejecta of supernova explosions; others are the escaped outer fringes of tenuous, overblown stars. Whatever their origins, these clouds can become the raw stuff from which a new generation of stars and planets can be made.

This mixture of gas and dust is stirred into motion here and there by waves expanding from stellar explosions, or by the gravitational tugs and streams of radiation from nearby stars. The dust is carried along by the gas in the manner of "dust devils" produced by the wind on a dry day. Occasionally, gas clouds collide with each other, creating high temperatures and generating "shock waves" resembling the chaotic surf near an ocean beach.

For the most part, these clouds merely swirl through space, too hot and too agitated to condense into stars. But if a cloud becomes sufficiently large and dense, the mutual gravitational attraction of its parts will overcome the pressure and cause the cloud to start falling in on itself. As it falls inward, the gas gains energy, much the way a hammer gains energy when it falls toward the ground. Some of this energy is converted to heat, and this heat can raise the temperature and the pressure of the gas. This will tend to slow the collapse.

In fact, the collapse will be halted unless some of this energy can be absorbed or carried away, letting the cloud stay cool and keeping the pressure down. (See Fig. 12.3.) Radiation will not suffice to cool a large cloud, because the gas and dust reabsorb the radiation before it can escape. But once the internal pressure has built up – under the influence of gravity – the high temperature of the interior will generate wind currents that can

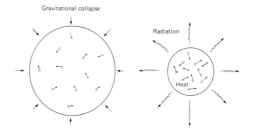

Figure 12.3. Gravitational collapse produces heat and radiation. Schematic diagram showing collapse and heating of an interstellar cloud of gas and dust. When the cloud shrinks, gravitational energy is converted into heat as the particles fall inward and collide with each other. This causes excitation of the gas and leads to radiation that can carry off some of the energy. The longer arrows in the collapsed cloud indicate the higher speeds of its atoms.

carry some of the heat toward the surface of the cloud. If the cloud has sufficient time to release a substantial part of its energy, it will cool and collapse like a pierced soufflé.

Despite astronomers' uncertainty about the actual mechanism of collapse, such collapsing clouds do exist. They have been discovered during the past decade in the form of giant molecular clouds with masses equivalent to many thousands of Suns. These clouds are in the process of forming stars in great clusters. Their interiors are cold (about 15 degrees Kelvin) and they are studded with dense concentrations of dust and molecular gas. These concentrations are much more massive than the Sun and they probably represent the earliest stages of star formation. The larger denser concentrations are already collapsing on themselves. Clouds of smaller mass have a more difficult time; and they probably require external compression to get the collapse underway.

Several processes can provide this external compression. Probably the most common is the formation of "density waves" in a rotating galaxy. These waves vaguely resemble the congestion of highway traffic at a roadblock. The interior of our galaxy is rotating at a speed that carries the Sun around once in 200 million years – the "cosmic year." The gravi-

Figure 12.4. Supernova remnant and young stars. A stellar explosion seems frozen in space as the star's scattered debris ploughs through interstellar gas and dust. A number of young stars lie along this expanding ring of gas, extending some 3 degrees across Canis Major. The shock of the stellar explosion triggered the birth of these stars. Yet, the chance of simultaneously witnessing a supernova remnant and the young stars that it spawns is very small because the remnant often expands away and dissipates into space in the time it takes for an interstellar cloud to collapse and become a star.

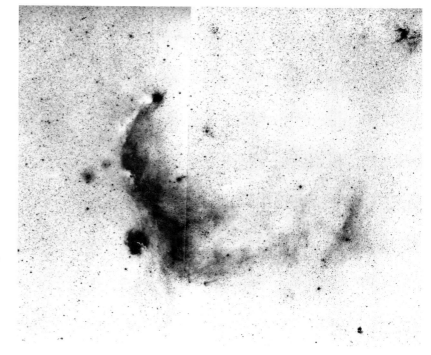

tational pull of an occasional clump of stars can act like a roadblock in this pattern of rotation, causing the gas to slow down and concentrate into a long spiralling wave. If the wave builds up sufficiently, it will compress the gas clouds, causing them to become the site of star formation. Such waves often appear as the brightly studded arms on the face of a spiral galaxy, where the young stars are strung like jewels in a necklace.

A more spectacular type of external compression is provided by the supernova (Fig. 12.4). When a massive star dies in a supernova explosion, it ejects a spherical shock wave that expands at a speed of 10 000 kilometers per second. (The wave produced by a nuclear explosion is a close analogy to the shock wave of a supernova.) Like a great snowplow, the shock wave pushes the interstellar gas and dust into a compacted shell, producing an extended region in which clumps of matter may further condense into stars. (See Fig. 12.5.)

The condensations in the giant molecular clouds are moving slowly among each other, and they are probably rotating gently. As each one condenses further, it will become separated from the chaotic background of

Figure 12.5. The Birth of the World. This painting by Joan Miró conveys an artist's impression of the formation of our solar system. By looking at the way the painting was built up from a structureless background, one gets a sense of the gradual development of structure in the planetary realm. (Courtesy of the Museum of Modern Art.)

the interstellar gas and dust, rotating faster and becoming flattened as it shrinks.

12.4 The original nebular hypothesis

There are two historical versions of the nebular hypothesis to describe subsequent stages in the formation of the Sun and planets. One proposed by the German philosopher–scientist, Immanuel Kant, in 1755, and the other by Laplace in 1796.

Kant suggested that the Sun formed at the center of a spinning nebula, while the planets formed from swirling condensations in a disc revolving around it. According to Laplace, the shrinking Sun shed a succession of gaseous rings, and each ring condensed into a planet (Fig. 12.6). Then each planet, in turn, became a small rotating nebula in which its own family of satellites was born.

This is the essence of the original nebular hypothesis, which explained qualitatively the fact that the planets and their moons all revolve in the plane that coincides with the equator of the rotating Sun. The highly regular pattern, which cannot be accidental, is a natural consequence of the rotation of a nebula composed of gas and dust from which the planets were then produced.

12.5 Questions raised by the original nebular hypothesis

The original nebular hypothesis greatly enriched the field of astronomy by incorporating new physical processes in the approach to cosmogony, the

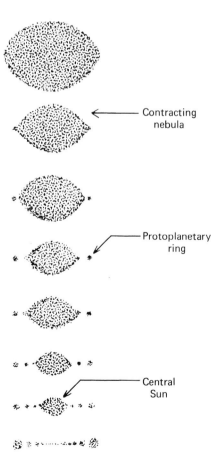

Figure 12.6. Laplace's version of the nebular hypothesis. This cross-sectional representation shows a contracting, rotating nebula that assumes a lenticular shape and successively sheds a concentric system of gaseous rings during its contraction. Here, the order of evolution is from top to bottom; the large central Sun has been omitted from the lower diagrams to reveal the proto-planetary rings. Laplace argued that Saturn's rings illustrate this nebular process in action, and mimic the earliest stages in the formation of the much larger planetary system.

Contracting nebula

Protoplanetary ring

Central Sun

theory of origins. But it raised problems that remained unanswered until recently. In the process of finding answers to these questions, astronomers have drastically modified the theory.

A successful theory must answer the following questions:

(a) Why does the Sun rotate so slowly?

The Sun's rotation period is about 27 days, far longer than the rotation periods of most of the planets. (Jupiter has the shortest period of any planet, 9 hours 55 minutes.) According to the law of conservation of angular momentum, the rotation of a shrinking object should speed up as the radius decreases, just as a skater will when her arms are pulled in. Objects of similar density, such as the Sun and Jupiter, ought to have similar rotation periods if they formed under similar conditions. On this basis, the period of the Sun ought to be no longer than about 10 hours, one percent of its actual value.

In fact, it is a wonder that stars can form at all, because the rotation of a collapsing cloud will accelerate as the star condenses. The resulting centrifugal force will soon inhibit further contraction. To see the severity of the problem of spin, we may assume that an interstellar gas cloud was initially turning with the period of rotation of the Milky Way, 200 million years. (It is hard to imagine a slower rotation!) By the time it has condensed to the density of the Sun, the cloud ought to be rotating twice a second if its angular momentum were perfectly conserved. But this is impossible; the Sun could never have collapsed that far, because the centrifugal force of such rotation would long ago have stopped the contraction. A star cannot be formed by gravitational collapse alone; some other phenomenon must slow the spin, perhaps mass-loss or magnetic braking.

In addition to the rotation problem, a heat problem must be overcome if the planets are to form. The interstellar gas must not only reduce its rotation, it must also lose heat and reduce its temperature.

To see the nature of the problem, suppose that a small pebble were to fall into the Sun from a great distance, say the orbit of Pluto or beyond. By the time it reached the surface of the Sun, the pebble would be falling inward at about 600 kilometers per second. If the pebble is to stay in the vicinity of the Sun, it must be brought to rest, and the energy created by its falling motion must be converted into heat by friction. (In the same way, a speeding bullet warms the air by friction, and the bullet becomes very hot when it strikes a solid surface.)

The result is a rise in temperature, to millions of degrees. At these temperatures, the matter of the pebble would vaporize into a gas and there would be no dust.

The continual inward rain of matter that built up the Sun must have brought a great quantity of heat. Only if this heat can be absorbed or carried away will the gas remain cool enough to continue collapsing and draw near to the Sun.

(b) Why are there two types of planets?

One of the most intriguing features of the solar system is the existence of four rocky planets in the inner region (Mercury through Mars) and four giant, icy planets (Jupiter through Neptune). These two planetary domains are separated by a region of asteroids. A satisfactory theory must explain this dichotomy.

12.6 Modern versions of the nebular hypothesis

As we shall see, the modern versions of the nebular hypothesis are quite different from their ancestors. Astronomers have not yet come to a consen-

Top view

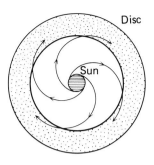

Side view

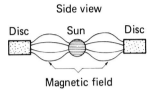

Figure 12.7. Magnetic brakes. Schematic top and side views of an ionized disc connected to a central Sun by a magnetic field. This slowly revolving ionized ring and the rapidly rotating Sun twist the magnetic field into a spiral shape; thereby providing a magnetic brake on the Sun's rotation and transferring angular momentum from the Sun to the proto-planetary disc.

sus concerning all the details, but recent computations have shown promise of explaining the observed patterns in the structures of the planets.

(a) Rotation and the nebular environment

If a star is to form, the rotation of the original gas must be slowed down. One way this may occur – if conditions are right – is through the action of magnetic brakes. Magnetic fields are threaded among the stars and they pervade the interstellar medium. The giant molecular clouds probably have magnetic fields, and if the gas is electrical, it will carry the magnetic field with it as it rotates. During the earliest stages, the gas is probably very cold and not sufficiently electrical to make the magnetic field effective. When the condensation has progressed, the inner region will become hot and the gas will become electrified. Now the magnetic field will act like a network of elastic cords and will tie together the various regions of the nebula. The inner regions will be moving faster, and the magnetic field will accelerate the slower moving outer regions of the nebula and slow down the inner regions. In effect, the outer regions will be swung around, and the magnetic field will transport some of the rotation from the inner region outward (Fig. 12.7).

In addition to the magnetic field, there are occasional hot eddies of gas, resembling bubbles, that surge outward carrying excess heat. Occasionally a colder eddy will sink toward the center of the disc. These turbulent motions carry heat from the interior of the nebula, aiding the collapse, and they also imitate internal friction by carrying momentum from one part of the gas to another. This friction tends to slow the inner regions and speed up the outer regions. As these outer regions become subject to greater centrifugal force, they tend to move further out and the nebula becomes flattened into a disc.

Recent studies of star-forming regions reveal energetic outflows in two opposite directions, as though following the shape of a magnetic dipole. This flow may help brake the rotation and carry away excess energy.

Magnetic braking, turbulent friction, and mass outflow gradually carry away the rotation of the core and produce a nebula with a slowly rotating core. The core is surrounded by a disc that is rotating with speeds that are determined by the gravitational attraction of the core. This pattern of a slow core surrounded by a disc is similar to the pattern of the solar system, where the Sun rotates relatively slowly, and most of the angular momentum is found in the orbital motions of the planets, which are spread in a disc.

(b) One alternative: collapse of a massive disc

Some astronomers suppose that the disc at this stage has only enough matter to form the planets – a few percent of the Sun's mass. Others suppose that the disc is much more massive – holding perhaps as much mass as the Sun itself. The two types of discs would behave quite differently. The more massive disc might be gravitationally unstable and break up into proto-planets whose subsequent development would lead to the planets as we see them today. Rocky material might have migrated inside these proto-planets, forming rocky cores enveloped by hydrogen in the form of ice and a thick atmosphere. With little change, the outer proto-planets could have then evolved into today's Jovian planets.

A major difficulty of this theory is the necessary removal of the large amount of debris after the formation of the planets. About 99 percent of the massive disc would have to be cleared away to leave one percent for the planets. Any process that would achieve 99 percent removal might also be expected to clean out the remaining one percent – leaving the Sun without any planets!

(c) Another alternative: formation and breakup of a dusty disc

According to the alternative theory, the disc has a mass only a few percent of the Sun's. Such a disc would be gravitationally stable and would not immediately break into protoplanets. In this case, the story might go as follows. Grains of dust would be carried about by the gas. Occasionally, two grains would strike each other and bounce apart like billiard balls. But then another pair would collide and this time they would stick together, forming a larger body. For the most part these developing grains were probably quite fragile, perhaps like a dust-ball or a tangled spider-web (Fig. 12.8). The larger grains are more massive per unit area and they would be less easily buffeted about by the turbulent motions of the gas. Like sand falling through water to the bottom of a lake, the grains would gradually fall to the central plane of the disc. They would accumulate in increasing numbers, forming a thin disc of particles. Calculations have shown that, when the density of such a disc becomes sufficiently high, the disc becomes unstable against gravitational forces. It breaks into rings, and the rings quickly bend and break into clumps. These clumps form planetesimals that move in roughly circular orbits about the center of the nebula. These planetesimals form the nuclei for the next stage: the formation of the solid cores of the planets.

(d) The planetesimals and proto-planets

Here again, astronomers have divided opinions. Some hold that the planetesimals move freely in gravitational orbits about the central core that will become the Sun; others hold that there is a residual of gas that affects the motions of the planetesimals. In both theories, the planetesimals and the gas collide and coalesce with each other and gradually grow to the size of planetary cores.

Planetesimals that collide with speeds greater than a few centimeters per second will either bounce or break each other apart, so only the gentlest collisions can lead to the growth of planetesimals. The more massive planetesimals would tend to stay in their own orbit, sweeping debris from around them, so they would have acted like cosmic vacuum-cleaners growing into the planets by accumulating innumerable tiny crystals and dust grains. The increased spacing between the outer planets may simply be a result of the need for more and more space to accumulate sufficient matter.

The distinction between the rocky inner planets and the icy giant planets may be understood as the result of two processes, perhaps working together. In the first place, the neighborhood of the inner planets would have been warmer than the outer solar system, so the inner planets would have been less likely to accumulate the easily vaporized ices that predominate the outer planets. Also, according to some astronomers, the cores of the giant planets were in a region of relatively high density and they produced gravitational collapse of the gas in their neighborhood, consisting largely of hydrogen and helium. This led to the creation of massive mantles around the rocky cores. Jupiter and Saturn, having more gas to draw in, became largest. Uranus and Neptune were in a region of lower density, so they developed more slowly and accumulated less gas. It is also possible that the hydrogen and helium in their neighborhood was swept away before Uranus and Neptune had become as large as Jupiter or Saturn. This would account for their having relatively less ice than Jupiter and Saturn.

According to this picture, the inner planets, being closer to the Sun, did not wrap themselves in massive mantles, but merely developed thick primordial atmospheres of hydrogen and helium.

And what about the asteroids? Can it be an accident that the region of the asteroids forms a division between the terrestrial planets and the giant planets? The asteroids do not presently contain enough material to form a large moon, let alone a planet, but perhaps the asteroid belt is a region of debris between two distinctly different types of planet-forming regions. It seems likely that Jupiter's immense gravity perturbed the asteroids and prevented them from continuing on the road to planethood.

(e) Clearing out the gas

As the final step in the drama, the Sun ignited its nuclear reactor and became a star. Its turbulent outer layers may have generated a gaseous outflow from its outer atmosphere, a primitive solar wind that would have cleaned the left-over dust and gas out of the solar system. The wind may also have drained most of the remaining rotational motion from the Sun.

The recent discovery of powerful winds from young stars implies that the Sun may have continued to shed its rotation by throwing its outer layers into space, spinning as they went.

The primordial atmospheres of the inner planets – if such atmospheres existed – may have been swept away by the intense wind of the young Sun. These planets may have then developed secondary atmospheres by outgassing from their hot interiors after the solar wind had calmed down. The wind would have been weaker at great distances from the Sun, and it may have left the atmospheres of the outer planets relatively unaffected.

This scenario is only an hypothesis (Fig. 12.9). So far we have been able to observe only one planetary system – ours, and it is in an advanced state of evolution. No one has yet claimed to have observed Earth-like planets outside our solar system, and reported detections of giant planets near other stars are controversial.

If the theories are to be completely trusted they must first be compared with observations of other planetary systems in various stages of evolution. The search is underway and may have already begun to bear fruit.

12.7 The search for other planetary systems

Evidence for solid material orbiting nearby stars was obtained by the Infrared Astronomical Satellite. The star Vega (fifth brightest in the sky, in the constellation Lyra) was found to be about five times brighter in the infrared than expected.

The infrared portion of the spectrum is the most likely to reveal solid objects such as dust and planets, because such objects will be relatively cool and will emit most of their radiation as low-energy infrared photons. Most of the brighter stars, on the other hand, will radiate primarily higher energy photons in the visible and ultraviolet portions of the spectrum. Thus, planets and dust clouds will be relatively brighter than the stars when observed in the infrared.

It was natural, therefore, to suspect that the excess infrared brightness of Vega was caused by a dust disc. From its temperature and brightness, astronomers have estimated that the disc is roughly twice the diameter of our own planetary system. (It is far too bright to be a single planet.) Since that initial observation, more than two dozen other stars have also been found to emit about 100 times more infrared radiation than expected, suggesting that the phenomenon is not extremely uncommon.

The disc itself has been mapped around the star Beta Pictoris. By masking the brilliant light of the star, astronomers found a faint streak of light on either side (Fig. 12.10). The streaks are produced by an edge-on view of starlight scattered from a circumstellar disc of dust. The disc is enormous – more than twenty times the size of Pluto's orbit. This size, and the fact that Beta Pictoris is more massive and younger than the Sun (with

Figure 12.8. Number 3, 1949: Tiger. This painting by Jackson Pollock, created by dripping paint onto a horizontal canvas, resembles the tangled, filamentary web of dust and gas that may have dominated the early stages of the solar system, particularly in the outer regions where the comets formed. (Courtesy of the Hirshhorn Museum and Sculpture Garden, Smithsonian Institution.)

an age of about one billion years), suggest that the disc around Beta Pictoris may be a protoplanetary system in the process of forming planets. The variation of color with wavelength indicates that the particles are larger than typical interstellar dust grains; they are at least the size of sand grains, and perhaps even larger. We may be witnessing the early coalescence of

*Figure 12.9. Planetary
growth in the solar nebula.
In this cross-sectional
representation, cosmic dust
and ice first rained down
into a thin equatorial plane.
Only dust could condense in
the vicinity of the hot central
regions, but both ice and
dust accumulated in the
colder outer regions. As a
result, the outer planets had
more massive cores that
permitted them to capture the
gas component of the solar
nebula. The rocky inner
planets could not hold on to
this gas because of their low
mass. The newly formed Sun
then generated winds that
swept the remaining gas out
of the solar nebula.*

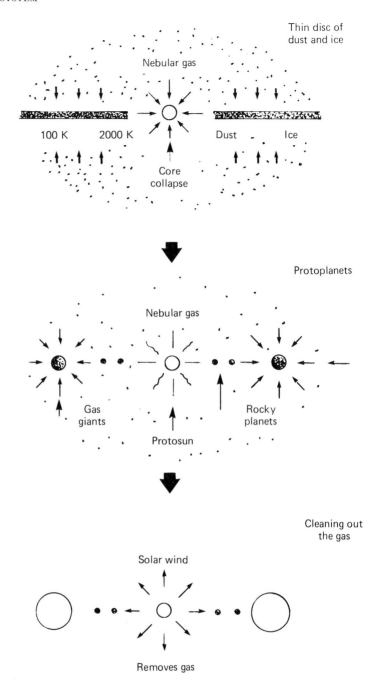

larger planets, and there may already be planets around Beta Pictoris.
Unfortunately, such planets would be lost in the glare of the dust and
invisible from Earth.

Radio astronomers have found another type of evidence for a cool
circumstellar disc that might be the birthplace of planets. Around the star
HL Tauri, thought to be about half a billion years old, they find the radio
emissions of carbon monoxide out to a distance 1500 times the Earth's

Figure 12.10. Circumstellar disc. Light extending from the star Beta Pictoris could represent an early stage of planet formation. By blocking the bright light of the central star with a coronagraph, astronomers were able to obtain this image of the scattered starlight from a circumstellar disc. It is seen edge on as it circles the central star, extending some 500 astronomical units (A.U.) to either side of it. (Pluto's average distance from the Sun is only about 40 A.U., but our comet cloud could extend out for hundreds of A.U.) The solid circumstellar particles that orbit Beta Pictoris are at least the size of sand grains, perhaps representing an early stage of accretion in a stellar nebula. (Courtesy of Richard Terrile, Jet Propulsion Laboratory.)

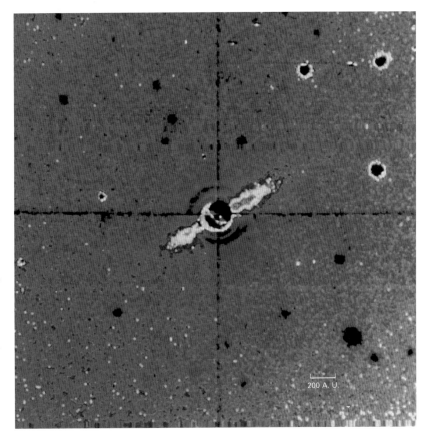

200 A. U.

distance from the Sun. The gas nearer the star is revolving about the star more rapidly than the outer gas, in accord with Kepler's third law of planetary motion. Thus, it is bound to the star and it may eventually condense to planets.

Many of the youngest T Tauri stars (which are thought to be in the early stages of formation) also glow prominently in infrared light, suggesting that they are surrounded by dust discs. However, these discs are not detected around the older T Tauri stars. Perhaps they have been swept up by newly formed planets.

These tantalizing observations suggest that the formation of planetary systems is a common event in our Galaxy, and that we may be on the threshold of detecting planets outside the solar system. Given that there are one hundred billion stars in our Milky Way, we must expect numerous planetary systems. Nevertheless, the possibility of extraterrestrial life is still pure speculation. We do not know whether conditions appropriate to life have arisen elsewhere.

However, one of the many lessons of recent space travel – and an implication of recent theories of planet formation – is that when a system of planets does form, it displays an amazing variety of conditions. Many astronomers find it hard to believe that ours is the only planet that could support life. Nevertheless, from the scientific standpoint, there is a possibility that we are alone among the wanderers in space – that our heritage is unique!

Appendix A1 Planetary properties

Mean orbital elements of the planets

Planet	Semi-major axis (A.U.)	Semi-major axis (10^6 km)	Sidereal period (tropical years)	Sidereal period (days)	Synodic period (days)
Mercury	0.387 099	57.9	0.240 85	87.969	115.88
Venus	0.723 332	108.2	0.615 21	224.701	583.92
Earth	1.000 000	149.6	1.000 04	365.256	
Mars	1.523 688	227.9	1.880 89	686.980	779.94
Jupiter	5.202 833 481	778.3	11.862 23	4 332.589	398.88
Saturn	9.538 762 055	1427.0	29.457 7	10 759.22	378.09
Uranus	19.191 391 28	2869.6	84.013 9	30 685.4	369.66
Neptune	30.061 069 06	4496.6	164.793	60 189	367.49
Pluto	39.529 402 43	5900	247.7	90 465	366.73

Planet	Mean orbital Velocity, V (km/s)	Mean daily motion, n (degrees)	Eccentricity, e	Inclination to the ecliptic, i (degrees)	Longitude of ascending node, Ω (degrees)	Longitude of perihelion, ω (degrees)
Mercury	47.89	4.0923	0.2056	7.00	47.7	76.7
Venus	35.03	1.6021	0.0068	3.39	76.2	130.9
Earth	29.79	0.9856	0.0167	0.01	174.4	102.1
Mars	24.13	0.5240	0.0933	1.85	49.2	335.1
Jupiter	13.06	0.0831	0.048	1.31	99.8	13.3
Saturn	9.64	0.0335	0.056	2.49	113.5	91.5
Uranus	6.81	0.0117	0.046	0.77	73.7	172.1
Neptune	5.43	0.0060	0.010	1.77	131.2	38
Pluto	4.74	0.0040	0.248	17.15	109.7	223

Physical elements of the inner planets

	Mercury	Venus	Earth	Mars
Equatorial radius, R_e (km)	2439	6051	6378.140	3397
Polar radius, R_p (km)	2439	6051	6356.775	3376
Oblateness, $(R_e - R_p)/R_e$	0.0	0.0000	0.003 352 9	0.0059
Equatorial radius ($R_E = 1.0$)	0.382	0.949	1.000	0.533
Angular diameter at closest approach (″)	10.90	61.0		17.88
Angular diameter at 1 A.U. (″)	6.74	16.92	17.60	9.36
Reciprocal mass ($M_\odot/M_P$)	6 023 600	408 525.1	332 946.043	3 098 710
Mass, M_P (grams)	3.3022×10^{26}	4.8690×10^{27}	5.9742×10^{27}	6.4191×10^{26}
Mass, M_P ($M_E = 1.0$)	0.055 27	0.814 99	1.000 00	0.107 45
Mean density (g/cm^3)	5.43	5.25	5.52	3.93
Sidereal rotation period	58.6462 Earth days	243.01 Earth days (retrograde)	23 hours 56 minutes 4.099 seconds	24 hours 37 minutes 22.66 seconds
Inclination of equator to orbit (degrees)	7.0	177.4	23.45	23.98
Magnetic moment (gauss $R^3{}_P$)	0.0035	<0.0003	0.31	≤0.0006
Tilt angle of magnetic axis (degrees)	<10		11.5	
Solar wind stagnation point			11	
Equatorial acceleration of gravity (cm/s^2)	370	887	980	371
Equatorial escape velocity (km/s)	4.25	10.36	11.18	5.02
Surface temperature (K)	100 to 700	730 ± 5	288 to 298	183 to 268
Surface pressure (bars)		90± 2	1.0	0.007 to 0.010

Physical elements of the outer planets

	Jupiter	Saturn	Uranus	Neptune	Pluto
Equatorial radius, R_e (km)	71 492	60 268	25 559	24 760	1123
Polar radius, R_p (km)	66 854	54 364	24 973		
Oblateness, $(R_e - R_p)/R_e$	0.064 87	0.097 96	0.022 93		
Equatorial radius ($R_E = 1.0$)	11.19	9.46	3.98	3.81	0.176
Angular diameter at closest approach (")	46.86	19.52	3.60	2.12	0.08
Angular diameter at 1 A.U. (")	196.74	165.6	65.8	33.9	2.9
Reciprocal mass ($M_\odot/M_P$)	1047.3492	3497.91	22 902.94	19 434	13×10^7
Mass, M_P (grams)	1.8992×10^{30}	5.6865×10^{29}	8.6849×10^{28}	1.0235×10^{29}	1.36×10^{25}
Mass, M_P ($M_E = 1.0$)	317.894	95.1843	14.5373	17.1321	0.025 61
Mean density (g/cm³)	1.33	0.71	1.24	1.67	1.89 to 2.14
Sidereal rotation period	9 hours 55 minutes 29.7 seconds	10 hours 39 minutes 22.4 seconds	17.24 ± 0.01 hours	15 ± 3 hours	6.38718 days
Inclination of equator to orbit	3.08	26.73	97.92	28.8	≥50

Magnetic moment (gauss R_P^3)				1.27
Tilt angle of magnetic axis (degrees)	9.6	0,8	60.0	72
Solar wind stagnation point (R_P)	70	22	18	40 to 60
Equatorial acceleration of gravity (cm/s²)	2312	896	869	1100
Equatorial escape velocity (km/s)	59.54	35.49	21.29	23.71
Effective temperature (K)	124 ± 0.3	95.0 ± 0.4	58 ± 2	55.5 ± 2.3
Temperature at one-bar level (K)	165 ± 5	134 ± 4	76 ± 2	
Ratio of emitted energy to incident energy	1.2 ± 0.2	2.2 ± 0.7		2.1 ± 0.5

Appendix A2 Satellites of the planets

Mean orbital elements for the satellites

Satellite		Distance from planet center (10^3 km)	Distance from planet center (R_P)	Orbital period (days)	Eccentricity	Inclination (degrees)
Earth						
Moon		384.4	60.2	27.3217	0.05490	18.2 to 28.6
Mars						
M1	Phobos	9.37	2.76	0.3189	0.0150	1.1
M2	Deimos	23.52	6.90	1.262	0.0008	0.9 to 2.7
Jupiter						
J14	Adrastea	128	1.80	0.295	~ 0.0	~ 0.0
J16	Metis	128	1.80	0.295	~ 0.0	~ 0.0
J5	Amalthea	181	2.55	0.489	0.003	0.4
J15	Thebe	221	3.11	0.675	~ 0.0	~ 0.0
J1	Io	422	5.95	1.769	0.004	0.0
J2	Europa	671	9.47	3.551	0.000	0.5
J3	Ganymede	1 070	15.1	7.155	0.001	0.2
J4	Callisto	1 880	26.6	16.69	0.010	0.2
J13	Leda	11 110	156	240	0.146	26.7
J6	Himalia	11 470	161	251	0.158	27.6
J10	Lysithea	11 710	164	260	0.130	29.0
J7	Elara	11 740	165	260	0.207	24.8
J12	Ananke	20 700	291	617	0.17	147
J11	Carme	22 350	314	692	0.21	164
J8	Pasiphae	23 300	327	735	0.38	145
J9	Sinope	23 700	333	758	0.28	153

Satellite		Distance from planet center (10^3 km)	Distance from planet center (R_P)	Orbital period (days)	Eccentricity	Inclination (degrees)
Saturn						
S17	Atlas	137.7	2.276	0.602	0.002	0.3
S16	Prometheus	139.4	2.310	0.613	0.004	0.0
S15	Pandora	141.7	2.349	0.629	0.004	0.1
S10	Janus	151.4	2.510	0.694	0.009	0.3
S11	Epimetheus	151.5	2.511	0.695	0.007	0.1
S1	Mimas	186	3.08	0.942	0.020	1.5
S2	Enceladus	238	3.95	1.370	0.004	0.0
S3	Tethys	295	4.88	1.888	0.000	1.1
S13	Telesto	295	4.88	1.888		
S14	Calypso	295	4.88	1.888		
S4	Dione	377	6.26	2.737	0.002	0.0
S12	Helene	377	6.26	2.737	0.005	0.2
S5	Rhea	527	8.73	4.518	0.001	0.4
S6	Titan	1 222	20.3	15.95	0.029	0.3
S7	Hyperion	1 481	24.6	21.28	0.104	0.4
S8	Iapetus	3 561	59	79.33	0.028	14.7
S9	Phoebe	12 954	215	550	0.163	150
Uranus						
U13	Cordelia	49.7	1.94	0.333		
U14	Ophelia	53.8	2.10	0.375		
U15	Bianca	59.2	2.32	0.433		
U9	Cressida	61.8	2.42	0.463		
U12	Desdemona	62.7	2.45	0.475		
U8	Juliet	64.6	2.53	0.492		
U7	Portia	66.1	2.58	0.513		
U10	Rosalind	69.9	2.73	0.558		
U11	Belinda	75.3	2.94	0.621		
U6	Puck	86.0	3.36	0.763		
U5	Miranda	129.9	5.08	1.413	0.017	3.4
U1	Ariel	190.9	7.47	2.521	0.0028	
U2	Umbriel	266.0	10.41	4.146	0.0035	
U3	Titania	436.3	17.07	8.704	0.0024	
U4	Oberon	583.4	22.82	13.463	0.0007	
Neptune						
N1	Triton	354	14.6	5.877	0.00	160.0
N2	Nereid	5 510	227	365.2	0.75	27.6

Physical elements for the principal satellites

Satellite		Radius (km)	Mass (10^{23} g)	Density (g/cm^3)	Visual magnitude opposition, V_{opp}	Visual magnitude unit distance $V(1,0)$	Geometrical visual albedo (ϱ_v)
Earth							
	Moon	1738	735	3.34		+ 0.21	0.12
Mars							
M1	Phobos	14 × 10	9.6×10^{-5}	≤ 2	11.6	+11.9	0.06
M2	Deimos	8 × 6	2.0×10^{-5}	≤ 2	12.7	+13.0	0.07
Jupiter							
J3	Ganymede	2631	1490	1.93	4.6	− 2.09	0.43
J4	Callisto	2400	1075	1.83	5.6	− 1.05	0.17
J1	Io	1815	892	3.55	5.0	− 1.68	0.63
J2	Europa	1569	487	3.04	5.3	− 1.41	0.64
Saturn							
S6	Titan	2575	1346	1.88	8.4	− 1.20	0.21
S5	Rhea	765	24.9	1.33	9.7	+ 0.16	0.60
S8	Iapetus	730	18.8	1.15	10.2 – 11.9	+ 1.6	0.12
S4	Dione	560	10.52	1.41	10.4	+ 0.88	0.60
S3	Tethys	530	7.55	1.20	10.3	+ 0.7	0.8
S2	Enceladus	250	0.74	1.13	11.8	+ 2.2	1.0
S1	Mimas	196	0.455	1.44	12.9	+ 3.3	0.7
Uranus							
U3	Titania	790	34.8	1.68	14.0	+ 1.3	0.28
U4	Oberon	762	29.2	1.58	14.2	+ 1.5	0.24
U2	Umbriel	586	12.7	1.51	15.3	+ 2.6	0.19
U1	Ariel	579	13.5	1.66	14.4	+ 1.7	0.40
U5	Miranda	236	0.8	1.35	16.5	+ 3.8	0.34
Neptune							
N1	Triton	1360	216	2.05	13.6	− 1.2	
N2	Nereid	150			18.7	+ 4.0	

Bibliography

Books about the solar system mainly for general audiences

Baugher, Joseph F.: *The Space-Age Solar System*, John Wiley & Sons, New York, 1988.

An introductory text that describes the modern exploration of our solar system, with an emphasis on the internal structures, surfaces and histories of the planets and their satellites. This is a complete textbook that should be read by any serious student of astronomy. The appendices contain interesting data on the planets and satellites, a listing of lunar and interplanetary space probes from 1950 to 1990, and a good technical bibliography.

Beatty, J. Kelly, Brian O'Leary and Andrew Chaikin (editors): *The New Solar System*, Cambridge University Press, New York, 1982.

A comprehensive collection of articles written for the well-educated layman by outstanding planetary scientists. There are excellent accounts of the atmospheres and surfaces of the terrestrial planets, asteroids, meteorites, planetary rings, the Galilean satellites, Titan, Jupiter, Saturn, comets and the origin of the solar system.

Chapman, Clark R.: *Planets of Rock and Ice*, Scribners, New York, 1982.

An eminent planetologist discusses the space-age exploration of the Moon, Mercury, Venus, Mars, Jupiter and Saturn in an engaging style.

Cole, G.H.A.: *Inside a Planet*, Hull University Press, England, 1986.

A good, non-technical introduction to planetary interiors, including the outgassing of planetary atmospheres, planetary magnetism, the detailed internal structure of the Earth and Moon, and the general properties of the other planets and satellites. The origin of the solar system and the question of life on planets are also discussed.

Couper, Heather and Nigel Henbest: *New Worlds – In Search of the Planets*, Addison-Wesley, Reading, Mass., 1986.

This entertaining description balances historical background with the latest space-age discoveries; authored by two well-known popular writers during the preparation of a British television series on the planets.

Elliot, James and Richard Kerr: *Rings – Discoveries from Galileo to Voyager*, The MIT Press, Cambridge, Mass., 1984.

An outstanding account of how modern scientists conduct astronomical research, from the fortuitous discovery of rings around Uranus in 1977 to the Voyager spacecraft's detection of the Jovian ring.

Greeley, Ronald: *Planetary Landscapes*, Allen and Unwin, Winchester, Mass., 1985.

An elementary, well-illustrated survey of the surface features of the Moon, Mercury, Venus, Mars and the large satellites of Jupiter and Saturn; surface-forming processes including tectonism, volcanism, erosion-sedimentation and impact cratering provide the focal points.

Guest, John, Paul Butterworth, John Murray and William O'Donnell: *Planetary Geology*, John Wiley, New York, 1979.
Half of this fascinating book is devoted to a description of the surfaces of the Moon and Mercury, whereas the remaining half deals largely with Martian surface features. The effective format places a one-page essay next to a page-length photographic illustration.

Hartmann, William K.: *Moons and Planets* (second edition), Wadsworth Publishing Co., Belmont, CA, 1983.
An introductory textbook filled with many facts, speculations and insights about the moons and planets. It also has a good treatment of asteroids, meteorites, comets and the formation of the solar system.

Kivelson, Margaret G. (editor): *The Solar System – Observations and Interpretations*, Prentice-Hall, Englewood Cliffs, New Jersey, 1986.
A mixed collection of seventeen lectures designed for upper-division undergraduate students. Expert scientists discuss the atmospheres, interiors and magnetic fields of the planets, as well as their origin and destiny, but the satellites are overlooked and the comet material was written before the spacecraft rendezvous with Halley's comet.

Lang, Kenneth R. and Owen Gingerich: *A Source Book in Astronomy and Astrophysics, 1900–1975*, Harvard University Press, Cambridge, Mass., 1979.
Historical essays and classical papers are combined to highlight the major advances in twentieth-century astronomy including: the atmospheres of the planets, continental drift, the hot surface of Venus, the slow rotation of Venus and Mercury, space-age exploration of the Moon and Mars, the origin and nature of comets, the precession of Mercury's perihelion, and tests of Einstein's new theory of gravity.

Littmann, Mark: *Planets Beyond – Discovering the Outer Solar System*, John Wiley & Sons, New York, 1988.
This interesting account of Uranus, Neptune and Pluto combines a long history of their discoveries with the Voyager 2 investigations of Uranus, expectations for its encounter with Neptune and the possibility of a currently unknown Planet X. An appendix contains a chronology of events linked with discoveries in the outer solar system.

Morrison, David and Tobias Owen: *The Planetary System*, Addison-Wesley, New York, 1988.
An introductory textbook written by two scientists with extensive research experience in the field. The text contains all the pedagogical tools, including review questions, key terms, summaries, and suggestions for additional reading. The order of presentation emphasizes the processes that formed and continue to influence the planetary system.

Murray, Bruce, Michael C. Malin and Ronald Greeley: *Earthlike Planets*, W.H. Freeman and Co., San Francisco, 1981.
This book emphasizes the differences and similarities of Mercury, Venus, the Earth, the Moon and Mars; it provides an excellent description of impact cratering and volcanism on the surfaces of the Moon, Mercury and Mars.

Preiss, Byron (editor): *The Planets*, Bantam Books, New York, 1985.
Scientific essays about each of the planets are followed by fictional speculation about future life on them. The contributors to this stimulating combination are well-known planetary scientists and eminent science-fiction writers.

Sagan, Carl: *Cosmos*, Random House, 1980.
A highly personal, best-selling introduction to contemporary astronomy woven into a set of themes that include the space-age exploration of the planets, planetary evolution, stellar evolution, the evolution of intelligence, and the possibilities of life elsewhere in the universe.

Shipman, Harry L.: *Space 2000 – Meeting the Challenge of a New Era*, Plenum Press, New York, 1987.
A comprehensive assessment of the U.S. space program, including its past achievements and future prospects. Communication and weather satellites, planetary exploration, and the search for extraterrestrial life are included. Space activities in the next century might include permanent stations in space or on the Moon, and human settlements on Mars.

Smoluchowski, Roman: *The Solar System*, W.H. Freeman, New York, 1984.

A clear, informative popular-level glance at the Sun, Moon, planets and the possibility of extraterrestrial life, but a short account that does not add much beyond other general books about the solar system.

Representative books, articles and technical papers arranged by chapters

Chapter 1. Worlds in motion

Drake, Stillman: *Discoveries and Opinions of Galileo*, Doubleday Anchor Books, New York, 1957.
A translation of Galileo's *The Starry Messenger* and three of his letters dealing with astronomical themes.

Galilei, Galileo: *Dialogue Concerning the Two Chief World Systems – Ptolemaic and Copernican*, translated by Stillman Drake, University of California Press, Berkeley, 1970.
Astronomical and philosophical arguments are presented for and against the motion of the Earth. A foreword by Albert Einstein relates Galileo's work to that of his predecessors and to our own scientific age.

Koestler, Arthur: *The Sleepwalkers*, Grosset and Dunlap, New York, 1973.
A classic account of the great astronomers from the early Greeks through Kepler and Galileo, and of the history of man's changing vision of the universe.

Whitney, Charles A.: *Whitney's Star Finder – A Field Guide to the Heavens*, Alfred A. Knopf, New York, 1989.
An account of back-yard astronomy including observations of eclipses, the Moon and the planets.

Will, Clifford M.: "Gravitation theory," *Scientific American* **231**, 25–33 (1974) – May.
A variety of experimental tests confirm Einstein's new theory of gravity.

Chapter 2. The Moon: stepping stone to the planets

Beatty, J. Kelly: "The making of a better Moon," *Sky and Telescope* **72**, 558–9 (1986) – December.
Various theories for the origin of the Moon are presented.

Cadogan, Peter: *The Moon – Our Sister Planet*, Cambridge University Press, New York, 1981.
A somewhat technical account of the Moon that is partially based on Stuart Ross Taylor's book on lunar science.

French, Bevan M.: *The Moon Book*, Penguin Books, New York, 1977.
A very readable non-technical account with many scientific results, but it does not go beyond Apollo 12.

Goldreich, Peter: "Tides and the Earth–Moon system," *Scientific American* **226**, 42–52 (1972) – April.
Tidal friction causes the length of the day to increase and the Moon to recede from the Earth.

Hartmann, William K.: "Cratering in the solar system," *Scientific American* **236**, 84–99 (1977) – January.
A heavy bombardment in the inner solar system was associated with the formation of the Moon, Mercury and Mars, but cratering by meteoritic impact continues at a reduced rate today.

Kitt, Michael T.: "Sculpting the Moon," *Astronomy* **15**, 82–7 (1986) – February.
Excavated by violent impacts, layered with flows of lava, and softened by a relentless rain of meteorites, the lunar surface displays a geological past over 4 billion years.

Kitt, Michael T.: "Eight lunar wonders," *Astronomy* **17**, 66–71 (1989) – March.

How to use a small telescope to investigate eight geological oddities on the Moon and ponder the mysteries of their formation.

Moore, Patrick: *The Moon*, Rand McNally, New York, 1981.

A well-written historical account and scientific discussion precedes colorful photographs and detailed maps of the lunar surface.

Price, Fred W.: *The Moon Observer's Handbook*, Cambridge University Press, New York, 1989.

An excellent guide for observing the Moon's surface features, including craters, maria, mountains and valleys.

Robinson, Leif J.: "The big one is coming!" *Sky and Telescope* **77**, 134–9 (1989) – February.

A description of the total solar eclipse that will be viewed from Hawaii, Mexico and Brazil on July 11, 1991.

Rubin, Alan E.: "Whence came the Moon?" *Sky and Telescope* **68**, 389–93 (1984) – November.

A fine summary of the competing theories for the origin of the Moon. There are attractive aspects and obstacles to each of the three main theories.

Runcorn, S.K.: "The Moon's ancient magnetism," *Scientific American* **257**, 60–8 (1987) – December.

The Moon is now a dead body, but it seems once to have generated its own magnetic field. Since then it has been shifted with respect to its spin axis – perhaps by collisions with Moon-orbiting satellites.

Taylor, Stuart Ross: *Planetary Science – A Lunar Perspective*, Lunar and Planetary Institute, Houston, 1982.

Although scientific details and technical terms keep the reader from obtaining a general perspective, all aspects of lunar science are treated in a remarkably comprehensive and thorough way.

Wahr, John: "The Earth's inconstant rotation," *Sky and Telescope* **71**, 545–9 (1986) – June.

A good review of recent measurements of changes in the length of the Earth's day and polar wandering.

More technical works

Apollo missions to the Moon

The preliminary examinations of the rocks returned from the Moon by Apollo 11, 12, 14, 15, 16 and 17 are respectively discussed in *Science* **165**, 1211–27 (1969), **167**, 1325–47 (1970), **173**, 681–93 (1971), **175**, 363–75 (1972), **179**, 23–34 (1973), and **182**, 659–71 (1973). Reviews of lunar science based upon the Apollo results are given by D.S. Burnett, "Lunar science: the Apollo legacy," *Review of Geophysics and Space Physics* **13**, 13–34 (1975), Farouk El-Baz, "The Moon after Apollo," *Icarus* **25**, 495–537 (1975), and James W. Head, "Lunar volcanism in space and time," *Review of Geophysics and Space Physics* **14**, 265–300 (1976). Science reports for the Apollo missions are also given in *NASA reports* **SP-214** (1969), **SP-235** (1970), **SP-272** (1971), **SP-289** (1972), **SP-315** (1972) and **SP-330** (1973), Superintendent of Documents, U.S. Government Printing Office, Washington D.C.

Boss, Alan P.: "The origin of the Moon," *Science* **231**, 341–5 (1986).

This article discusses the theory that the Moon formed after a Mars-sized body impacted the proto-Earth.

Hartmann, William K., R.J. Phillips and G.J. Taylor (editors): *Origin of the Moon*, Lunar and Planetary Institute, Houston, 1986.

A collection of 33 papers that discuss both the historical background to lunar origin theories and the dynamical, geochemical, and geophysical constraints on these theories.

Massey, Sir Harry *et al.* (editors): "The Moon – a new appraisal from space missions and laboratory analysis," *Philosophical Transactions of the Royal Society (London)* **A285**, 1–606 (1977).

A collection of 66 technical papers discuss the composition, evolution and interior of the Moon.

Proceedings of the Lunar and Planetary Science Conferences

Book-length collections of technical articles published between 1970 and the 1980s.

Chapter 3. Mercury: a battered world

Davies, Merton E., Stephen E. Dwornik, Donald E. Gault and Robert G. Strom: *Atlas of Mercury – NASA SP-423*, U.S. Government Printing Office, Washington, D.C., 1978.

A pictorial summary of the Mariner 10 photographs of Mercury.

Dunne, James A. and Eric Burgess: *The Voyage of Mariner 10 – NASA SP-424*, U.S. Government Printing Office, Washington, D.C., 1978.

A well-illustrated description of the Mariner 10 Venus–Mercury mission.

Gault, Donald E., Joseph A. Burns, Patrick Cassen and Robert G. Strom: "Mercury," *Annual Reviews of Astronomy and Astrophysics* **15**, 97–126 (1977).

A post Mariner 10 account of Mercury's size, mass, orbit, rotation, atmosphere, magnetic field, magnetosphere, surface features, interior and geologic history.

Kunzig, Robert: "Iron planet," *Discover* **10**, 66–9 (1989) – February.

Tiny Mercury may once have been blindsided by a Moon-size object and blown apart. In the early days of the solar system, that sort of thing may have happened a lot.

Murray, Bruce C.: "Mercury," *Scientific American* **233**, 59–68 (1975) – September.

A discussion of Mercury's Moon-like surface, Earth-like interior, and magnetic field.

Stewart, Glen R.: "A violent birth for Mercury," *Nature* **335**, 496–7 (1988).

One way to produce an iron-rich, high-density body like Mercury is to strip the mantle from a larger, differentiated planet by a catastrophic collision with another slightly smaller one.

Strom, Robert G.: *Mercury – The Elusive Planet*, Smithsonian Institution Press, Washington, D.C., 1987.

This entertaining book includes a history of Earth-based observations of Mercury, an account of the flight of Mariner 10, a description of the planet's surface features and physical properties, and an explanation for its likely origin and geological evolution.

More technical works

Mariner 10 mission to Mercury.

The preliminary scientific results of the Mariner 10 encounters with Mercury may be found in *Science* **185**, 141–80 (1974). More thorough discussions of these results are given in the *Journal of Geophysical Research* **80**, 2341–2514 (1975) and *Icarus* **28**, 429–609 (1976). Also see Robert G. Strom: "Mercury: a post Mariner 10 assessment," *Space Science Reviews* **24**, 3–70 (1979).

Vilas, Faith, Clark R. Chapman and Mildred S. Matthews (editors): *Mercury*, University of Arizona Press, Tucson, 1988.

Twenty-three technical articles take into account findings from the Mariner 10 spacecraft and Earth-based observations. Topics include Mercury's atmosphere, craters, evolution, geology, plains, magnetic field, magnetosphere, metal-rich composition, slow rotation, surface composition, and tectonic history.

Chapter 4. Venus: the veiled planet

Bazilevskiy, Aleksandr T.: "The planet next door," *Sky and Telescope* **77**, 360–8 (1989) – April.

An excellent description of Venus' topographic details revealed by the Pioneer Venus orbiter and the Venera 15 and 16 spacecraft, including the plains of Venus, their age, Venus' mountains, and its volcanic and tectonic zones.

Burgess, Eric: *Venus – An Errant Twin*, Columbia University Press, New York, 1985.

A simple, non-scientific look at Venus.

Head, James W., Sandra E. Yuter and Sean C. Solomon: "Topography of Venus and Earth – a test for the presence of plate tectonics," *American Scientist* **69**, 614–23 (1981).

Radar topography indicates that Venus has many landforms that are different from those on Mars or Mercury and these landforms might be due to Earth-like plate tectonics.

Head, James W. and L.S. Crumpler: "Evidence for divergent plate boundary characteristics and crustal spreading – Aphrodite Terra, Venus," *Science* **238**, 1380–5 (1987).

This paper discusses similarities between Aphrodite Terra, Venus and terrestrial fracture zones and mid-ocean spreading centers.

Pettengill, Gordon H., Donald B. Campbell, and Harold Masursky: "The surface of Venus," *Scientific American* **243**, 54–65 (1980) – August.

Radar probes from the Earth and from the Pioneer Venus Orbiter indicate that Venus has an exceptionally flat surface, but that elevated plateaus and volcanoes also occur.

Pieters, Carle M. *et al.*: "The color of the surface of Venus," *Science* **234**, 1379–83 (1986).

Computer processing of Venera 13 images indicate that the surface of Venus is dark without significant color at visible wavelengths, but the high reflectivity at infrared wavelengths suggests oxidized surface material.

Prinn, Ronald G.: "The volcanoes and clouds of Venus," *Scientific American* **252**, 46–53 (1985) – March.

Radar maps of Venus and chemical analysis of its atmosphere and crust imply the existence of active volcanoes. The sulfur gases they release form a global cover of sulfuric acid clouds.

Rasool, S. Ichtiaque: "Venus, star of sweet confidences," *Natural History* **78**, 52–7 (1969) – June/July.

A fine discussion of why the Earth became a prolific haven for life while Venus became a sterile, lifeless world.

Schubert, Gerald and Curt Covey: "The atmosphere of Venus," *Scientific American* **245**, 66–74 (1981) – July.

This article emphasizes the circulation of Venus' atmosphere.

Taylor, Harry A. and Paul A. Cloutier: "Venus, dead or alive?" *Science* **234**, 1087–93 (1986).

Pioneer Venus Orbiter observations indicate that certain electrical events occur in the planet's ionosphere. This very controversial paper suggests that they are not due to lightning and that they do not suggest currently active volcanoes.

Wall, Stephen D.: "Venus unveiled," *Astronomy* **17**, 26–32 (1989) – April.

Piercing Venus' cloud cover with radar, the orbiting spacecraft Magellan will produce the most detailed map yet of the planet's mountains and valleys. It will take pictures over at least 70 percent of the surface with enough resolution to identify objects as small as half a mile across.

Young, Andrew and Louise Young: "Venus," *Scientific American* **233**, 71–8 (1975) – September.

A well-written account of the sulfuric acid clouds, the high surface temperature, the greenhouse effect, and the origin of Venus' massive carbon dioxide atmosphere.

More technical works

Crumpler, L.S., James W. Head and Donald B. Campbell: "Orogenic belts on Venus," *Geology* **14**, 1031–4 (1986).

This paper presents evidence that the mountains around Lakshmi Planum mark the location of concentrated horizontal compressional deformation.

Hunten, Donald M. *et al.* (editors): *Venus*, University of Arizona Press, Tucson, 1983.

An excellent collection of technical papers and major reviews that describe American and Soviet investigations of the atmosphere and surface of Venus.

Mariner missions to Venus

The preliminary scientific results of the Mariner 2, 5 and 10 encounters with Venus are respectively found in *Science* **139**, 900–11 (1963), **158**, 1665–90 (1967) and **183**, 1289–320 (1974). Additional details may be found in *Mariner-Venus 1962 NASA SP-59*, *Mariner-Venus 1967 NASA SP-190* and *The Voyage of Mariner 10 NASA SP-424*, U.S. Government Printing Office, Washington, D.C.

Pioneer missions to Venus

The preliminary results of the two Pioneer Venus missions are presented in *Nature* **279**, 577, 613–20 (1979) and *Science* **203**, 743–808, **205**, 41–121 (1979). More detailed scientific results can be found in the *Journal of Geophysical Research* **85**, 7575–8337 (1980) and *Icarus* **51**, 167–461, **52**, 209–365 (1982).

Vega balloon mission to Venus

The balloon results for the composition, dynamics, structure and winds of the clouds and atmosphere of Venus are given in *Soviet Astronomy Letters* **12**, 1–176 (1986) – January/February. The tracking of the balloons from a network of observing stations on Earth and the resulting meteorologic and atmospheric data are given in *Science* **231**, 1349, 1369, 1407–25 (1986).

Venera missions to Venus

Summary articles on the Soviet series of Venus entry probes, Venera 4–8, are given in the *Journal of Atmospheric Science* **25**, 533–4 (1968), **27**, 561–79 (1970), **28**, 263–4 (1971), **30**, 1210–14 (1973), **30**, 1215–18 (1973) and *Icarus* **20**, 407–21 (1973). Results of the Venera 9 and 10 orbiter/lander missions are discussed in *Cosmic Research* **14**, no. 5, 573–701 (1976) and *Icarus* **30**, 605–25 (1977).

Chapter 5. The restless Earth

Anderson, Don L.: "Where on Earth is the crust?" *Physics Today* **42**, 38–46 (1989) – March.

Mineralogical data and comparative planetology suggest a controversial hypothesis – that much of the Earth's crust may be buried within the mantle.

Anderson, Don L. and Adam M. Dziewonski: "Seismic tomography," *Scientific American* **251**, 60–8 (1984) – October.

By analyzing many earthquake waves with this technique, it is now possible to map the Earth's mantle in three dimensions. The maps throw light on the convective flow that propels the crustal plates.

Ballard, Robert D.: *Exploring Our Living Planet*, National Geographic Society, Washington, D.C., 1983.

This beautifully illustrated description of the changing Earth includes the birth of oceans, mountains beneath the sea, volcanic hotspots, grinding plates, the ring of fire, and mountain building by colliding plates.

Berner, Robert A. and Antonio C. Lasaga: "Modeling the geochemical carbon cycle," *Scientific American* **260**, 74–81 (1989) – March.

Natural geochemical processes that result in the slow buildup of atmospheric carbon dioxide may have caused past geologic intervals of global warming through the greenhouse effect.

Bonatti, Enrico: "The rifting of the continents," *Scientific American* **256**, 96–103 (1987) – March.

It begins above a hot zone in the mantle. Upwelling molten rock "underplates" and weakens the continental crust, piercing it at discrete points and finally rifting it in two. An ocean is born.

Burchfield, B. Clark: "The continental crust," *Scientific American* **249**, 130–41 (1983) – September.

An excellent account of the way in which the continental and ocean crust are respectively created at converging and diverging plate boundaries.

Burke, Kevin C. and J. Tuzo Wilson: "Hot spots on the Earth's surface," *Scientific American* **235**, 46–57 (1976) – August.
Regions of isolated volcanic activity, called hot spots, are anchored deep within the Earth's interior; they create chains of volcanic islands and help fracture continents.

Carter, William E. and Douglas S. Robertson: "Studying the Earth by very-long-baseline interferometry," Scientific American **255**, 46–54 (1986) – November.
In which radio signals emitted by quasars billions of light-years away serve as benchmarks for measuring the Earth's wobble, tiny changes in its spin rate and the imperceptible drift of its plates.

Cattermole, Peter and Patrick Moore: *The Story of the Earth*, Cambridge University Press, New York, 1985.
A clear, accessible and beautifully illustrated account of the evolution of the Earth from its beginning in the primeval solar nebula, through its bombardment by cosmic particles, continental drifting and the formation of mountains and oceans, the ice ages, the rise of man and his influence on the planet.

Covey, Curt: "The Earth's orbit and the ice ages," *Scientific American* **250**, 58–66 (1984) – February.
Periodic variations in the geometry of the Earth's orbit have long been considered a possible cause of the ice ages. The idea is now supported by a more reliable chronology of glaciations.

Dewey, John F.: "Plate tectonics," *Scientific American* **226**, 56–68 (1972).
The Earth's surface is divided into a mosaic of moving plates that are related to sea-floor spreading, continental drift, earthquakes, volcanoes, and the creation of mountains and oceans.

Francis, Peter and Stephen Self: "Collapsing volcanoes," *Scientific American* **256**, 91–7 (1987) – June.
In the life cycle of many volcanoes a catastrophic collapse is a "normal" event. The details of the process are revealed in the deposits left by the devastating avalanches of debris.

Frohlich, Cliff: "Deep earthquakes," *Scientific American* **260**, 48–55 (1989) – January.
They have posed a fruitful puzzle since their discovery 60 years ago. How can rock fail at the temperatures and pressures that prevail hundreds of kilometers down?

Garwin, Laura: "Of impacts and volcanoes," *Nature* **336**, 714–16 (1988) – December.
Recent compelling evidence indicates at least one large-body impact wiped out the dinosaurs.

Gordon, Arnold L. and Josefino C. Comiso: "Polynyas in the Southern Ocean," *Scientific American* **258**, 90–7 (1988) – June.
They are vast gaps in the sea ice around Antarctica. By exposing enormous areas of seawater to frigid air, they help to drive the global heat engine that couples the ocean and the atmosphere.

Hekinian, Roger: "Undersea volcanoes," *Scientific American* **251**, 46–55 (1984) – July.
New techniques for exploring the ocean floor are yielding a remarkably detailed picture of the volcanic processes responsible for generating the bulk of the Earth's crust.

Heppenheimer, T.A.: "Journey to the center of the Earth," *Discover* **8**, 86–95 (1987) – November.
Geologists are now probing the Earth's deepest secrets. The center of its core seems to be hotter than the surface of the Sun.

Hoffman, Kenneth A.: "Ancient magnetic reversals – clues to the geodynamo," *Scientific American* **258**, 76–83 (1988) – May.
Is the Earth headed for a reversal of its magnetic field? No one can answer this question yet, but rocks magnetized by ancient fields offer clues to the underlying reversal mechanism in the Earth's core.

Hones, Edward W.: "The Earth's magnetotail," *Scientific American* **254**, 40–7 (1986) – March.
The solar wind sweeps the Earth's magnetic field into a vast tail. Disruptions of the tail generate bright auroras at the Earth and propel great bodies of magnetized gas into interplanetary space.

Houghton, Richard A. and George M. Woodwell, "Global climatic change," *Scientific American* **260**, 36–44 (1989) – April.
Evidence suggests that production of carbon dioxide and methane from human activities has already begun to change the climate, and that radical steps must be taken to halt any further change.

Ingersoll, Andrew P.: "The atmosphere," *Scientific American* **249**, 162–74 (1983) – September.
This article describes the climate, composition and weather of the Earth's dynamic Sun-driven atmosphere.

Jeanloz, Raymond: "The Earth's core," *Scientific American* **249**, 56–66 (1983) – September.
Motions in the outer liquid core generate the Earth's magnetic field.

Jordan, Thomas H. and J. Bernard Minster: "Measuring crustal deformation in the American West," *Scientific American* **259**, 48–58 (1988) – August.
Continental crust is actively deforming as the Pacific and North America plates slide past each other. Direct measurements of the process rely on extraterrestrial reference points such as quasars.

Lanzerotti, Louis J. and Chanchal Uberoi: "Earth's magnetic environment," *Sky and Telescope* **76**, 360–2 (1988) – October.
An up-to-date report on Earth's magnetosphere, magnetotail, Van Allen belts, auroras, whistlers, micropulsations and practical effects of the magnetosphere.

Marvin, Ursula B.: *Continental Drift – The Evolution of a Concept*, Smithsonian Institution Press, Washington, D.C., 1973.
A fine book that traces the historical development of the concepts of continental drift, sea-floor spreading and plate tectonics.

McKenzie, D.C.: "The Earth's mantle," *Scientific American* **249**, 67–78 (1983) – September.
Internal heat energizes massive convection currents that drive the Earth's plates along like a conveyor belt.

Mohnen, Volker A.: "The challenge of acid rain," *Scientific American* **259**, 30–8 (1988) – August.
Acid rain's effects in soil and water leave no doubt about the need to control its causes. Now advances in technology have yielded environmentally and economically attractive solutions.

Molnar, Peter: "The structure of mountain ranges," *Scientific American* **255**, 70–9 (1986) – July.
What holds mountains up? Some stand on plates of strong rock; others are buoyed by crustal roots reaching deep into the mantle. The latter may collapse when their flanks are not pushed together.

Muecke, Gunter K. and Peter Moller: "The not-so-rare earths," *Scientific American* **258**, 72–7 (1988) – January.
The rare-earth elements, on which electronic, metallurgical and glass industries depend, are not all that scarce in minerals. The elemental abundances reveal the geochemistry that leads to a mineral's formation.

Mutter, John C.: "Seismic images of plate boundaries," *Scientific American* **254**, 66–75 (1986) – February.
By bouncing sound off rock layers under the sea floor and recording the reflections with many detectors, structural images of the crust can be made at the boundaries where plates collide and rift apart.

Nance, R. Damian, Thomas R. Worsley and Judith B. Moody: "The supercontinent cycle," *Scientific American* **259**, 72–9 (1988) – July.
Several times in Earth history the continents have joined to form one body, which later broke apart. The process seems to be cyclic; it may shape geology and climate and thereby influence biological evolution.

Parker, Eugene N.: "Magnetic fields in the cosmos," *Scientific American* **249**, 44–54 (1983) – August.
The dynamo mechanism that generates the Earth's changing magnetic field can also explain magnetic fields in the Sun and the Galaxy.

Revkin, Andrew C.: "Endless summer – living with the greenhouse effect," *Discover* **9**, 50–61 (1988) – October.
Global warming has begun, and we had better start preparing for the dramatic changes to come.

Schneider, Stephen H.: "Climate modeling," *Scientific American* **256**, 72–80 (1987) – May.
Will the "greenhouse effect" bring on another dust bowl? Would nuclear war mean "nuclear winter"? Computer models of the Earth's climate yield clues to its future as well as to its checkered past.

Sclater, John G. and Christopher Tapscott: "The history of the Atlantic," *Scientific American* **240**, 156–74 (1979) – June.
The Atlantic Ocean was created about 165 million years ago when the continents that now border it were joined together.

Stolarski, Richard S.: "The Antarctic ozone hole," *Scientific American* **258**, 30–6 (1988) – January.
Each spring for the past decade the ozone layer in the atmosphere has thinned at the South Pole. Is the loss an anomaly, or is it a sign that the ultraviolet-absorbing layer is in jeopardy globally?

Taubes, Gary: "Made in the shade? No way," *Discover* **8**, 62–71 (1987) – August.
Whatever the implications of the Antarctic ozone hole are, the ozone layer is being ominously depleted – and Sun hats aren't the solution.

Toon, Owen B. and Steve Olson: "The warm Earth," *Science* **85**, 50–7 (1985) – October.
Our planet didn't boil like Venus and it didn't freeze like Mars. This popular article discusses what has kept the Earth liveable.

Van Andel, Tjeerd H.: *New Views on an Old Planet – Continental Drift and the History of the Earth*, Cambridge University Press, New York, 1985.
A geologist's history of the dynamic, ever-changing Earth, including drifting continents, ancient oceans, mountain building, changing climates and the evolution of life on Earth.

Vink, Gregory E., W. Jason Morgan and Peter R. Vogt: "The Earth's hot spots," *Scientific American* **252**, 50–7 (1985) – April.
These plumes of hot rock welling up from deep in the mantle are a key link in the plate-tectonic cycle. The marks they leave on passing plates include volcanoes, swells and midocean plateaus.

Weiner, Jonathan: *Planet Earth*, Bantam Books, New York, 1986.
This companion volume to a PBS television series about our home planet focuses on topics that tell how the Earth works, including plate tectonics, sea-floor mountains with exotic inhabitants, climatic change, comparisons with the other planets, the solar connection, the Earth's resources, and the origin, evolution, and fate of life on Earth.

Wesson, Robert L. and Robert E. Wallace: "Predicting the next great earthquake in California," *Scientific American* **252**, 35–43 (1985) – February.
On part of the San Andreas fault an earthquake of magnitude 8 in the next 30 years is assigned

a probability of 50 percent. Precise predictions await better knowledge of how quakes are triggered.

Chapter 6. Mars: the red desert

Carr, Michael H.: "The volcanoes of Mars," *Scientific American* **234**, 32–43 (1976) – January.
A good discussion of the old and young volcanic features on Mars.

Carr, Michael H.: *The Surface of Mars*, Yale University Press, New Haven, 1981.
An excellent, well-illustrated overview of nearly everything that is known about Mars.

Carr, Michael H.: "Water on Mars," *Nature* **326**, 30–5 (1987).
Estimates for the amounts of water at the surface of Mars and in its atmosphere can be reconciled if the planet lost part of its atmosphere.

Carroll, Michael: "The changing face of Mars," *Astronomy* **15**, 8–22 (1987) – March.
Evidence for water on Mars is changing our perceptions of that lifeless, desert world.

Chandler, David L.: *Life on Mars*, E.P. Dutton, New York, 1979.
A good discussion of the probable origin of life on Earth, the possibility of microbes surviving the harsh Martian environment, and our current knowledge of the surface and atmosphere of Mars.

Cooper, Henry S.F.: *The Search for Life on Mars – Evolution of an Idea*, Holt, Rinehart and Winston, New York, 1980.
Beginning with a revealing profile of Carl Sagan, the author places the search for life on Mars within a personal context that is designed for the layman.

Haberle, Robert M.: "The climate of Mars," *Scientific American* **254**, 54–62 (1986) – May.
The climate of Mars may have started out much like the Earth's early climate, but it evolved differently. Once warm enough to support flowing water, Mars is now so cold that carbon dioxide freezes at the poles every winter.

Horowitz, Norman H.: *To Utopia and Back – The Search for Life in the Solar System*, W.H. Freeman, New York, 1986.
A lucid discussion of the possibility of life in the solar system written by one of the participants in the Viking search for life on Mars. The author concludes that our small planet is unique, and that the Earth is the only life-bearing planet in the solar system.

Horowitz, Norman H.: "The search for life on Mars," *Scientific American* **237**, 52–61 (1977) – November.
A good review of the unsuccessful tests for life on Mars written by one of the scientists involved.

Leovy, Conway B.: "The atmosphere of Mars," *Scientific American* **237**, 34–43 (1977) – July.
An excellent summary of the composition, pressure, temperature, variations and winds of the Martian atmosphere.

Kasting, James F., Owen B. Toon and James B. Pollack: "How climate evolved on the terrestrial planets," *Scientific American* **258**, 90–7 (1988) – February.
Why is Mars too cold for life, Venus too hot and the Earth just right? Distance from the Sun is not the full answer. The Earth cycles carbon dioxide into its atmosphere to promote greenhouse warming when its surface cools. Mars lacks that ability. Venus has the opposite problem; it cannot drain carbon dioxide from its atmosphere. Planets with temperate, Earth-like climates were once thought to be rare in our Galaxy. Mathematical models now suggest that if planets do exist outside the solar system, many of them might be habitable.

Mariner 9 television and science teams: *Mars as Viewed by Mariner 9 – NASA SP-329*, U.S. Government Printing Office, Washington, D.C., 1974.

Schultz, Peter H.: "Polar wandering on Mars," *Scientific American* **253**, 94–102 (1985) – December.

Regions at the planet's equator seem once to have been near a pole; possibly because the entire lithosphere has shifted in relation to the axis of spin. This theory explains many puzzling features and processes.

Viking lander imaging team: *The Martian Landscape* – NASA SP-425, U.S. Government Printing Office, Washington, D.C., 1978.

Viking orbiter imaging team: *Viking Orbiter Views of Mars* – NASA SP-441, U.S. Government Printing Office, Washington, D.C., 1980.

More technical works

Baker, Victor R.: *The Channels of Mars*, University of Texas Press, Austin, Texas, 1982.

A well-written, beautifully illustrated discussion of water-related processes on Mars.

Mariner missions to Mars

The preliminary scientific results for the Mariner 4, 6, 7, and 9 encounters with Mars are respectively given in *Science* **149**, 1226–48 (1965), **166**, 49–68 (1969) and **175**, 293–323 (1972). More thorough discussions of the discoveries of the Mariner 9 orbiter can be found in the *Journal of Geophysical Research* **78**, 4009–440 (1973). A number of papers resulting from the Mariner 9 and the Soviet Mars 3 are collected in *Icarus* **17**, 289–327 (1972).

Symposium on Mars: *Evolution of Its Climate and Atmosphere*, Lunar and Planetary Institute, Houston, Texas, 1986.

Synder, Conway W.: "The planet Mars as seen at the end of the Viking mission," *Journal of Geophysical Research* **84**, 8487–519 (1979) – December.

Viking missions to Mars

The preliminary scientific results of the Viking missions to Mars are given in *Science* **193**, 759–815 and **194**, 57–104 (1976). More detailed discussions are given in *Science* **194**, 1274–353 (1976) and the *Journal of Geophysical Research* **82**, 3959–4681 (1977), **84**, 2793–3007 (1979), and **84**, 7909–8519 (1979). A special issue devoted to dust on Mars is in *Icarus* **66**, 1–142 (1986).

Chapter 7. *Asteroids, meteors and meteorites*

Alvarez, Luis W.: "Mass extinctions caused by large bolide impacts," *Physics Today* **40**, 24–33 (1987) – July.

Evidence indicates that the collision of Earth and a large piece of solar system debris such as a meteoroid, asteroid or comet caused the great extinctions of 65 million years ago, leading to the transition from the age of the dinosaurs to the age of the mammals.

Asimov, Isaac: *A Choice of Catastrophes – The Disasters that Threaten Our World*, Simon and Schuster, New York, 1979.

A novel compilation of world-threatening disasters such as the dying Sun, drifting continents, ice ages, changing magnetism, diseases, wars, resource depletion, population growth and the bombardment of Earth by asteroids, comets and meteorites.

Burke, John G.: *Cosmic Debris – Meteorites In History*, University of California Press, Berkeley, 1986.

A thorough, scholarly history of meteorite studies from Aristotle to the present. An excellent source for references and changing theories for meteorite origins. The growth of large collections, folklore and myths are also documented.

Cameron, I.R.: "Meteorites and cosmic radiation," *Scientific American* **229**, 64–73 (1973) – July.

Meteorites have spent much of their lives embedded within asteroidal-sized parent bodies.

Chapman, Clark R.: "The nature of asteroids," *Scientific American* **232**, 24–33 (1975) – January.

The surface compositions of asteroids are inferred from measurements of their colors and absorption features. Mechanisms for gravitationally shuffling meteorites and asteroids into Earth-crossing orbits are also discussed.

Cunningham, Clifford J.: *Introduction to Asteroids – The Next Frontier*, Willmann-Bell, Inc., Richmond, Va., 1988.

An up-to-date account of the diverse aspects of asteroid study, including a historical background, observing methods, orbital characteristics, composition, rotation, sizes, shapes, families, satellites, Apollos, Amors, Trojans, mass extinctions, infrared detections and future space missions to the asteroids.

Dodd, Robert T.: *Thunderstones and Shooting Stars*, Harvard University Press, Cambridge, Mass., 1986.

A good summary of what we know and think about meteorites, including their clues to the origin of our solar system and their possible disruption of life on Earth.

Goldsmith, Donald: *Nemesis*, Walker and Co., New York, 1985.

A hypothetical death star, nemesis, might produce periodic comet showers in our solar system every 26 million years; other possible extraterrestrial causes of mass extinction are also discussed.

Gould, Stephen Jay: "The belt of an asteroid," *Natural History* **89**, 26–33 (1980).

An excellent article about mass extinctions written by an eminent biologist who advocates an extraterrestrial cause for the extinction of the dinosaurs.

Grossman, Lawrence: "The most primitive objects in the solar system," *Scientific American* **232**, 30–8 (1975) – February.

The class of meteorites called carbonaceous chondrites may be the least altered, or most primitive, objects in the solar system.

Hartmann, William K.: "The smaller bodies of the solar system," *Scientific American* **233**, 143–59 (1975) – September.

Many of the smaller bodies of the solar system have a common origin as the fragments of collisions between growing planetesimals.

Hutchinson, Robert: *The Search for Our Beginning*, Oxford University Press, Oxford, 1983.

An enquiry, based on meteorite research, into the origin of our planet and of life.

Kowal, Charles T.: *Asteroids – Their Nature and Utilization*, John Wiley & Sons, New York, 1988.

This short account highlights special topics like asteroids near the Earth, mining the asteroids, and the future of asteroid research. An appendix gives the orbital and physical data for 3445 numbered asteroids.

Marvin, Ursula B.: "Extraterrestrials have landed on Antarctica," *New Scientist* 710–15 (1983) – March.

An excellent, succinct description of unusual meteorites found in Antarctica, particularly those coming from the Moon and Mars.

McSween, Harry Y., Jr.: *Meteorites and Their Parent Planets*, Cambridge University Press, New York, 1987.

Meteorites are traced back to their origin in asteroidal parent bodies. Their chemical and physical properties contain information about the formation of our solar system. Meteorites are also placed in the overall scheme of space science.

McSween, Harry Y. and Edward M. Stolper: "Basaltic meteorites," *Scientific American* **242**, 54–63 (1980) – June.

A discussion of the rare meteorites that contain volcanic rock is placed within the larger context of volcanism within the solar system.

Morrison, David: "Asteroids," *Astronomy* **4**, 6–17 (1976) – June.

A very good summary that combines a capsule history of the discovery of asteroids with a review of their reflectivity and surface composition.

Russell, Dale A.: "The mass extinctions of the Late Mesozoic," *Scientific American* **246**, 58–65 (1982) – January.

A colliding asteroid wiped out many of the existing plants and animals, including the dinosaurs, at the end of the Cretaceous Age 65 million years ago.

Wasson, John T.: *Meteorites – Their Record of Early Solar-System History*, W.H. Freeman and Co., New York, 1985.

This text describes meteoritic evidence about the formation and early evolution of the solar system.

Wetherill, George W.: "Apollo objects," *Scientific American* **240**, 54–65 (1979) – March.

A comprehensive account of asteroids that cross the Earth's orbit, including their collision rate with the Earth.

More technical works

Chapman, Clark R., James G. Williams and William K. Hartmann, "The asteroids," *Annual Reviews of Astronomy and Astrophysics* **16**, 33–75 (1978).

A thorough review of observations, physical properties, rotations and surface compositions of the asteroids.

Delsemme, Armand H. (editor): *Comets, Asteroids, Meteorites – Interrelations, Evolution and Origins*, University of Toledo, Toledo, Ohio, 1977.

As the title suggests, this collection of 75 technical articles stresses the interrelations between asteroids, meteorites and comets; it contains good discussions about the physical nature and orbital evolution of asteroids.

Gehrels, Tom (editor): *Asteroids*, University of Arizona Press, Tucson, 1979.

Experts have contributed 45 articles on topics that include the composition, diameters, origin and evolution of asteroids; an extensive index of asteroid data is included.

Sears, D.W.: *The Nature and Origin of Meteorites*, Oxford University Press, Oxford, 1978.

This attractive volume begins with an interesting historical introduction; it also includes discussions of meteorite falls, terrestrial impact craters, and the classification, composition, physical properties, and origin of meteorites.

Chapter 8. Jupiter: a giant primitive world

Burgess, Eric: *By Jupiter – Odysseys to a Giant*, Columbia University Press, New York, 1982.

A non-technical account of the Pioneer 10 and 11 and Voyager 1 and 2 missions written for a general audience.

Hunt, Garry E. and Patrick Moore: *Jupiter*, Rand McNally, New York, 1981.

This concise, informative book includes a short historical introduction and thorough discussions of Jupiter's atmosphere, clouds, interior, magnetosphere, radiation and satellites.

Ingersoll, Andrew P.: "The meteorology of Jupiter," *Scientific American* **234**, 45–56 (1976) – March.

A good survey article describing the circulation, radiation, structure and temperature of the Jovian atmosphere.

Ingersoll, Andrew P.: "Jupiter and Saturn," *Scientific American* **245**, 90–108 (1981) – December.

Two competing models may describe the atmospheric circulation on Jupiter and Saturn.

Johnson, Torrence V. and Laurence A. Soderblom: "Io," *Scientific American* **249**, 56–67 (1983) – December.

A thorough review of volcanic activity on Io, including eruptive plumes, volcanic calderas, and lava lakes.

Johnson, Torrence V., *et al.*: "Volcanic hotspots on Io – stability and longitudinal distribution," *Science* **226**, 134–7 (1984).
A discussion of ground-based infrared observations of volcanic activity on Io.

Krimigis, Stamatios Mike: "A post-Voyager view of Jupiter's magnetosphere," *Endeavor* **5**, 50–60 (1981).
Jupiter's magnetosphere is compared to those of Mercury, Earth and Saturn.

Lanzerotti, Louis J.: "The planets' magnetic environment," *Sky and Telescope* **77**, 149–52 (1989) – February.
A discussion of the magnetic fields and magnetospheres of Jupiter, Saturn and Uranus.

Morrison, David: "The enigma called Io," *Sky and Telescope* **69**, 198–205 (1985) – March.
A good colorful account of volcanic activity on Io, also including Io's torus and its orbital resonance with Europa and Ganymede.

Morrison, David and Jane Samz: *Voyage to Jupiter – NASA SP-439*, National Aeronautics and Space Administration, Washington, D.C., 1981.
A well-illustrated, complete description of the Voyager 1 and 2 investigations of Jupiter.

Peek, Bertram M.: *The Planet Jupiter*, Faber and Faber, London, 1958.
A thorough description of telescopic observations of Jupiter before the late 1950s.

Pollack, James B. and Jeffrey N. Cuzzi: "Rings in the solar system," *Scientific American* **245**, 104–29 (1981) – November.
The similarities and differences of the rings of Saturn, Jupiter and Uranus are placed within the context of the complex forces that give them their shape.

Washburn, Mark: *Distant Encounters – The Exploration of Jupiter and Saturn*, Harcourt Brace Jovanovich, New York, 1983.
A fine story that places the scientific results of the Voyager missions to Jupiter and Saturn within the context of space-age politics, interstellar messages, scientific showmanship and flashy journalism.

More technical works
Burns, Joseph A. and Mildred Shapley Matthews (editors): *Satellites of the Solar System*, University of Arizona Press, Tucson, 1985.
A very good collection of technical articles written by experts in the field.

Gehrels, Tom (editor): *Jupiter*, University of Arizona Press, Tucson, 1976.
This collection of 44 technical articles provides a comprehensive summary of our post-Pioneer 10 and 11, but pre-Voyager, knowledge of the Jovian realm.

Morrison, David (editor): *Satellites of Jupiter*, University of Arizona Press, Tucson, 1982.
A post-Voyager summary of the Jovian satellites is given in a series of technical articles.

Pioneer missions to Jupiter
The preliminary results of the Pioneer 10 and 11 encounters with Jupiter can be found in *Science* **183**, 301–24 (1974) and *Science* **188**, 445–77 (1975), respectively. More thorough discussions of the respective missions can be found in the *Journal of Geophysical Research* **79**, 3522–64 (1974) and *Icarus* **30**, 97–274 (1977).

Voyager missions to Jupiter
The preliminary scientific results of the Voyager 1 encounter with Jupiter may be found in *Science* **204**, 945–1008 (1979) and *Nature* **280**, 725–806 (1979), while those of the Voyager 2 encounter are in *Science* **206**, 925–96 (1979). More thorough discussions of the respective missions can be found in *Geophysical Research Letters* **7**, 1–963 (1980) and the *Journal of Geophysical Research* **86**, 8123–841 (1981). Various aspects of the Voyager missions are discussed in *Space Science Reviews* **21**, 75–376 (1977); observations of the atmosphere and magnetosphere of Jupiter are treated in *Geophysical Research Letters* **7**, 1–68 (1980). The Jovian atmosphere is reviewed in *Icarus* **65**, 159–466 (1986), while discussions of the giant planet's satellites are

found in *Icarus* **44**, 225–547 (1980) and *Icarus* **58**, 135–329 (1984). The magnetic environment of Jupiter is discussed in *Physics of the Jovian Magnetosphere*, edited by A.J. Dessler, Cambridge University Press, New York, 1982.

Chapter 9. Saturn: lord of the rings

Cooper, Henry S.F.: *Imaging Saturn – The Voyager Flights to Saturn*, Holt, Rinehart and Winston, New York, 1982.
A dramatic moment-by-moment account of the Voyager encounters with Saturn written by a *New York Times* journalist who covered the events.

Cuzzi, Jeffrey N.: "Ringed planets – still mysterious," *Sky and Telescope* **68**, 511–15 (1984) – December, **69**, 19–23 (1985) – January.
A good description of resonance effects – including ring gaps and spiral density waves, shepherding moonlets, viscous instabilities, meteorite bombardment and the origin of planetary rings.

Elliot, James and Richard Kerr: *Rings – Discoveries from Galileo to Voyager*, The MIT Press, Cambridge, Mass., 1984.
Rings within rings in Saturn's realm and elsewhere.

Morrison, David: *Voyages to Saturn – NASA SP-451*, National Aeronautics and Space Administration, Washington, D.C., 1982.
A well-written, beautifully illustrated book about the two Voyager encounters with Saturn.

Owen, Tobias: "Titan," *Scientific American* **246**, 98–109 (1982) – February.
This excellent article describes the nitrogen-rich atmosphere on Titan.

Soderblom, Laurence A. and Torrence V. Johnson: "The moons of Saturn," *Scientific American* **246**, 100–16 (1982) – January.
The surfaces of Saturn's six medium-sized satellites display a surprising range of geological evolution.

Washburn, Mark: *Distant Encounters – The Exploration of Jupiter and Saturn*, Harcourt Brace Jovanovich, New York, 1983.
The scientific results of the Voyager missions to Jupiter and Saturn are placed within the context of space-age politics, scientific showmen and flashy journalism.

More technical works

Burns, Joseph A. and Mildred Shapley Matthews (editors): *Satellites of the Solar System*, University of Arizona Press, Tucson, 1985.
A very good collection of technical articles written by experts in the field.

Gehrels, Tom (editor): *Saturn*, University of Arizona Press, Tucson, 1983.
A technical summary of our space-age understanding of Saturn's atmosphere, composition, evolution, interior, magnetosphere, rings and satellites.

Pioneer mission to Saturn
The preliminary scientific results of the Pioneer 11 encounter with Saturn are given in *Science* **207**, 400–53 (1980). A more thorough discussion of these results may be found in the *Journal of Geophysical Research* **85**, 5651–6082 (1980).

Voyager missions to Saturn
The preliminary scientific results of the Voyager 1 encounter with Saturn may be found in *Science* **212**, 159–243 (1981) and in *Nature* **292**, 675–755 (1981), while those of the Voyager 2 encounter can be found in *Science* **215**, 499–594 (1982). A thorough discussion of the Voyager 1 and 2 technical results can be found in *Icarus* **53**, 163–387 (1983) and the *Journal of Geophysical Research* **88**, 8625–9018 (1983).

Chapter 10. Frozen worlds: Uranus, Neptune and Pluto

Beatty, J. Kelly: "A place called Uranus," *Sky and Telescope* **71**, 333–7 (1986) – April.
A fine description of the Voyager 2 encounter with Uranus including the planet's atmosphere, tilted magnetic field, electroglow and rings.

Beatty, J. Kelly: "Discovering Pluto's atmosphere," *Sky and Telescope* **76**, 624–30 (1988) – December.

Pluto's occultation of a star indicates that the distant planet has a cold, tenuous atmosphere of methane. Also see Hubbard *et al.*, *Nature* **336**, 452–4 (1988).

Burgess, Eric: *Uranus and Neptune – The Distant Giants*, Columbia University Press, New York, 1988.

A good description of our current knowledge about Uranus, its interior, its rings, and its satellites as gleaned from the Voyager 2 encounter. A short conclusion described expectations for the future Voyager approach to Neptune.

Chaikin, Andrew: "Voyager among the ice worlds," *Sky and Telescope* **71**, 338–43 (1986) – April.

Voyager 2 images of Uranus' five largest satellites reveal a striking variety of surface sculpture.

Croswell, Ken: "Pluto – enigma on the edge of the solar system," *Astronomy* **14**, 6–22 (1986) – July.

A good summary of our recent knowledge about Pluto and Charon.

Cuzzi, Jeffrey N. and Larry W. Esposito: "The rings of Uranus," *Scientific American* **257**, 52–66 (1987) – July.

Why are they so narrow and dark? Findings from the Voyager 2 encounter suggest that the austere ring system may be only a fleeting stage in a continuing saga of creation and destruction.

Drake, Stillman and Charles T. Kowal: "Galileo's sighting of Neptune," *Scientific American* **243**, 74–9 (1980).

An examination of Galileo's notebooks indicates that he observed Neptune in 1612 and 1613 when he noticed its apparent motion with respect to the stars.

Gold, Michael: "Voyager to the seventh planet," *Science 86*, **7**, 32–9 (1986) – May.

In less than six hours Voyager 2 examined Uranus' backward weather, unusual magnetic field and exotic moons.

Grosser, Morton: *The Discovery of Neptune*, Dover Publications, New York, 1979.

A good historical account of the prediction and discovery of Neptune.

Hoyt, William Graves: *Planets X and Pluto*, University of Arizona Press, Tucson, 1980.

This excellent book conveys the mystery and excitement of the discovery of Pluto.

Ingersoll, Andrew P.: "Uranus," *Scientific American* **256**, 38–45 (1987) – January.

The giant blue-green planet has one pole pointing toward the Sun, and its magnetic field is askew. Its atmosphere is dense and icy, yet its winds resemble the Earth's.

Johnson, Torrence V., Robert Hamilton Brown and Laurence A. Soderblom: "The moons of Uranus," *Scientific American* **256**, 48–60 (1987) – April.

Voyager 2 photographed the five major moons of Uranus at close range. All have icy surfaces, but they are darker and rockier than Saturn's moons. Early in their history three were geologically vigorous.

Laeser, Richard P., William I. McLaughlin and Donna M. Wolff: "Engineering Voyager 2's encounter with Uranus," *Scientific American* **255**, 36–44 (1986) – November.

Difficult problems posed by vast distances, low light levels, aging equipment and mechanical breakdowns were solved by radio control from the ground as the Voyager 2 spacecraft hurtled toward Uranus.

McKinnon, William B. and Steve Mueller: "Pluto's structure and composition suggest origin in the solar, not a planetary, nebula," *Nature* **335**, 240–3 (1988).

Pluto is not just another ice ball at the edge of the solar system, but instead a denser, rock-rich object.

Tombaugh, Clyde W. and Patrick Moore: *Out of the Darkness – The Planet Pluto*, Stackpole Books, Harrisburg, Pa., 1981.
A popular account of planetary discoveries including Uranus, the asteroids, Neptune, Pluto and Charon.

More technical works
Beebe, Reta F. and Herbert A. Beebe (editors): "Pluto – the ninth planet's golden year," *Icarus* **44**, 1–83 (1980).
A collection of papers presented at a scientific meeting in honor of the 50th anniversary of Pluto's discovery.

Goldreich, Peter and Scott Tremaine: "Towards a theory for the Uranian rings," *Nature* **277**, 97–9 (1979).
A theoretical prediction of small shepherding satellites that gravitationally control the Uranian rings, confining them within narrow lanes.

Hunt, Garry E. (editor): *Uranus and the Outer Planets*, Cambridge University Press, New York, 1982.
This book begins with a reproduction of William Herschel's original paper on Uranus' discovery; five historical articles are then followed by technical accounts of the planet's rotation, internal structure, composition, satellites and rings.

Jankowski, David G. and Steven W. Squyres: "Solid-state ice volcanism on the satellites of Uranus," *Science* **241**, 1322–5 (1988).
Flow features detected on the surfaces of the Uranian satellites Ariel and Miranda were detected in Voyager images; this technical account explains them in terms of solid-state ice volcanism.

Voyager mission to Uranus
The preliminary scientific results of the Voyager 2 encounter with Uranus are given in *Science* **233**, 39–109 (1986). A complete discussion of the technical results can be found in the *Journal of Geophysical Research* **92**, A13, 14873–15376 (1987).

Voyager mission to Neptune
The preliminary scientific results of the Voyager 2 encounter with Neptune are given in *Science*, **246**, 1417–1501 (1989).

Chapter 11. Comets: icy wanderers

Balsiger, Hans, Hugo Fechtig and Johannes Geiss: "A close look at Halley's comet," *Scientific American* **259**, 96–103 (1988) – September.
The armada of spacecraft that flew by the comet provided spectacular images and data that continue to yield quantitative information about the nature of the faithful visitor.

Beatty, J. Kelly: "An inside look at Halley's comet," *Sky and Telescope* **71**, 438–43 (1986) – May.
A summary of the Vega and Giotto investigations of the nucleus of Comet Halley.

Berry, Richard: "Search for the primitive," *Astronomy* **15**, 7–22 (1987) – June.
Comets are fluffy bodies made of recycled interstellar grains.

Berry, Richard and Richard Talcott: "What have we learned from Halley's comet?" *Astronomy* **14**, 6–22 (1986) – September.
New insights to comets have been provided by spacecraft encounters with Comet Halley.

Brandt, John C. (editor): *Comets – Readings from Scientific American*, W.H. Freeman, San Francisco, 1981.

A collection of articles about comets and related phenomena that have appeared in the magazine *Scientific American*.

Brandt, John C. and Robert D. Chapman: *Introduction to Comets*, Cambridge University Press, New York, 1981.
This is an excellent introduction to comets that will appeal to a wide range of readers including students, the educated layperson, and professional scientists.

Brandt, John C. and M.B. Niedner: "The structure of comet tails," *Scientific American* **254**, 49–56 (1986) – January.
A comet's plasma tail forms and disconnects in response to the solar wind and its magnetic field.

Brownlee, Donald E.: "Cosmic dust," *Natural History* **90**, 73–7 (1981) – April.
The porous, fragile cosmic dust that is collected from the stratosphere is cometary debris that may predate the origin of the solar system.

Calder, Nigel: *The Comet is Coming – The Feverish Legacy of Mr. Halley*, Penguin Books, New York, 1982.
The return of Halley's comet is used as an underlying theme for this excellent popular account of comets.

Delsemme, Armand H.: "Whence come comets?" *Sky and Telescope* **77**, 260–64 (1989) – March.
Despite their generally isotropic distribution, long-period comets do not approach the Sun from all directions equally; there are three zones depleted of cometary aphelia, perhaps as the result of galactic tides. Comets may be brought into the planetary system, and thus to visibility, by the fast passage of a star through the Oort cloud, the slow passage of a brown dwarf through the cloud, and by vertical galactic tides of a large invisible mass hidden in the galactic disc.

Knacke, Roger: "Sampling the stuff of a comet," *Sky and Telescope* **73**, 246–50 (1987) – March.
The gas and dust in Halley's coma hint at the solar system's origin.

Maffei, Paolo: *Monsters in the Sky*, MIT Press, Cambridge, Mass., 1980.
This book places comets within the perspective of other active astronomical objects like supernovae, black holes, active galaxies and quasars.

Maran, Stephen P.: "Where do comets come from?" *Natural History* **91**, 80–3 (1982) – May.
A good, concise description of the vast reservoir of comets that are stored in the deep freeze of outer space.

Sagan, Carl and Ann Druyan: *Comet*, Random House, New York, 1985.
A thorough, well-written description of the nature of comets, their origins and their fates, but written before the most recent return of Halley's comet.

Sagdeev, Roald Z. and Albert A. Galeev: "Comet Halley and the solar wind," *Sky and Telescope* **73**, 252–5 (1987) – March.
Spacecraft found a variety of electromagnetic activity inside Halley's coma.

Seargent, David A.: *Comets – Vagabonds of Space*, Doubleday and Co., New York, 1982.
This fine, short book places an emphasis on observations of comets, including descriptions of famous ones.

Weissman, Paul R.: "Realm of the comets," *Sky and Telescope* **73**, 238–40 (1987) – March.
The distant and unseen Oort cloud sends regular reminders that it exists.

Whipple, Fred L.: "The nature of comets," *Scientific American* **230**, 48–57 (1974) – February.

An excellent overview of cometary research written for the layperson; abundant evidence is provided for the author's dirty snowball model of the nucleus.

Whipple, Fred L.: *The Mystery of Comets*, Smithsonian Institution Press, Washington, D.C., 1985.

A fine summary of the growth of cometary science through the ages.

Whipple, Fred L.: "An inside look at Halley's comet," *Sky and Telescope* **71**, 438–43 (1986) – May.

A colorful description of the Vega 1, Vega 2 and Giotto encounters with Halley's comet.

Whipple, Fred L.: "The black heart of Comet Halley," *Sky and Telescope* **73**, 241–5 (1987) – March.

A discussion of spacecraft investigations of the comet's nucleus.

More technical works

Delsemme, Armand H. (editor): *Comets, Asteroids, Meteorites – Interrelations, Evolutions, and Origins*, University of Toledo Press, Toledo, Ohio, 1977.

A collection of technical papers that include the physical nature, orbital evolution and origin of comets.

Delsemme, Armand H.: "The chemistry of comets," *Philosophical Transactions of the Royal Society (London)* **A 325**, 509–23 (1988).

A review of the atomic, chemical and molecular composition of comets with a stress on spacecraft investigations of Halley's comet.

Kronk, Gary W.: *Comets – A Descriptive Catalog*, Enslow Publishers, Hillside, New Jersey, 1985.

A fantastic compilation of the physical descriptions of every comet observed from the year −371 to 1982.

Kronk, Gary W.: *Meteor Showers – A Descriptive Catalog*, Enslow Publishers, Hillside, New Jersey, 1988.

This catalog includes positions, durations, maximum dates and orbital elements for 112 meteor showers together with interesting historical or observational details.

Lust, Rhea: "Chemistry in comets," in *Topics in Current Chemistry – Cosmochemistry and Geochemistry*, Springer-Verlag, New York, 1981.

A technical review of the molecular chemistry of interstellar clouds, comets and planetary atmospheres.

Marsden, Brian G.: "Comets," *Annual Review of Astronomy and Astrophysics* **12**, 1–21 (1974).

A complete review that discusses the decay, orbits, structure and origin of comets.

Marsden, Brian G.: *Catalog of Cometary Orbits*, Enslow Publishers, Hillside, New Jersey, 1983.

A comprehensive catalog of orbital data for over one thousand apparitions of 710 comets.

Mendis, D.A.: "A postencounter view of comets," *Annual Reviews of Astronomy and Astrophysics* **26**, 11–49 (1988).

A report on spacecraft encounters with Halley's comet, including direct observations of its nucleus, cometary dust, ions and molecules, and the comet's interaction with the solar wind.

Oort, J.H.: "The origin and dissolution of comets," *Observatory* **106**, 186–93 (1986) – December.

A good description of the origin of comets by the man who proposed the existence of the comet cloud that now bears his name.

Smoluchowski, Roman, John N. Bahcall, and Mildred S. Matthews (editors): *The Galaxy and the Solar System*, University of Arizona Press, Tucson, 1986.

Twenty-two technical articles place the solar system within its galactic environment and discuss expected perturbations of the outer solar system by passing stars, by molecular clouds, by gravitational fields and by a hypothetical companion star that might produce showers of comets and periodic extinctions of species on the Earth.

Spacecraft encounter with Comet Giacobini-Zinner

The first scientific results from the International Cometary Explorer Mission to Comet Giacobini-Zinner are given in *Science* **232**, 297–428 (1986).

Spacecraft encounters with Comet Halley

The first results of spacecraft encounters during the 1986 perihelion passage of Comet Halley are presented in *Nature* **321**, 259–365 (1986); fantastic close-up images of the comet's nucleus are given by H.U. Keller, R. Kramm and N. Thomas: "Surface features on the nucleus of Comet Halley," *Nature* **321**, 227–31 (1988) and the volume's cover. Comprehensive details are found in over 160 papers collected in *Halley's Comet*, an entire issue of *Astronomy and Astrophysics* **187**, 1–936 (1987); these papers stress the Giotto space probe that sampled the cometary gas (hydrogen, oxygen, hydrogen cyanide, carbon, water vapor, etc.), plasma (electrons and ions), dust, magnetic field and solar wind interaction as the probe passed through the coma at a short distance from the comet's nucleus.

Wilkening, Laurel L.: *Comets*, University of Arizona Press, Tucson, 1982.

A comprehensive technical reference for nearly everything we knew about comets, before the spacecraft encounter with Comet Halley.

Chapter 12. Birth of the solar system

Chaisson, Eric: *Cosmic Dawn*, Little Brown, Boston, 1981.

This book places the origin and evolution of life within the more general context of the evolution of matter.

Fisher, David E.: *The Birth of the Earth – A Wanderlied Through Space, Time and the Human Imagination*, Columbia University Press, New York, 1987.

The contents and origin of the solar system are described in a conversational and historical style. Topics include the Earth's size and age, understanding planetary orbits, stellar encounters, supernova-induced collapse, protoplanets, accretion, differentiation, attempts to detect planets of other stars, and the radio search for extraterrestrial intelligence.

Lewis, John S.: "Chemistry of the solar system," *Scientific American* **230**, 50–65 (1974) – March.

A good discussion of how the density and chemical composition of the solid material in the solar system were determined by the temperature gradient in the primeval solar nebula, but some of the technical details are now out of date.

Lewis, Roy S. and Edward Anders: "Interstellar matter in meteorites," *Scientific American* **249**, 66–77 (1983) – August.

Carbonaceous chondrites contain a number of isotopic anomalies that originated outside the solar system, perhaps from the explosion of a nearby supernova.

More technical works

Black, David C. and Mildred Shapley Matthews (editors): *Protostars and Planets II*, University of Arizona Press, Tucson, 1985.

An excellent collection of technical articles on all aspects of planetary formation and evolution.

Cameron, Alastair G.W.: "Origin of the solar system," *Annual Reviews of Astronomy and Astrophysics* **26**, 441–72 (1988).

A technical review of current origin concepts including interstellar clouds and their collapse, the primitive solar nebula, the formation of the Sun, planetesimal and planetary accumulation, major planetary collisions, and the origin of comets.

Runcorn, S.K. (editor): *The Physics of the Planets – Their Origin, Evolution and Structure*, John Wiley & Sons, New York, 1988.

Twenty-six technical articles written by experts in the field are grouped together in four parts – structure and composition, planetary dynamics, celestial mechanics, and the origin of the solar system.

Index